Microarray Data Analysis

METHODS IN MOLECULAR BIOLOGY™

John M. Walker, SERIES EDITOR

METHODS IN MOLECULAR BIOLOGY™

Microarray Data Analysis

Methods and Applications

Edited by

Michael J. Korenberg

Department of Electrical and Computer Engineering
Queen's University, Kingston, Ontario, Canada

HUMANA PRESS ✳ TOTOWA, NEW JERSEY

© 2007 Humana Press Inc.
999 Riverview Drive, Suite 208
Totowa, New Jersey 07512

www.humanapress.com

This publication is printed on acid-free paper. ∞
ANSI Z39.48-1984 (American Standards Institute) Permanence of Paper for Printed Library Materials.

Cover design by Nancy K. Fallatt

Cover illustration: Support Vector Machine analysis constructs planes in multidimensional space such that sets of genes separate into distinct classes based on an iterative training algorithm (Fig. 6, Chapter 2; *see* complete caption on p. 32 and discussion on pp. 31–32).

For additional copies, pricing for bulk purchases, and/or information about other Humana titles, contact Humana at the above address or at any of the following numbers: Tel.: 973-256-1699; Fax: 973-256-8341; E-mail: orders@humanapr.com; or visit our Website: www.humanapress.com

Photocopy Authorization Policy:

Printed in the United States of America. 10 9 8 7 6 5 4 3 2 1

ISSN 1064-3745

E-ISBN 1-59745-390-0

Library of Congress Catloging-in-Publication Data

Microarray data analysis : methods and applications / edited by Michael J. Korenberg.
    p. ; cm. -- (Methods in molecular biology ; 377)
  Includes bibliographical references and index.
  ISBN-13: 978-1-58829-540-8 (alk. paper)
  ISBN-10: 1-58829-540-0 (alk. paper)
  1. DNA microarrays. 2. Gene expression. I. Korenberg, Michael J. II.
Series: Methods in molecular biology (Clifton, N.J.) ; v. 377.
  [DNLM: 1. Microarray Analysis--methods. 2. Gene Expression Profil-
ing. W1 ME9616J v. 377 2007 / QU 450 M6256 2007]
  QP624.5.D726M512 2007
  572.8'636--dc22
                                              2006037730

To my Mother and Father,

and to June

Preface

When the series editor, Prof. John Walker, asked me to edit a book on microarray data analysis, I began by writing to a number of researchers whose work I admired. Many of them agreed to contribute chapters. One of them, Dr. Orly Alter, suggested several others to me, and I am very grateful to her. The contributed chapters speak for themselves. They indeed cover a wide range of topics in both methods and applications; I found them fascinating, and thank the authors for all their work. I am very fortunate to have dealt with such an elite group.

Michael J. Korenberg

Contents

Contributors

ORLY ALTER • *Department of Biomedical Engineering, Institute for Cellular and Molecular Biology and Institute for Computational Engineering and Sciences, University of Texas at Austin, Austin, TX*

DAVID ATKINS • *Veridex LLC, a Johnson and Johnson Company, San Diego, CA*

GÁBOR BALÁZSI • *Department of Molecular Therapeutics, University of Texas M. D. Anderson Cancer Center, Houston, TX*

SOUMYAROOP BHATTACHARYA • *Center for Biomedical Informatics, Pittsburgh PA*

GHISLAIN BIDAUT • *Center for Bioinformatics, Department of Genetics, University of Pennsylvania School of Medicine, Philadelphia, PA*

SVEN BILKE • *Oncogenomics Section, Pediatric Oncology Branch, Advanced Technology Center, National Cancer Institute, Gaithersburg, MD*

JAMES L. BOLEN • *Dynamics Group, Department of Biology, Beckman Research Institute of the City of Hope Medical Center, Duarte CA*

CLAUDIA BREDEL • *Division of Oncology, Center for Clinical Sciences Research, Stanford University School of Medicine, Stanford, CA*

MARKUS BREDEL • *Department of Neurosurgery and Division of Oncology, Center for Clinical Sciences Research, Stanford University School of Medicine, Stanford, CA*

DEBOPRIYA DAS • *Lawrence Berkeley National Laboratory, Berkeley, CA*

PEDRO FARINHA • *Department of Pathology, British Columbia Cancer Agency, Vancouver, British Columbia, Canada*

JOHN FOEKENS • *Department of Medical Oncology, Erasmus Medical Center, Daniel den Hoed Cancer Center, Rotterdam, The Netherlands*

RANDY D. GASCOYNE • *Department of Pathology, British Columbia Cancer Agency, Vancouver, British Columbia, Canada*

BRADEN GREER • *Oncogenomics Section, Pediatric Oncology Branch, Advanced Technology Center, National Cancer Institute, Gaithersburg, MD*

DEJAN JURIC • *Division of Oncology, Center for Clinical Sciences Research, Stanford University School of Medicine, Stanford, CA*

JAVED KHAN • *Oncogenomics Section, Pediatric Oncology Branch, Advanced Technology Center, National Cancer Institute, Gaithersburg, MD*

WARREN A. KIBBE • *Robert H. Lurie Comprehensive Cancer Center, Northwestern University, Chicago, IL*

ROBERT R. KLEVECZ • *Dynamics Group, Department of Biology, Beckman Research Institute of the City of Hope Medical Center, Duarte, CA*

JAN KLIJN • *Department of Medical Oncology, Erasmus Medical Center, Daniel den Hoed Cancer Center, Rotterdam, The Netherlands*

xi

MICHAEL J. KORENBERG • *Department of Electrical and Computer Engineering, Queen's University, Kingston, Ontario, Canada*
ANDREW KOSSENKOV • *Fox Chase Cancer Center, Philadelphia, PA*
CAROLINE M. LI • *Dynamics Group, Department of Biology, Beckman Research Institute of the City of Hope Medical Center, Duarte, CA*
SIMON M. LIN • *Robert H. Lurie Comprehensive Cancer Center, Northwestern University, Chicago, IL*
JAMES LYONS-WEILER • *Center for Biomedical Informatics, Benedum Center for Oncology Informatics/Center for Pathology Informatics, and University of Pittsburgh Medical Center/Cancer Institute, Pittsburgh, PA*
TOBEY J. MACDONALD • *Center for Cancer and Immunology Research, Children's Research Institute, Department of Hematology-Oncology, Children's National Medical Center, Washington, DC*
MARCELO O. MAGNASCO • *Center for Studies in Physics and Biology, The Rockefeller University, New York, NY*
OLGA MODLICH • *Institute of Chemical Oncology, University of Düsseldorf, Düsseldorf, Germany*
MARC MUNNES • *Bayer Healthcare AG, Diagnostic Research Germany, Leverkusen, Germany*
MICHAEL F. OCHS • *Fox Chase Cancer Center, Philadelphia, PA*
HIDEHO OKADA • *Departments of Neurosurgery and Pathology, Cancer Institute Brain Tumor Center , University of Pittsburgh Medical Center and Children's Hospital of Pittsburgh, Pittsburgh, PA*
ZOLTÁN N. OLTVAI • *Department of Pathology, University of Pittsburgh, Pittsburgh, PA*
JOHN D. OSBORNE • *Robert H. Lurie Comprehensive Cancer Center, Northwestern University, Chicago, IL*
AIDAN J. PETERSON • *Fox Chase Cancer Center, Philadelphia, PA*
IAN F. POLLACK • *Departments of Neurosurgery and Pathology, Cancer Institute Brain Tumor Center, University of Pittsburgh Medical Center and Children's Hospital of Pittsburgh, Pittsburgh, PA*
MANOJ PRATIM SAMANTA • *Systemix Institute, Cupertino, CA*
BRANIMIR I. SIKIC • *Division of Oncology, Center for Clinical Sciences Research, Stanford University School of Medicine, Stanford, CA*
VIKTOR STOLC • *Systemix Institute, Cupertino, CA*
MAYTE SUÁREZ-FARIÑAS • *Center for Studies in Physics and Biology, The Rockefeller University, New York, NY*
DENNIS D. TAUB • *Laboratory of Immunology, National Institutes of Health, National Institute on Aging, Gerontology Research Center, Baltimore, MD*
WARAPORN TONGPRASIT • *Systemix Institute, Cupertino, CA*
YIXIN WANG • *Veridex LLC, a Johnson and Johnson Company, San Diego, CA*
ASHANI T. WEERARATNA • *Laboratory of Immunology, National Institutes of Health, National Institute on Aging, Gerontology Research Center, Baltimore, MD*

MICHAEL Q. ZHANG • *Cold Spring Harbor Laboratory, Cold Spring Harbor, NY*
YI ZHANG • *Veridex LLC, a Johnson and Johnson Company, San Diego, CA*
LIHUA (JULIE) ZHU • *Robert H. Lurie Comprehensive Cancer Center, Northwestern University, Chicago, IL*

1

Microarray Data Analysis
An Overview of Design, Methodology, and Analysis

Ashani T. Weeraratna and Dennis D. Taub

Summary

Microarray analysis results in the gathering of massive amounts of information concerning gene expression profiles of different cells and experimental conditions. Analyzing these data can often be a quagmire, with endless discussion as to what the appropriate statistical analyses for any given experiment might be. As a result many different methods of data analysis have evolved, the basics of which are outlined in this chapter.

Key Words: Microarray data analysis; MIAME; clustering.

1. Introduction

Microarray technology is widely used to examine the gene expression profiles of a multitude of cells and tissues. This technology is based on the hybridization of RNA from tissues or cells to either cDNA or oligonucleotides immobilized on a glass chip or, in increasingly rare cases, on a nylon membrane. One of the first experiments in which cDNA clones were arrayed onto a filter, and then hybridized with cell lysates, analyzed the gene expression profiles of colon cancer, and examined the expression of 4000 genes therein *(1)*. Since then, the identification of genes by the Human Genome Project *(2)* has allowed for the expansion of the number of cDNA clones or oligonucleotides spotted on a single slide. Today, the average commercial microarray contains roughly 20,000 clones or oligonucleotides, many of which are unique. Some companies, such as Agilent Technologies, also make a slide that encompasses genes from the whole genome with over 44,000 genes spotted on their arrays. Obviously, the analysis of so many data can prove quite overwhelming and labor intensive. The purpose of this chapter is to outline the available techniques for microarray data analysis.

From: *Methods in Molecular Biology, vol. 377, Microarray Data Analysis: Methods and Applications*
Edited by: M. J. Korenberg © Humana Press Inc., Totowa, NJ

2. Experimental Design

Successful data analysis begins with a good experimental design, and often, one of the most crucial and most overlooked parts of performing an informative array experiment is designating an appropriate reference, or standard. For example, when analyzing a given disease, it is useful to assign a "control" or "frame-of-reference" sample that can be used as a comparison for all states of that disease. This could be a sample such as a normal, nonmalignant tissue of origin when analyzing cancer, or resting T-cells as compared with those activated through the T-cell or cytokine receptors. It is, however, often difficult to determine what "normal" tissue or cell is best to use, and what exactly defines normal. Many users prefer to utilize universal RNA, so that comparisons can be made between several different gene expression profiles that may not have a common normal counterpart. To assess what constitutes a good reference for an experiment, the researchers must first have a clear idea of what precise questions they want to answer. Often, researchers fall into the trap of comparing experimental and control conditions directly to each other, when a slightly more complex experiment using a common reference for both experimental and control conditions may provide a more sophisticated analysis of the data. For example, when treating cancer cell lines with a drug, it is tempting to simply compare treated to untreated cell lines. However, more information could potentially be gathered by comparing both treated and untreated cell lines to a normal, untreated control cell line (e.g., melanocytes vs melanomas treated with different agents or vehicle controls). Ultimately, the more complex statistical analyses that can be performed on these types of data may reveal more subtle, but equally important, gene expression patterns.

3. Minimal Information About a Microarray Experiment

In an effort to standardize the thousands of array experiments, the Microarray Gene Expression Database (MIAME) society established guidelines that require researchers to conform to MIAME guidelines *(3)*. MIAME describes the minimal information about a microarray experiment that is required to interpret the results of the experiment, and compare it with other experiments from other groups. The checklist for complying with the MIAME guidelines is quite extensive and can be found at http://www.mged.org/Workgroups/MIAME/miame_checklist.html

In brief, these guidelines include:

1. Array design: information regarding the platform of the array, description of the clones and oligomers, and catalog numbers for commercial arrays. This also should include the location of each feature as well as the explanations of feature annotation.
2. Experimental design: a description and the goals of the experiment, rationale for cells/tissues and treatment used, quality control steps, and links to any public databases necessary.

3. Sample selection: criteria for the selection of samples, description of the procedures used for RNA extraction, and sample labeling.
4. Hybridization: conditions of hybridization, including blocking and washing of slides.
5. Data analysis: description of the raw data, as well as of the original images, hardware, and software used, and also the criteria used for processing and normalization of data.

In addition to the obvious benefits of standardizing microarray data, many of the top journals in the field currently require researchers to comply with these guidelines, so it is worth examining your selected array format for MIAME compliance prior to starting a microarray experiment.

4. Image Acquisition and Analysis

Once the RNA has been isolated and hybridized to the chip, the first stage of data analysis begins. This requires successful acquisition of the fluorescent or radioactive signal bound to the chip or membrane. With radioactive membranes, it is standard procedure to expose the membrane several times and then take an educated average of the best exposures *(4)*. With fluorescent dyes, it is essential to utilize a high-resolution scanner and that the first scan be performed as quickly and accurately as possible, as the dyes are quickly bleached and multiple scans are not possible. Some salient points of image acquisition are outlined next.

4.1. Quality of Scanner

It is important to use a scanner that can detect at a resolution of 10 microns or greater. In addition, the scanner must be able to excite and detect Cy3 (532 nm) and Cy5 fluorescence (633 nm). An adjustable photomultiplier tube to ensure equal scanning, while reducing as much bleaching as possible, is also ideal. Typically, the settings for the photomultiplier tube are around 30%.

4.2. Orientation of Image

The orientation of the image becomes particularly important when combining arrays from one company with a scanner from a different company as images may be inverted depending on the scanner being used. Thus, it is crucial that the array include "landing lights"—control cDNAs or oligonucleotides spotted on the arrays that yield a distinct pattern when the array is in the correct orientation **(Fig. 1A)**.

4.3. Spot Recognition

Often referred to as "gridding," this is the process used to identify each spot on the array prior to extracting information from it. When purchasing arrays and scanners from commercial sources, programs for spot recognition and information extraction are often included. Agilent and Affymetrix both have their own

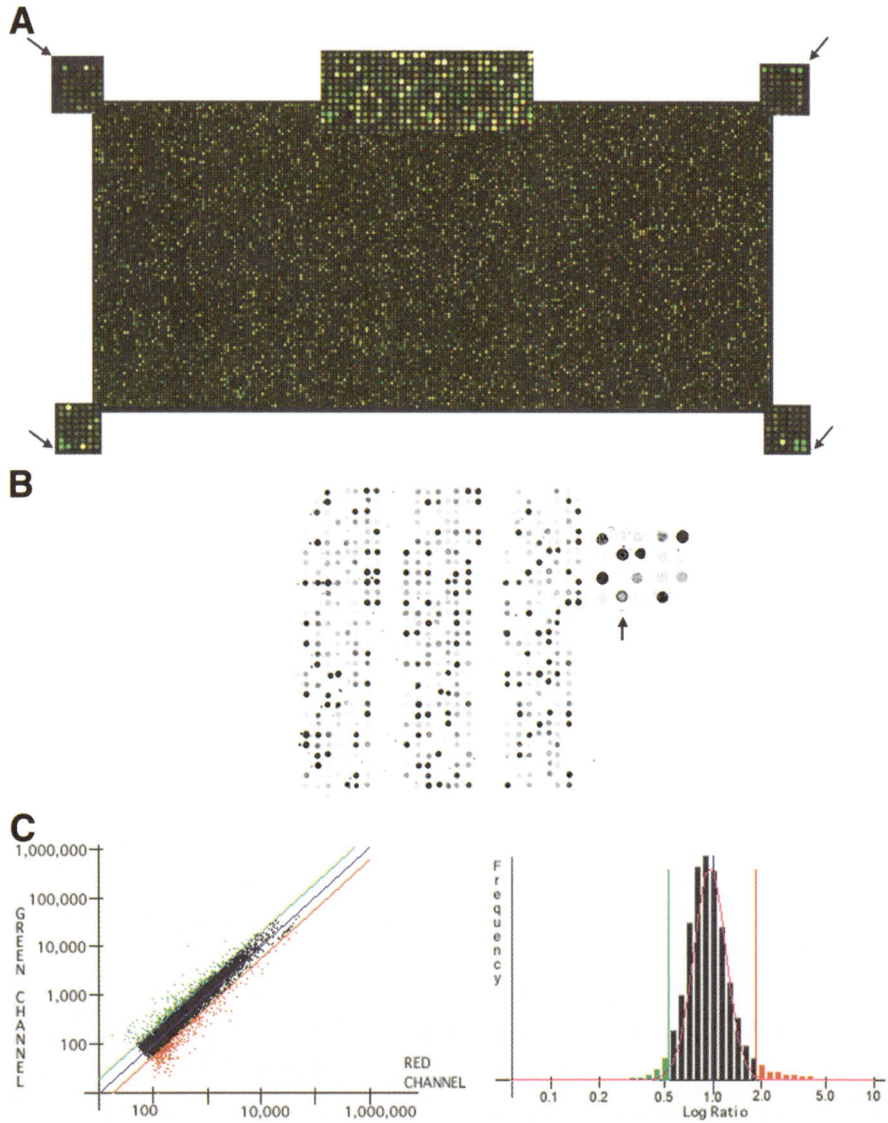

Fig. 1. Image analysis: (**A**) Image acquisition. Shown here is a scanned microarray slide from Agilent Technologies. Note the four corner features that show one, two, three, and four green dots, respectively (arrows), allowing for orientation of the slide by the user. The center blow-up demonstrates the green, zig-zag pattern of the control features on the array. (**B**) Spot recognition. A clip of a microarray experiment showing a single dye channel, prior to gridding of spots. The blow-up shows a variety of good spots, and bad spots, including blanks, donuts, and one spot that has a highly intense outer rim, and center, but low signal in between (arrow). (**C**) Data normalization. Data in an array experiment was normalized using internal targets for calibration, and the ratio distribution was extracted from the experiment in both a scatter plot and histogram form.

feature extractor software, which uses control spots on the array for automated spot recognition and feature extraction. Many other programs require that the user intervene and flag "bad" spots, and realign grids to fit the spots.

4.4. Segmentation

Once grids have been placed, information as to the pixel intensity within the spots must be extracted. This process is known as segmentation. Various methods exist to perform this including fixed circle segmentation, adaptive circle segmentation, fixed shape segmentation, adaptive shape segmentation, and seeded region growing method (also known as the histogram-based method).

1. Fixed circle segmentation: assumes that spots are circular, with a fixed radius—all information is extracted from within this fixed radius.
2. Adaptive circle segmentation: allows for radius to be adapted to the spot.
3. Adaptive shape segmentation-seeded region growing method: the foreground and background intensities are adapted from two initial growing seeds.
4. Histogram-based segmentation: uses a target mask that is larger than the spot, and calculates intensity from both foreground and background using given threshold values from the masked areas.

Lately, an approach that utilizes model-based recognition of spots, based on Bayesian information criterion has greatly improved this process, making the commonly seen "donuts," scratches, and blank spots (**Fig. 1B**) not addressed by the above methods much easier to recognize and remove from the analysis *(5)*. This method combines a histogram-based spot recognition, using a flexible adaptive shape segmentation approach with finding the large spatially connected components (>100 pixels) within each cluster of pixels, and may soon be available commercially. Finally, experimentation using DAPI to stain the spots on the array has been quite successful in removing limitations of these types of algorithmic approaches *(6)*. It has been suggested that this approach may lead to fully automated image analysis but has not as yet entered into the general mainstream of array data analysis. Ultimately, the goal of all these methods is to subtract background intensity from foreground intensity and give spot intensity for each dye channel, while reducing misinformation from contaminants, such as dust and scratches.

4.5. Analysis of the Quality of the Hybridization

All of these imaging parameters can then be used to analyze the quality of the microarray experiment. Intensities in each channel should ultimately cluster around a central norm in a Gaussian distribution (**Fig. 1C**). Background intensity abnormalities can be calculated statistically by computing the average background intensity and using the standard deviation among this intensity to calculate a confidence interval, the upper limit of which is used to assume background correction.

4.6. Data Normalization

In order to normalize the information received from a microarray experiment, several methods have been designed and are outlined next.

4.6.1. Housekeeping Genes

The use of housekeeping genes to normalize array data assumes that there is a set of standard genes whose expression does not change with experimental condition, or sample type, thus providing a basis for comparison between samples. However, as commonly used housekeeping genes such as GAPDH and actin can indeed change from one condition to another, it is sometimes dangerous to base calculations on this assumption.

4.6.2. Control Targets

Many arrays, especially commercial arrays, have targets for control features printed onto the chip. These targets are often DNA sequences that are designed to hybridize to positive control sequences on the chip. With Agilent chips, for example, the control nucleotides (Cy3-TAR25_C and Cy5-TAR25_C) are already labeled with Cy-3 or Cy-5 and are added to the solution just prior to hybridization. These targets hybridize to control features, Pro25+, on the array, which are arranged in a specific pattern. These control features can also serve as "landing lights" to help the user orient the slide image.

4.6.3. Global Normalization Techniques

Global normalization assumes that the majority of genes on the array are non-differentially expressed between the Cy-3 and Cy-5 channels, and that the number of genes expressed preferentially in one channel is equal to that of the genes expressed preferentially in the other. Thus, several algorithms can be used. Integral balance analysis assumes constant mRNA for all samples, whereas linear regression methods assume constant expression among most genes, regardless of experimental conditions *(7,8)*. Regression methods can account for intensity and spatial dependence on dye bias variables *(9,10)*. In both types of normalization, a best-fit equation is used and the normalization signal becomes either the logarithmic or linear mean of expression intensity, or expression intensity ratios. The pitfall of this type of analysis is that when the reference RNA is significantly different from the experimental RNA, or when intensities vary significantly, the assumptions may be invalid. Newly available methods attempt to address these discrepancies. In a recent paper by Zhao et al. *(11)*, a mixture model-based normalization method was used to analyze dual channel (fluorescent) experiments. As with all other parts of microarray data analysis, the normalization method selected should be tailored to the experiment and biological samples in question.

4.7. Data Transformation

After background correction has been performed, the data must be transformed for statistical analysis. The analyses applied to the data (e.g., parametric vs nonparametric) determine the type of transformation that must be performed. Parametric tests are the most commonly utilized, as these tests are much more sensitive and require the data to be normally distributed. This is often achieved by using log transformation of the spot intensities to achieve a Gaussian distribution of the data. However, log transformation is not recommended for all types of downstream analysis, as some analyses rely on a distance measure (*see* **Subheadings 5.2.1.** and **5.2.2.**).

5. Differential Gene Expression

Differential gene expression is often measured by the ratio of intensity (as a measure of expression level) between two samples. Many early microarray experiments assigned a fold-change cutoff, and considered genes above this fold-change significant. However, this treatment of the data does not take into account interexperimental variability and requires that a few replicates of the arrays be performed. Recently, several model-based techniques have been developed, the newest of which assumes multiplicative noise, and eliminates statistically significant outliers from the data *(12)*. In addition, several statistical analyses can be utilized including maximum-likelihood analysis, F-statistic, ANOVA (analysis of variance), and *t*-tests. The results of these tests can often be improved by log transformation of data as mentioned previously, and by random permutations of the data. Nonparametric tests used to analyze microarray data include Mann–Whitney tests and Kruskal–Williams rank analysis.

5.1. Reducing Error Rate: False-Positives and False-Negatives

Ultimately, all of the statistical tests calculate significance values for gene expression, most commonly as a "*p*-value." *P*-values are then compared to α-levels, which determine the false-positive and false-negative rates by setting a predetermined acceptance level for the *p*-value. False-negative rates depend not only on α-levels, as do false-positive rates, but also on the number of replicates, the population effect size, and random errors of measurement. These methods calculate the overall chance that at least one gene is a false-positive or -negative, i.e., the family-wise error rate *(13)*. Another method for discovering false positive/negative data is the Bonferoni approach, a stringent analysis that uses multiple tests. This linear step-up approach multiplies the uncorrected *p*-value by the number of genes tested treating each gene as an individual test, which can significantly increase specificity by reducing the number of false-positives identified, but unfortunately leads to a decrease in sensitivity by increasing the number of false-negatives. A modification of the Bonferoni approach, the false-discovery rate, uses random permutation while assuming each gene is an independent test, and bootstrapping approaches can improve significantly on the Bonferoni approach, as they are less stringent *(14)*. Resampling-based false discovery rate-controlling procedures can also be used *(15)*, and software to perform this analysis is available at www.math.tau.ac.il/~ybenja.

5.2. Pattern Discovery

Often called exploratory or unsupervised data analysis, this approach can encompass a number of different techniques listed next that allow for a global view of the data. These methods often rely on clustering techniques that allow for quick viewing of distinct gene expression patterns within a dataset. Cluster analysis is available free of charge as part of the gene expression omnibus, a site that attempts to catalog gene expression data *(16)*, providing a valuable data mining resource (http://www.ncbi.nlm.nih.gov/geo/). Dimension reduction techniques such as principal component analysis (PCA) and multidimensional scaling analysis can often be used in conjunction with other supervised techniques such as artificial neural networks to provide even more robust data analysis.

5.2.1. PCA

PCA can analyze multivariate data by expressing the maximum variance as a minimum number of principal components. Redundant components are eliminated, thus reducing the dimensions of the input vectors. For information on the mathematical origins of this equation, *see* http://www.cis.hut.fi/~jhollmen/dippa/node30.html.

5.2.2. Multidimensional Scaling

This analysis is often based on a pair-wise correlation coefficient and assesses the similarities and dissimilarities between samples and assigns the difference as a "distance" between samples, such that the more similar two samples are, the closer they are together, and vice versa **(Fig. 2A)**. The multi- as opposed to two-dimensional analysis comes into play when not only the degree of difference (distance) but also the spatial relationship of three or more samples to each other (direction) is taken into account. For further mathematical description of this process, *see* http://www.statsoft.com/textbook/stmulsca.html.

5.2.3. Singular Value Decomposition

Singular value decomposition (SVD) treats microarray data as a rectangular matrix, A, which is composed of n rows (genes) by p columns (experiments). SVD is represented by the mathematical equation, with U being the gene coefficient vectors, S the mode amplitudes, and V^T the expression level vectors.

$$A_{nxp} = U_{nxn} \, S_{nxp} \, V^T_{pxp}$$

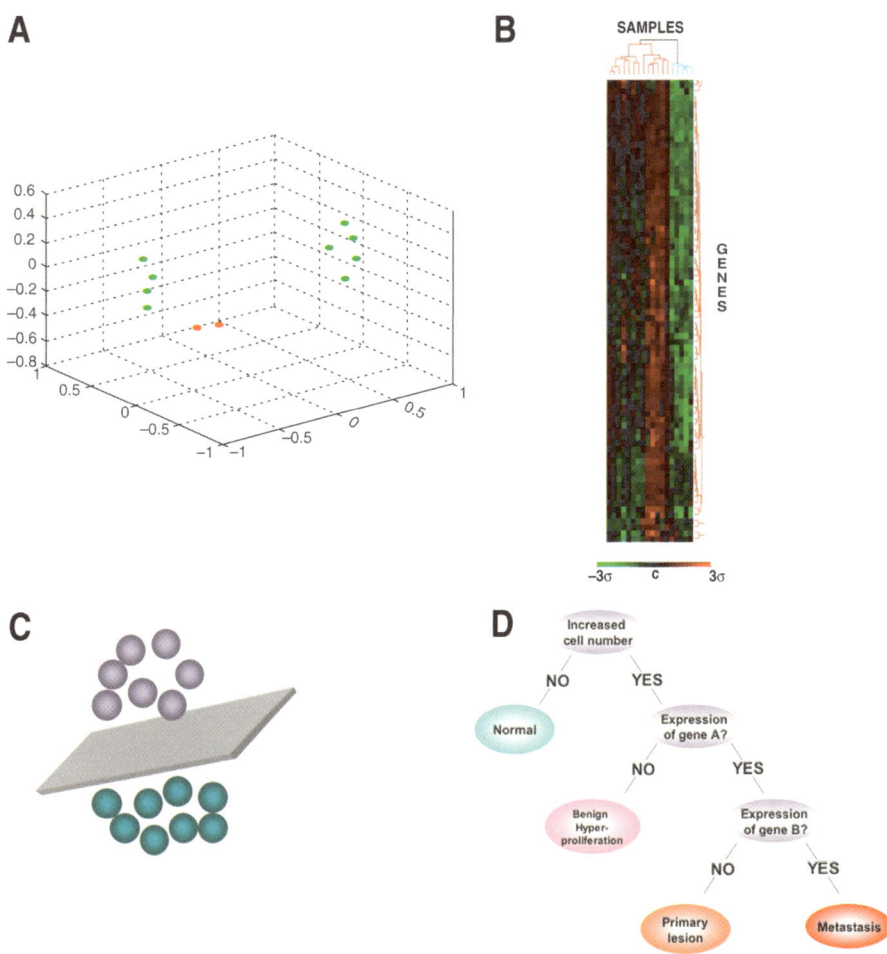

Fig. 2. Data analysis. Unsupervised clustering algorithms include (**A**) multidimensional scaling, and (**B**) hierarchical clustering. Supervised methods include (**C**) support vector machines and (**D**) decision trees. *See* **Subheadings 5.2.** and **5.3.** for more details.

For those readers interested in solving for the SVD equation, an excellent description of the problem can be found online at http://web.mit.edu/be.400/www/SVD/Singular_Value_Decomposition.htm.

5.2.4. Hierarchical Clustering

Perhaps the most familiar to biologists, hierarchical clustering presents the data as a gene list organized into a dendrogram, and is a bottom-up analysis. This is obtained by assigning a similarity score to all gene pairs, calculating the Pearson's correlation coefficient, and then building a tree of genes by replacing the two most similar genes with a node that contains the average, then repeating the process for the next closest pair of data points, and then the next. This process is repeated several times (iterative process) to generate the dendrogram or Treeview, as well as heat maps that represent a two-color checkerboard view of the data (**Fig. 2B**) *(17)*.

5.2.5. K-Means Clustering

K-means clustering is a top-down technique that groups a collection of nodes into a fixed number of clusters (k) that are subjected to an iterative process. Each class must have a center point that is the average position of all the distances in that class (representative element), and each sample must fall into the class to which its center is closest. Fuzzy k-means is performed by "soft" assignment of genes to these clusters *(17)*.

5.2.6. Self-Organizing Maps

These maps are basically two-dimensional grids containing nodes of genes in "K"-dimensional space. These can be represented by sample and weight vectors, which are composed of the data and their natural location. Weight vectors are initialized, and then sample vectors are randomly selected to determine which weight best represents that sample, and these are used to map the nodes into K-dimensional space into which the gene expression data falls. Like the previously mentioned methods, this is also iterative and is often repeated more than 1000 times, and these methods can often be used in combination to generate the best overview of the data *(18)*.

5.3. Class Prediction

Class prediction is based on supervised data analysis methods that impose known groups on datasets. First, a training set is identified—this is a group of genes with a known pattern of expression that is used to "train" a dataset, by comparing the data to the training set and thus classifying it *(19)*. This particular method is very useful in the subclassification of similar samples *(20)*, cancer diagnosis *(21)*, or to predict cell or patient response to drug therapy *(22,23)*. In some cases, this type of analysis has also been used to predict patient outcome *(24)*, allowing for a very clinically relevant use of microarray data. Importantly, gene selection by these methods relies on the assignment of discriminatory weights to these genes, i.e., how often a single gene correlates to a given class or phenotype, often calculated using random permutation tests. Random permutation tests are also used to calculate p (probability the weight can be obtained by chance) and α (probability of high weight resulting from random classification) values for these weights. Many different statistical methods can be used to find discriminant genes.

5.3.1. Fisher Linear Discriminant Analysis

This theory assumes that a random vector x has a multivariate normal distribution between each defined class or group, and the covariance within each group is identical for all the groups. This makes the optimal decision function for the comparison of data a linear transformation of x (25). Variations on this theme include quadratic discriminant analysis, flexible discriminant analysis, penalized discriminant analysis, and mixture discriminant analysis.

5.3.2. Nearest-Neighbor Classification

These methods are based on a measure of distance (e.g., Euclidean distance) between two gene expression profiles. Observations are given a value (x) and the number of observations (k) closest to x is used to choose the class. The value of k can be determined by using cross-validation techniques (26).

5.3.3. Support Vector Machines

This type of analysis is based on constructing planes in a multidimensional space that separate the different classes of genes, and set decision boundaries using an iterative training algorithm (27). Data is mapped into the higher dimensional space from its original input space, and a nonlinear decision boundary is assigned (Fig. 2C). This plane is known as the maximal margin hyperplane, and can be located by the use of a kernel function (a nonparametric weighting function). For further mathematical description, see http://www.statsoft. com/textbook/stsvm.html.

5.3.4. Artificial Neural Networks

Neural networks, or perceptrons, another machine-learning technique, are so named because they model the human brain—they learn by experience. Multilayer perceptrons can be used to classify samples based on their gene expression (28,29). Gene expression data for a sample are input into the model, and a response is generated in the next layer, ultimately triggering a response in the output layer. This output perceptron should represent the class to which the sample belongs.

5.3.5. Decision Trees

These are built by using criteria to divide samples into nodes. Samples are divided recursively until they either fall into partitions, or until a termination condition is met (30). Ultimately the intermediate nodes represent splitting points or partitioning criteria, and the leaf nodes represent those decisions (Fig. 2D).

6. Pathway Analysis Tools

Once all the genes in an experiment have been analyzed, the next step is to biologically interpret the data. The use of gene ontology programs, such as those listed next, take the gene lists identified by the experiment and compare the patterns therein to the available literature, and thus extract information about potentially important pathways affected by the experiment. All of these programs are available online, but only a few are freely available.

6.1. GoMiner

GoMiner maps lists of genes to functional categories using a tree view. This program also links to PubMed, and LocusLink. In addition it provides biological molecular interaction map and signaling pathway packages for more detailed analysis *(31)*.

6.2. Database for Annotation, Visualization, and Integrated Discovery (DAVID)

DAVID is available at http://www.david.niaid.nih.gov; this program has four components *(32)*.

1. Annotation tool: annotates the gene lists by adding gene descriptions from public databases.
2. GoCharts: functionally categorizes genes based on user-selected classifications and term specificity level.
3. KeggCharts: assigns genes to the Kyoto Encyclopedia of Genes and Genomes (KEGG) metabolic processes and enables users to view genes in the context of biochemical pathway maps.
4. DomainCharts: groups genes according to conserved protein domains.

6.3. PATIKA: Pathway Analysis Tool for Integration and Knowledge Acquisition

Patika is a multi-user tool that is composed of a server-side, scalable, object-oriented database and client-side. As with the other programs, there is pathway layout, functional computation support, advanced querying, and a user-friendly graphical interface *(33)*.

6.4. Ingenuity Pathway Analysis

Of all the above programs, Ingenuity pathway analysis is perhaps the most efficient at analyzing multiple datasets across different experimentation platforms. Like GOMiner, Ingenuity can identify key functional pathways *(34)*. It is currently the largest curated database that comprises individually modeled relationships between proteins, genes, complexes, cells, tissues, drugs, and diseases, and provides a large variety in the presentation of the data.

7. Data Validation

As complex and robust as the available analyses for microarray data currently are, there is always room for error, and many inherent problems in the experimental technique. Thus, it is critical that researchers validate their data before drawing any firm biological conclusions from the data. One of the most common techniques for validating array data is the use of real-time PCR *(35)*. Real-time PCR effectively quantitates differences in transcript levels between different samples *(36)*, but it must be remembered that the ratios acquired from a microarray experiment are quite likely to be much lower than fold changes seen in real-time PCR, as this method is much more sensitive.

Ultimately, protein expression is of course the final confirmation, as most gene expression-profiling experiments, whether of a classifier or exploratory nature seek protein markers, and this is most often confirmed using immunohistochemistry. As such, tissue microarrays have become an important companion to DNA microarrays. These are slides that contain small punches of paraffin-embedded tissue, often up to 500 sections on one slide *(37)*. Tissue arrays often encompass all the stages of a disease being studied or can be made from animal tissues, as confirmation for in vivo mouse experiments, for example. The current large whole-genome arrays pose a problem when it comes to this aspect, as the actual rate of antibody production for all these novel proteins, many of which are hypothetical, lags far behind the rate of gene discovery. One can only hope that soon this will catch up with the available genomic data, leaving us with valuable tools to identify markers and pathways, and that truly take us from bench to bedside.

8. Future of Microarray Analysis and Technology

Over the last decade, microarray analysis has been utilized almost exclusively as a research tool that requires significant effort and computer time by trained individuals to prepare high-quality RNA, label and hybridize the arrays, and analyze the data. As evidenced by the recent surge of microarray use in the medical literature over the past 5 yr, this technique has become increasingly popular in comparing "normal" to "diseased" tissues or "treated" to "untreated" cells or clinical samples derived from various conditions. Despite this recent use in clinical studies, several significant hurdles need to be overcome to optimize it for routine clinical lab use. Considerable improvements are required to optimize microarray fabrication, hybridization methodology, and analysis that will permit a great deal of these processes to become fully automated and thus increase the reproducibility within and across experiments. New technologies, such as the use of carbon nanotubules to produce microarray-like devices, may increase the use, automation, accuracy, and throughput in the study of gene expression within research, clinical, and diagnostic samples. Moreover, continual advances in the field of proteomics, in combination with microarray technology, should greatly enhance our ability to identify proteins and antigens for therapeutic use. Several commercial software vendors have already initiated modifications in their data-mining software to link the nucleotide and protein databases and analysis tools to permit the examination of an individual gene transcription and translation. With the advent of new technologies and more rapid methods of analysis, the microarray technique will most likely become a more commonplace and invaluable tool not only for basic research studies but also for clinical analysis and diagnosis.

Acknowledgments

We thank Dr. Kevin Becker for helpful comments on the manuscript.

References

1. Augenlicht, L. H., Wahrman, M. Z., Halsey, H., Anderson, L., Taylor, J., and Lipkin, M. (1987) Expression of cloned sequences in biopsies of human colonic tissue and in colonic carcinoma cells induced to differentiate in vitro. *Cancer Res.* **47,** 6017–6021.
2. Lander, E. S., Linton, L. M., Birren, B., et al. (2001) Initial sequencing and analysis of the human genome. *Nature* **409,** 860–921.
3. Brazma, A., Hingamp, P., Quackenbush, J., et al. (2001) Minimum information about a microarray experiment (MIAME)-toward standards for microarray data. *Nat. Genet.* **29,** 365–371.
4. Dodson, J. M., Charles, P. T., Stenger, D. A., and Pancrazio, J. J. (2002) Quantitative assessment of filter-based cDNA microarrays: gene expression profiles of human T-lymphoma cell lines. *Bioinformatics* **18,** 953–960.
5. Li, Q., Fraley, C., Bumgarner, R.E., Yeung, K.Y., and Raftery, A.E. (2005) In: "Technical Report no. 473" (http://www.stat.washington.edu/www/research/reports/2005/tr473.pdf, Ed.), University of Washington, Seattle.
6. Jain, A. N., Tokuyasu, T. A., Snijders, A. M., Segraves, R., Albertson, D. G., and Pinkel, D. (2002) Fully automatic quantification of microarray image data. *Genome Res.* **12,** 325–332.
7. Quackenbush, J. (2002) Microarray data normalization and transformation. *Nat Genet* **32 (Suppl),** 496–501.
8. Zien, A., Aigner, T., Zimmer, R., and Lengauer, T. (2001) Centralization: a new method for the normalization of gene expression data. *Bioinformatics* **17 (Suppl 1),** S323–S331.
9. Yang, Y. H., Dudoit, S., Luu, P., et al. (2002) Normalization for cDNA microarray data: a robust composite method addressing single and multiple slide systematic variation. *Nucleic Acids Res.* **30,** e15.
10. Kepler, T. B., Crosby, L., and Morgan, K. T. (2002) Normalization and analysis of DNA microarray data by self-consistency and local regression. *Genome Biol.* **3,** RESEARCH0037.

11. Zhao, Y., Li, M. C., and Simon, R. (2005) An adaptive method for cDNA microarray normalization. *BMC Bioinformatics* **6**, 28.
12. Sasik, R., Calvo, E., and Corbeil, J. (2002) Statistical analysis of high-density oligonucleotide arrays: a multiplicative noise model. *Bioinformatics* **18**, 1633–1640.
13. Li, H., Wood, C. L., Getchell, T. V., Getchell, M. L., and Stromberg, A. J. (2004) Analysis of oligonucleotide array experiments with repeated measures using mixed models. *BMC Bioinformatics* **5**, 209.
14. Meuwissen, T. H., and Goddard, M. E. (2004) Bootstrapping of gene-expression data improves and controls the false discovery rate of differentially expressed genes. *Genet. Sel. Evol.* **36**, 191–205.
15. Reiner, A., Yekutieli, D., and Benjamini, Y. (2003) Identifying differentially expressed genes using false discovery rate controlling procedures. *Bioinformatics* **19**, 368–375.
16. Barrett, T., Suzek, T. O., Troup, D. B., et al. (2005) NCBI GEO: mining millions of expression profiles—database and tools. *Nucleic Acids Res.* **33 Database Issue,** D562–D566.
17. Sherlock, G. (2000) Analysis of large-scale gene expression data. *Curr. Opin. Immunol.* **12**, 201–205.
18. Wang, J., Delabie, J., Aasheim, H., Smeland, E., and Myklebost, O. (2002) Clustering of the SOM easily reveals distinct gene expression patterns: results of a reanalysis of lymphoma study. *BMC Bioinformatics* **3**, 36.
19. Dharmadi, Y., and Gonzalez, R. (2004) DNA microarrays: experimental issues, data analysis, and application to bacterial systems. *Biotechnol. Prog.* **20**, 1309–1324.
20. Bittner, M., Meltzer, P., Chen, Y., et al. (2000) Molecular classification of cutaneous malignant melanoma by gene expression profiling. *Nature* **406**, 536–540.
21. Ramaswamy, S., Tamayo, P., Rifkin, R., et al. (2001) Multiclass cancer diagnosis using tumor gene expression signatures. *Proc. Natl. Acad. Sci. USA* **98**, 15,149–15,154.
22. Cunliffe, H. E., Ringner, M., Bilke, S., et al. (2003) The gene expression response of breast cancer to growth regulators: patterns and correlation with tumor expression profiles. *Cancer Res.* **63**, 7158–7166.
23. Burczynski, M. E., Oestreicher, J. L., Cahilly, M. J., et al. (2005) Clinical pharmacogenomics and transcriptional profiling in early phase oncology clinical trials. *Curr. Mol. Med.* **5**, 83–102.
24. Nutt, C. L., Mani, D. R., Betensky, R. A., et al. (2003) Gene expression-based classification of malignant gliomas correlates better with survival than histological classification. *Cancer Res.* **63**, 1602–1607.
25. Mendez, M. A., Hodar, C., Vulpe, C., Gonzalez, M., and Cambiazo, V. (2002) Discriminant analysis to evaluate clustering of gene expression data. *FEBS Lett.* **522**, 24–28.
26. Olshen, A. B., and Jain, A. N. (2002) Deriving quantitative conclusions from microarray expression data. *Bioinformatics* **18**, 961–970.
27. Brown, M. P., Grundy, W. N., Lin, D., et al. (2000) Knowledge-based analysis of microarray gene expression data by using support vector machines. *Proc. Natl. Acad. Sci. USA* **97**, 262–267.

28. Khan, J., Wei, J. S., Ringner, M., et al. (2001) Classification and diagnostic prediction of cancers using gene expression profiling and artificial neural networks. *Nat Med* **7**, 673–679.
29. Ringner, M., Peterson, C., and Khan, J. (2002) Analyzing array data using supervised methods. *Pharmacogenomics* **3**, 403–415.
30. Zhang, H., Yu, C. Y., Singer, B., and Xiong, M. (2001) Recursive partitioning for tumor classification with gene expression microarray data. *Proc. Natl. Acad. Sci. USA* **98**, 6730–6735.
31. Zeeberg, B. R., Feng, W., Wang, G., et al. (2003) GoMiner: a resource for biological interpretation of genomic and proteomic data. *Genome Biol.* **4**, R28.
32. Dennis, G., Jr., Sherman, B. T., Hosack, D. A., et al. (2003) DAVID: Database for Annotation, Visualization, and Integrated Discovery. *Genome Biol.* **4**, P3.
33. Demir, E., Babur, O., Dogrusoz, U., et al. (2002) PATIKA: an integrated visual environment for collaborative construction and analysis of cellular pathways. *Bioinformatics* **18**, 996–1003.
34. Raponi, M., Belly, R. T., Karp, J. E., Lancet, J. E., Atkins, D., and Wang, Y. (2004) Microarray analysis reveals genetic pathways modulated by tipifarnib in acute myeloid leukemia. *BMC Cancer* **4**, 56.
35. Jenson, S. D., Robetorye, R. S., Bohling, S. D., et al. (2003) Validation of cDNA microarray gene expression data obtained from linearly amplified RNA. *Mol. Pathol.* **56**, 307–312.
36. Winer, J., Jung, C. K., Shackel, I., and Williams, P. M. (1999) Development and validation of real-time quantitative reverse transcriptase-polymerase chain reaction for monitoring gene expression in cardiac myocytes in vitro. *Anal. Biochem.* **270**, 41–49.
37. Kononen, J., Bubendorf, L., Kallioniemi, A., et al. (1998) Tissue microarrays for high-throughput molecular profiling of tumor specimens. *Nat. Med.* **4**, 844–847.

2

Genomic Signal Processing: From Matrix Algebra to Genetic Networks

Orly Alter

Summary

DNA microarrays make it possible, for the first time, to record the complete genomic signals that guide the progression of cellular processes. Future discovery in biology and medicine will come from the mathematical modeling of these data, which hold the key to fundamental understanding of life on the molecular level, as well as answers to questions regarding diagnosis, treatment, and drug development. This chapter reviews the first data-driven models that were created from these genome-scale data, through adaptations and generalizations of mathematical frameworks from matrix algebra that have proven successful in describing the physical world, in such diverse areas as mechanics and perception: the singular value decomposition model, the generalized singular value decomposition model comparative model, and the pseudoinverse projection integrative model. These models provide mathematical descriptions of the genetic networks that generate and sense the measured data, where the mathematical variables and operations represent biological reality. The variables, patterns uncovered in the data, correlate with activities of cellular elements such as regulators or transcription factors that drive the measured signals and cellular states where these elements are active. The operations, such as data reconstruction, rotation, and classification in subspaces of selected patterns, simulate experimental observation of only the cellular programs that these patterns represent. These models are illustrated in the analyses of RNA expression data from yeast and human during their cell cycle programs and DNA-binding data from yeast cell cycle transcription factors and replication initiation proteins. Two alternative pictures of RNA expression oscillations during the cell cycle that emerge from these analyses, which parallel well-known designs of physical oscillators, convey the capacity of the models to elucidate the design principles of cellular systems, as well as guide the design of synthetic ones. In these analyses, the power of the models to predict previously unknown biological principles is demonstrated with a prediction of a novel mechanism of regulation that correlates DNA replication initiation with cell cycle-regulated RNA transcription in yeast. These models may become the foundation of a future in which biological systems are modeled as physical systems are today.

From: *Methods in Molecular Biology, vol. 377, Microarray Data Analysis: Methods and Applications*
Edited by: M. J. Korenberg © Humana Press Inc., Totowa, NJ

Key Words: Singular value decomposition (SVD); generalized SVD (GSVD); pseudoinverse projection; blind source separation (BSS) algorithms; genome-scale RNA expression and proteins' DNA-binding data; cell cycle; yeast *Saccharomyces cerevisiae;* human HeLa cell line; analog harmonic and digital ring oscillators.

1. Introduction

1.1. DNA Microarray Technology and Genome-Scale Molecular Biological Data

The Human Genome Project, and the resulting sequencing of complete genomes, fueled the emergence of the DNA microarray hybridization technology in the past decade. This novel experimental high-throughput technology makes it possible to assay the hybridization of fluorescently tagged DNA or RNA molecules, which were extracted from a single sample, with several thousand synthetic oligonucleotides *(1)* or DNA targets *(2)* simultaneously. Different types of molecular biological signals, such as DNA copy number, RNA expression levels, and DNA-bound proteins' occupancy levels, that correspond to activities of cellular systems, such as DNA replication, RNA transcription, and binding of transcription factors to DNA, can now be measured on genomic scales (e.g., **refs.** *3* and *4*). For the first time in human history it is possible to monitor the flow of molecular biological information, as DNA is transcribed to RNA, RNA is translated to proteins, and proteins bind to DNA, and thus to observe experimentally the global signals that are generated and sensed by cellular systems. Already laboratories all over the world are producing vast quantities of genome-scale data in studies of cellular processes and tissue samples (e.g., **refs.** *5–9*).

Analysis of these new data promises to enhance the fundamental understanding of life on the molecular level and might prove useful in medical diagnosis, treatment, and drug design. Comparative analysis of these data among two or more organisms promises to give new insights into the universality as well as the specialization of evolutionary, biochemical, and genetic pathways. Integrative analysis of different types of these global signals from the same organism promises to reveal cellular mechanisms of regulation, i.e., global causal coordination of cellular activities.

1.2. From Technology and Large-Scale Data to Discovery and Control of Basic Phenomena Using Mathematical Models: Analogy From Astronomy

Biology and medicine today, with these recent advances in DNA microarray technology, may very well be at a point similar to where physics was after the advent of the telescope in the 17th century. In those days, astronomers were compiling tables detailing observed positions of planets at different times for navigation. Popularized by Galileo Galilei, telescopes were being used in these sky surveys, enabling more accurate and more frequent observations of a growing number of celestial bodies. One astronomer, Tycho Brahe, compiled some of the more extensive and accurate tables of such astronomical observations. Another astronomer, Johannes Kepler, used mathematical equations from analytical geometry to describe trends in Brahe's data, and to determine three laws of planetary motion, all relating observed time intervals with observed distances. These laws enabled the most accurate predictions of future positions of planets to date. Kepler's achievement posed the question: why are the planetary motions such that they follow these laws? A few decades later, Isaac Newton considered this question in light of the experiments of Galileo, the data of Brahe, and the models of Kepler. Using mathematical equations from calculus, he introduced the physical observables mass, momentum, and force, and defined them in terms of the observables time and distance. With these postulates, the three laws of Kepler could be derived within a single mathematical framework, known as the universal law of gravitation, and Newton concluded that the physical phenomenon of gravitation is the reason for the trends observed in the motion of the planets *(10)*. Today, Newton's discovery and mathematical formulation of the basic phenomenon that is gravitation enables control of the dynamics of moving bodies, e.g., in exploration of outer space.

The rapidly growing number of genome-scale molecular biological datasets hold the key to the discovery of previously unknown molecular biological principles, just as the vast number of astronomical tables compiled by Galileo and Brahe enabled accurate prediction of planetary motions and later also the discovery of universal gravitation. Just as Kepler and Newton made their discoveries by using mathematical frameworks to describe trends in these large-scale astronomical data, also future predictive power, discovery, and control in biology and medicine will come from the mathematical modeling of genome-scale molecular biological data.

1.3. From Complex Signals to Simple Principles Using Mathematical Models: Analogy From Neuroscience

Genome-scale molecular biological signals appear to be complex, yet they are readily generated and sensed by the cellular systems. For example, the division cycle of human cells spans an order of one day only of cellular activity. The period of the cell division cycle in yeast is of the order of an hour.

DNA microarray data or genomic-scale molecular biological signals, in general, may very well be similar to the input and output signals of the central nervous system, such as images of the natural world that are viewed by the retina and the electric spike trains that are produced by the neurons in the visual cortex. In a series of classic experiments, the neuroscientists Hubel and Wiesel *(11)* recorded the activities of individual neurons in the visual cortex in response to different patterns of light falling on the retina. They showed that the visual cortex represents a spatial map of the visual field. They also discovered that there exists a class of neurons, which they called "simple cells," each of which responds selectively to a stimulus of an edge of a given scale at a given orientation in the neuron's region of the visual field. These discoveries posed the question: what might be the brain's advantage in processing natural images with a series of spatially localized scale-selective edge detectors? Barlow *(12)* suggested that the underlying principle of such image processing is that of sparse coding, which allows only a few neurons out of a large population to be simultaneously active when representing any image from the natural world. Naturally, such images are made out of objects and surfaces, i.e., edges. Two decades later, Olshausen and Field *(13; see also* Bell and Sejnowski, **ref.** *14)* developed a novel algorithm, which separates or decomposes natural images into their optimal components, where they defined optimality mathematically as the preservation of a characteristic ensemble of images as well as the sparse representation of this ensemble. They showed that the optimal sparse linear components of a natural image are spatially localized and scaled edges, thus validating Barlow's postulate.

The sensing of the complex genomic-scale molecular biological signals by the cellular systems might be governed by simple principles, just as the processing of the complex natural images by the visual cortex appear to be governed by the simple principle of sparse coding. Just as the natural images could be represented mathematically as superpositions, i.e., weighted sums of images, which correlate with the measured sensory activities of neurons, also the complex genomic-scale molecular biological signals might be represented mathematically as superpositions of signals, which might correspond to the measured activities of cellular elements.

1.4. Matrix Algebra Models for DNA Microarray Data

This chapter reviews the first data-driven predictive models for DNA microarray data or genomic-scale molecular biological signals in general. These models use adaptations and generalizations of matrix algebra frameworks *(15)* in order to provide mathematical descriptions of the genetic networks that generate and sense the measured data. The singular value decomposition (SVD) model formulates a dataset as the result of a simple linear network (**Fig. 1A**): the measured gene patterns are expressed mathematically as superpositions of the effects of a few independent sources, biological or experimental, and the

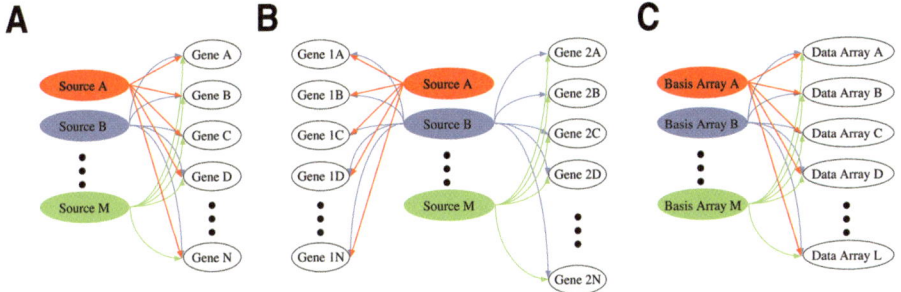

Fig. 1. The first data-driven predictive models for DNA microarray data. (**A**) The singular value decomposition (SVD) model describes the overall observed genome-scale molecular biological data as the outcome of a simple linear network, where a few independent sources, experimental or biological, and the corresponding cellular states, affect all the genes and arrays, i.e., samples, in the dataset. (**B**) The generalized SVD (GSVD) comparative model describes the two genome-scale molecular biological datasets as the outcome of a simple linear comparative network, where a few independent sources, some common to both datasets whereas some are exclusive to one dataset or the other, affect all the genes in both datasets. (**C**) The pseudoinverse projection integrative model approximates any number of datasets as the outcome of a simple linear integrative network, where the cellular states, which correspond to one chosen "basis" set of observed samples, affect all the samples, or arrays, in each dataset.

measured sample patterns, as superpositions of the corresponding cellular states *(16–18)*. The comparative generalized SVD (GSVD) model formulates two datasets, e.g., from two different organisms such as yeast and human, as the result of a simple linear comparative network (**Fig. 1B**): the measured gene patterns in each dataset are expressed mathematically simultaneously as superpositions of a few independent sources that are common to both datasets, as well as sources that are exclusive to one of the datasets or the other *(19)*. The integrative pseudoinverse projection model approximates any number of datasets from the same organism, e.g., of different types of data such as RNA expression levels and proteins' DNA-binding occupancy levels, as the result of a simple linear integrative network (**Fig. 1C**): the measured sample patterns in each dataset are formulated simultaneously as superpositions of one chosen set of measured samples, or of profiles extracted mathematically from these samples, designated the "basis" set *(20,21)*.

The mathematical variables of these models, i.e., the patterns that these models uncover in the data, represent biological or experimental reality. The "eigengenes" uncovered by SVD, the "genelets" uncovered by GSVD, and the pseudoinverse correlations uncovered by pseudoinverse projection, correlate with independent processes, biological or experimental, such as observed genome-wide effects of known regulators or transcription factors, the cellular elements that generate the genome-wide RNA expression signals most commonly measured by DNA microarrays. The corresponding "eigenarrays" uncovered by SVD and "arraylets" uncovered by GSVD, correlate with the corresponding cellular states, such as measured samples in which these regulators or transcription factors are overactive or underactive.

The mathematical operations of these models, e.g., data reconstruction, rotation, and classification in subspaces spanned by these patterns also represent biological or experimental reality. Data reconstruction in subspaces of selected eigengenes, genelets, or pseudoinverse correlations, and corresponding eigenarrays or arraylets, simulates experimental observation of only the processes and cellular states that these patterns represent, respectively. Data rotation in these subspaces simulates the experimental decoupling of the biological programs that these subspaces span. Data classification in these subspaces maps the measured gene and sample patterns onto the processes and cellular states that these subspaces represent, respectively.

Because these models provide mathematical descriptions of the genetic networks that generate and sense the measured data, where the mathematical variables and operations represent biological or experimental reality, these models have the capacity to elucidate the design principles of cellular systems as well as guide the design of synthetic ones (e.g., **ref. 22**). These models also have the power to make experimental predictions that might lead to experiments in which the models can be refuted or validated, and to discover previously unknown molecular biological principles *(21,23)*. Ultimately, these models might enable the control of biological cellular processes in real time and in vivo *(24)*.

Although no mathematical theorem promises that SVD, GSVD, and pseudoinverse projection could be used to model DNA microarray data or genome-scale molecular biological signals in general, these results are not counterintuitive. Similar and related mathematical frameworks have already proven successful in describing the physical world, in such diverse areas as mechanics and perception *(25)*.

First, SVD, GSVD, and pseudoinverse projection, interpreted as they are here as simple approximations of the networks or systems that generate and sense the processed signals, belong to a class of algorithms called blind source separation (BSS) algorithms. BSS algorithms, such as the linear sparse coding algorithm by Olshausen and Field *(13)*, the independent component analysis by Bell and Sejnowski *(14)* and the neural network algorithms by Hopfield *(26)*, separate or decompose measured signals into their mathematically defined optimal components. These algorithms have already proven successful in modeling natural signals and computationally mimicking the activity of the brain as it expertly perceives these signals, for example, in face recognition *(27,28)*.

Second, SVD, GSVD, and pseudoinverse projection can be also thought of as generalizations of the eigenvalue decomposition (EVD) and generalized EVD (GEVD) of Hermitian matrices, and inverse projection onto an orthogonal matrix, respectively. In mechanics, EVD of the Hermitian matrix, which tabulates the energy of a system of coupled oscillators, uncovers the eigenmodes and eigenfrequencies of this system, i.e., the normal coordinates, which oscillate independently of one another, and their frequencies of oscillations. One of these eigenmodes represents the center of mass of the system. GEVD of the Hermitian matrices, which tabulate the kinetic and potential energies of the oscillators, compares the distribution of kinetic energy among the eigenmodes with that of the potential energy. The inverse projection onto the orthogonal matrix, which tabulates the eigenmodes of this system, is equivalent to transformation of coordinates to the frame of reference, which is oscillating with the system (e.g., **ref. *29***). SVD, GSVD, and pseudoinverse projection are, therefore, generalizations of the frameworks that underlie the mathematical theoretical description of the physical world.

In this chapter, the mathematical frameworks of SVD, GSVD, and pseudoinverse projection are reviewed with an emphasis on the mathematical definition of the optimality of the components, or patterns, that each algorithm uncovers in the data. These models are illustrated in the analyses of RNA expression data from yeast and human during their cell cycle programs and DNA-binding data from yeast cell cycle transcription factors and replication initiation proteins. The correspondence between the mathematical frameworks and the genetic networks that generate and sense the measured data is outlined in each case, focusing on the correlations between the mathematical patterns and the observed cellular programs, as well as between the mathematical operations in subspaces spanned by selected patterns and the experimental observation of the cellular programs. Two alternative pictures of RNA expression oscillations during the cell cycle that emerge from these analyses are considered, and parallels between these pictures and well-known designs of physical oscillators, namely the analog harmonic oscillator and the digital ring oscillator, are drawn to convey the capacity of the models to elucidate the design principles of cellular systems, as well as guide the design of synthetic ones. Finally, the power of these models to predict previously unknown biological principles is demonstrated with a prediction of a novel mechanism of regulation that correlates DNA replication initiation with cell cycle-regulated RNA transcription in yeast.

2. SVD for Modeling DNA Microarray Data

This section reviews the SVD model for DNA microarray data (*16–18, 22–24*). SVD is a BSS algorithm that decomposes the measured signal, i.e., the measured gene and array patterns of, e.g. RNA expression, into mathematically decorrelated and decoupled patterns, the "eigengenes" and "eigenarrays." The correspondence between these mathematical patterns uncovered in the measured signal and the independent biological and experimental processes and cellular states that compose the signal is illustrated with an analysis of genome-scale RNA expression data from the yeast *Saccharomyces cerevisiae* during its cell cycle program *(6)*. The picture of RNA expression oscillations during the yeast cell cycle that emerges from this analysis suggests an underlying genetic network or circuit that parallels the analog harmonic oscillator.

2.1. Mathematical Framework of SVD

Let the matrix $\hat{e}$ of size N-genes $\times M$-arrays tabulate the genome-scale signal, e.g., RNA expression levels, measured in a set of M samples using M DNA microarrays. The vector in the mth column of the matrix $\hat{e}$, $|a_m\rangle \equiv \hat{e}|m\rangle$, lists the expression signal measured in the mth sample by the mth array across all N genes simultaneously. The vector in the nth row of the matrix $\hat{e}, \langle g_n| \equiv \langle n|\hat{e}$, lists the signal measured for the nth gene across the different arrays, which correspond to the different samples.*

SVD is a linear transformation of this DNA microarray dataset from the N-genes $\times M$-arrays space to the reduced L-eigenarrays $\times L$-eigengenes space (**Fig. 2**), where $L = \min\{M,N\}$,

$$\hat{e} = \hat{u}\hat{\varepsilon}\hat{v}^T. \tag{1}$$

In this space, the dataset or matrix $\hat{e}$ is represented by the diagonal nonnegative matrix $\hat{\varepsilon}$ of size L-eigenarrays $\times L$-eigengenes. The diagonality of $\hat{\varepsilon}$ means that each eigengene is decoupled of all other eigengenes, and each eigenarray is decoupled of all other eigenarrays, such that each eigengene is expressed only in the corresponding eigenarray.

The "fractions of eigenexpression" $\{p_l\}$ are calculated from the "eigenexpression levels" $\{\varepsilon_l\}$, which are listed in the diagonal of $\hat{\varepsilon}$,

$$p_l = \frac{\varepsilon_l^2}{\sum_{k=1}^{L}\varepsilon_k^2}. \tag{2}$$

These fractions of eigenexpression indicate for each eigengene and eigenarray their significance in the dataset relative to all other eigengenes and eigenarrays in terms of the overall expression information that they capture in the data. Note that each fraction of eigenexpression can be thought of as the probability for any given gene among all genes in the dataset to express the corresponding

---

*In this chapter, $\hat{m}$ denotes a matrix, $|v\rangle$ denotes a column vector, and $\langle u|$ denotes a row vector, such that, $\hat{m}|v\rangle$, $\langle u|\hat{m}$, and $\langle u|v\rangle$ all denote inner products and $|v\rangle\langle u|$ denotes an outer product.

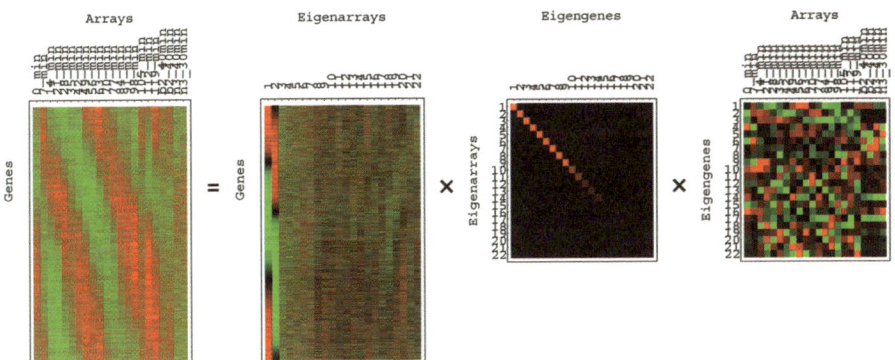

Fig. 2. Raster display of the SVD of the yeast cell cycle RNA expression dataset, with overexpression (red), no change in expression (black), and underexpression (green) around the steady state of expression of the 4579 yeast genes. SVD is a linear transformation of the data from the 4579-genes × 22-arrays space to the reduced diagonalized 22-eigenarrays × 22-eigengenes space, which is spanned by the 4579-genes × 22-eigenarrays and 22-eigengenes × 22-arrays bases.

eigengene, and at the same time, the probability for any given array among all arrays to express the corresponding eigenarray.

The "normalized Shannon entropy" of the dataset,

$$0 \leq d = -\frac{1}{L} \sum_{k=1}^{L} p_k \log(p_k) \leq 1, \tag{3}$$

measures the complexity of the data from the distribution of the overall expression information between the different eigengenes and corresponding eigenarrays, where $d = 0$ corresponds to an ordered and redundant dataset in which all expression is captured by one eigengene and the corresponding eigenarray, and $d = 1$ corresponds to a disordered and random dataset where all eigengenes and eigenarrays are equally expressed.

The transformation matrices $\hat{u}$ and $\hat{v}^T$ define the N-genes × L-eigenarrays and the L-eigengenes × M-arrays basis sets, respectively. The vector in the lth column of the matrix $\hat{u}$, $|\alpha_l\rangle \equiv \hat{u}|l\rangle$, lists the genome-scale expression signal of the lth eigenarray. The vector in the lth row of the matrix $\hat{v}^T$, $\langle\gamma_l| \equiv \langle l|\hat{v}^T$, lists signal of the lth eigengene across the different arrays. The eigengenes and eigenarrays are orthonormal superpositions of the genes and arrays, such that the transformation matrices $\hat{u}$ and $\hat{v}^T$ are both orthogonal,

$$\hat{u}^T \hat{u} = \hat{v}^T \hat{v} = \hat{I}, \tag{4}$$

where $\hat{I}$ is the identity matrix. The signal of each eigengene and eigenarray is, therefore, not only decoupled but also decorrelated from that of all other eigengenes and eigenarrays, respectively. The eigengenes and eigenarrays are unique up to phase factors of ±1 for a real data matrix $\hat{e}$, such that each eigengene and eigenarray captures both parallel and antiparallel gene and array expression patterns, except in degenerate subspaces, defined by subsets of equal eigenexpression levels. SVD is, therefore, data driven, except in degenerate subspaces.

2.2. SVD Analysis of Cell Cycle RNA Expression Data From Yeast

In this example, SVD is applied to a dataset that tabulates RNA expression levels of 4579 genes in 22 yeast samples, 18 samples of a time course monitoring the cell cycle in an α factor-synchronized culture, and two samples each of yeast strains where the genes *CLN3* and *CLB2*, which encode G_1 and G_2/M cyclins, respectively, are overexpressed or overactivated (*6*).

2.2.1. Significant Eigengenes and Corresponding Eigenarrays Correlate With Genome-Scale Effects of Independent Sources of Expression and Their Corresponding Cellular States

Consider the 22 eigengenes of the α factor, *CLB2*, and *CLN3* dataset (**Fig. 3A**). The first eigengene, which captures about 80% of the overall expression signal (**Fig. 3B**), and describes sample-invariant expression, is inferred to represent steady-state expression (**Fig. 3C**). The second and third eigengenes, which capture about 9.5% and 2% of the overall expression signal, respectively, describe initial transient increase and decrease in expression, respectively, superimposed on time-invariant expression during the cell cycle. These eigengenes are inferred to represent the responses to synchronization by the pheromone α factor. The fourth through ninth and 11th eigengenes, which capture together about 5% of the overall expression information, show expression oscillations of two periods during the α factor-synchronized cell cycle, and are inferred to represent cell cycle expression oscillations (**Fig. 3D–F**).

The corresponding eigenarrays are associated with the corresponding cellular states. An eigenarray is parallel and antiparallel associated with the most likely parallel and antiparallel cellular states, or none thereof, according to the annotations of the two groups of *n* genes each, with largest and smallest levels of signal, e.g., expression, in this eigenarray among all *N* genes, respectively. A coherent biological theme might be reflected in the annotations of either one of these two groups of genes. The *p*-value of a given association by annotation is calculated using combinatorics and assuming hypergeometric probability distribution of the *K* annotations among the *N* genes, and of the subset of $k \subseteq K$ annotations among the subset of $n \subseteq N$ genes,

$$P(k;n,N,K) = \binom{N}{n}^{-1} \sum_{l=k}^{n} \binom{K}{l}\binom{N-K}{n-l},$$

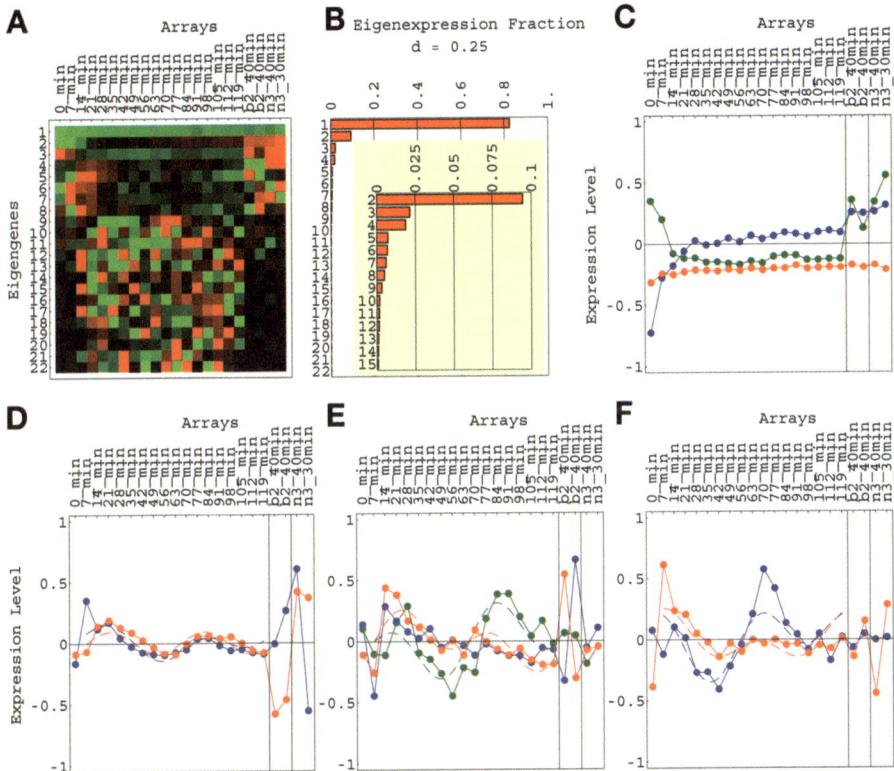

Fig. 3. The eigenegenes of the yeast cell cycle RNA expression dataset. (A) Raster display of the expression of 22 eigengenes in 22 arrays. (B) Bar chart of the fractions of eigenexpression, showing that the first eigengene captures about 80% of the overall relative expression. (C) Line-joined graphs of the expression levels of the first eigene-gene (red), which represents the steady expression state, and the second (blue) and third (green) eigengenes, which represent responses to synchronization of the yeast culture by α factor. (D) Expression levels of the fourth (red) and seventh (blue) eigengenes, (E) the fifth (red), eighth (blue), and 11th (green) eigengenes, and (F) the sixth (red) and ninth (blue) eigengenes, all fit dashed graphs of sinusoidal functions of two periods superimposed on sinusoidal functions of one period during the time course.

where
$$\binom{N}{n} = N!\,n!^{-1}(N-n)!^{-1}$$

is the Newton binomial coefficient *(30)*. The most likely association of an eigen-array with a cellular state is defined as the association that corresponds to the smallest *p*-value.

Following the p-values for the distribution of the 364 genes, which were microarray-classified as α factor regulated *(31)* and that of the 646 genes, which were traditionally or microarray-classified as cell cycle-regulated *(6)* among all 4579 genes and among each of the subsets of 200 genes with the largest and smallest levels of expression, respectively, the second and third eigenarrays are associated with the cellular states of the α factor response program, whereas the fourth through ninth and 11th eigenarrays are associated with the cellular states of the cell cycle program.

2.2.2. Filtering Out of Eigengenes and Eigenarrays Simulates the Experimental Suppression of the Cellular Processes and States That These Eigengenes and Eigenarrays Represent

Any eigengene $\langle\gamma_l|$ and corresponding eigenarray $|\alpha_l\rangle$ can be filtered out, without eliminating genes or arrays from the dataset, by setting their corresponding eigenexpression level in $\hat{e}$ to zero, $\varepsilon_l = 0$, and reconstructing the dataset according to **Eq. 1**, such that $\hat{e} \rightarrow \hat{e} - \varepsilon_l|\alpha_l\rangle\langle\gamma_l|$. The α factor, *CLB2*, and *CLN3* dataset is normalized by filtering out the first eigengene, which represents the additive steady-state expression level, the second and third eigengenes, which represent the α factor synchronization response, as well as the 10th and 12th through 22nd eigengenes. After filtering out the first eigengene, the expression pattern of each gene is approximately centered at its time-invariant level. Similarly, the expression of each gene is then approximately normalized by its steady scale of variance *(16,17)*. The normalized dataset tabulates for each gene an expression pattern that is of an approximately zero arithmetic mean, with a variance which is of an approximately unit geometric mean.

Consider the eigengenes of the normalized α factor, *CLB2*, and *CLN3* dataset (**Fig. 4A**). The first, second, and third normalized eigengenes, which are of similar significance, capture together about 60% of the overall normalized expression (**Fig. 4B**). Their time variations fit normalized sine and cosine functions of two periods superimposed on a normalized sine function of one period during the cell cycle (**Fig. 4C**). Although the first and third normalized eigengenes describe underexpression in both *CLB2*-overactive arrays, and overexpression in both *CLN3*-overactive arrays, the second normalized eigengene describes the antiparallel expression pattern of overexpression in both *CLB2*-overactive arrays and underexpression in both *CLN3*-overactive arrays. These normalized eigengenes are inferred to represent expression oscillations during the cell cycle superimposed on differential expression because of *CLB2* and *CLN3* overactivations. The corresponding eigenarrays are associated by annotation with the corresponding cellular states.

None of the significant eigengenes and eigenarrays of the normalized dataset represents either the steady-state expression or the response to the α factor

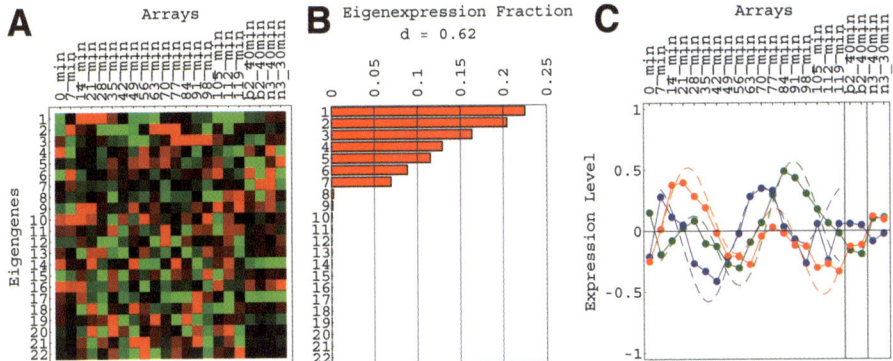

Fig. 4. The eigengenes of the normalized yeast cell cycle RNA expression dataset. (A) Raster display. (B) Bar chart of the fractions of eigenexpression, showing that the first, second, and third normalized eigengenes capture approximately 20% of the overall normalized expression information each, and span an approximately degenerate subspace. (C) Line-joined graphs of the expression levels of the first (red), second (blue), and third (green) normalized eigengenes, fit dashed graphs of two-period sinusoidal functions superimposed on one-period sinusoidal functions during the time course.

synchronization. The normalized dataset simulates an experimental measurement of only the cell cycle program and the differential expression in response to overactivation of *CLB2* and *CLN3*.

2.2.3. Rotation in an Almost Degenerate Subspace Simulates Experimental Decoupling of the Biological Programs the Subspace Spans

The almost degenerate subspaces spanned by the first, second, and third eigengenes and corresponding eigenarrays are approximated with degenerate subspaces, by setting each of the corresponding eigenexpression levels equal, $\varepsilon_1, \varepsilon_2, \varepsilon_3 \rightarrow \sqrt{(\varepsilon_1^2 + \varepsilon_2^2 + \varepsilon_3^2)/3}$, and reconstructing the dataset according to **Eq. 1**. With this approximation, the three eigengenes and corresponding eigenarrays can be rotated, such that the same expression subspaces that are spanned by these eigenegenes, and eigenarrays will be spanned by three orthogonal superpositions of these eigengenes and eigenarrays, i.e., by three rotated eigengenes and eigenarrays. Requiring two of these three rotated eigengenes to describe equal expression in the *CLB2*-overactive samples as in the *CLN3*-overactive samples, so that only the one remaining rotated eigengene captures the differential expression between these two sets of arrays, gives unique angles of rotations in the three-dimensional subspaces of eigengenes and eigenarrays, and therefore also unique rotated eigengenes and eigenarrays.

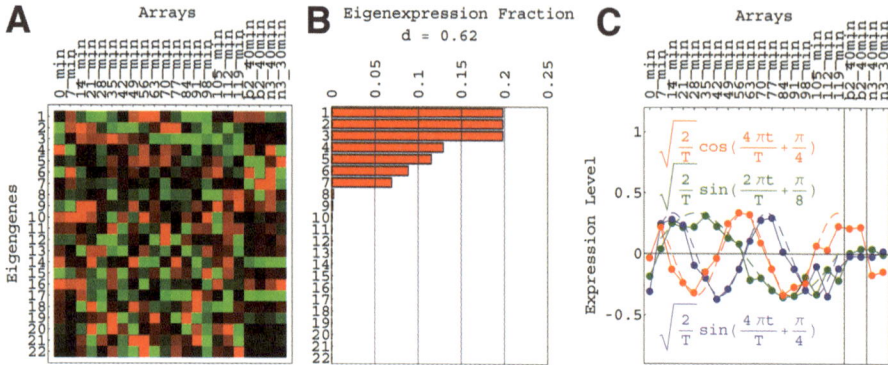

Fig. 5. The rotated eigengenes of the normalized yeast cell cycle RNA expression dataset. (A) Raster display. (B) Bar chart of the fractions of eigenexpression, showing that the first, second, and third rotated eigengenes span an exactly degenerate subspace. (C) Line-joined graphs of the expression levels of the first (red) and second (blue) rotated eigengenes fit normalized sine and cosine functions of two periods, and the third rotated eigengene (green) fits a normalized sine of one period during the time course.

Consider the eigengenes of the normalized and rotated α factor, *CLB2*, and *CLN3* dataset (**Fig. 5A**), where the first, second, and third fractions of eigenexpression are approximated to be equal (**Fig. 5B**). The time variations of the first and second rotated eigengenes fit normalized sine and cosine functions of two periods during the cell cycle (**Fig. 5C**). The time variation of the third rotated eigengene fits a normalized sine function of one period during the cell cycle, suggesting differences in expression between the two successive cell cycle periods, which may be due to dephasing of the initially synchronized yeast culture. Although the second and third rotated eigengenes describe steady-state expression in the *CLB2*- and *CLN3*-overactive arrays, the first rotated eigengene describes underexpression in the *CLB2*-overactive arrays and overexpression in the *CLN3*-overactive arrays. The first rotated eigengene, therefore, is inferred to represent cell cycle expression oscillations that are *CLB2*- and *CLN3*-dependent, whereas the second rotated eigengene is inferred to represent cell cycle expression oscillations that are *CLB2*- and *CLN3*-independent. The third rotated eigengene is inferred to represent variations in the cell cycle expression from the first period to the second, which also appear to be *CLB2*- and *CLN3*-independent. The first, second, and third rotated eigenarrays are associated by annotation with the corresponding cellular states.

The rotation of the data, therefore, simulates decoupling of the differential expression owing to *CLB2* and *CLN3* overactivation from at least one of the cell cycle stages. It also simulates decoupling of the variation between the first and the second cell cycle periods from the cell cycle stages and from the *CLB2* and *CLN3* overactivation.

2.2.4. Classification of the Normalized Yeast Data According to the Rotated Eigengenes and Eigenarrays Gives a Global Picture of the Dynamics of Cell Cycle Expression

Consider the normalized expression of the 22 α factor, *CLB2*, and *CLN3* arrays in the subspace spanned by the first and second rotated eigenarrays, which represents approximately all cell cycle cellular states (**Fig. 6A**). Sorting the arrays according to their correlations with the second rotated eigenarray along the *y*-axis, $\langle \alpha_2 | a_m \rangle / \sqrt{\langle a_m | a_m \rangle}$, vs that with the first rotated eigenarray along the *x*-axis, $\langle \alpha_1 | a_m \rangle / \sqrt{\langle a_m | a_m \rangle}$, reveals that all except for five arrays have at least 25% of their normalized expression in this subspace. This sorting gives an array order that is similar to that of the cell cycle time-points measured by the arrays, an order that describes the progression of the cell cycle from the M/G1 stage through G_1, S, S/G_2, and G_2/M and back to M/G_1 twice. The first rotated eigenarray is correlated with samples that probe the cellular state of cell cycle transition from G_2/M to M/G_1, which is simulated experimentally by *CLB2* overactivation. This eigenarray is also anticorrelated with the cellular state of transition from G_1 to S, which is simulated by *CLN3* overactivation. Similarly, the second rotated eigenarray is correlated with the transition from M/G_1 to G_1, and anticorrelated with S/G_2, both of which appear to be *CLB2* and *CLN3* independent.

Consider also the normalized expression of the 646 yeast genes in this dataset that were traditionally or microarray-classified as cell cycle regulated (**Fig. 6B**). Sorting the genes according to their correlations with the first and second rotated eigengenes reveals that 551 of these genes have at least 25% of their normalized expression in this subspace. This sorting gives a classification of these genes into the five cell cycle stages, which is in good agreement with both the traditional and microarray classifications. The first rotated eigengene is correlated with the observed expression pattern of *CLB2* and its targets, genes for which expression peaks at the transition from G_2/M to M/G_1. This eigengene is also anticorrelated with the observed expression of *CLN3* and its targets, genes for which expression peaks at the transition from G_1 to S. The second rotated eigengene is correlated with the cell cycle oscillations, which peak at the transition from M/G_1 to G_1 and anticorrelated with these which peak at S/G_2, both of which appear to be independent of the genome-scale effects of *CLB2* and *CLN3*.

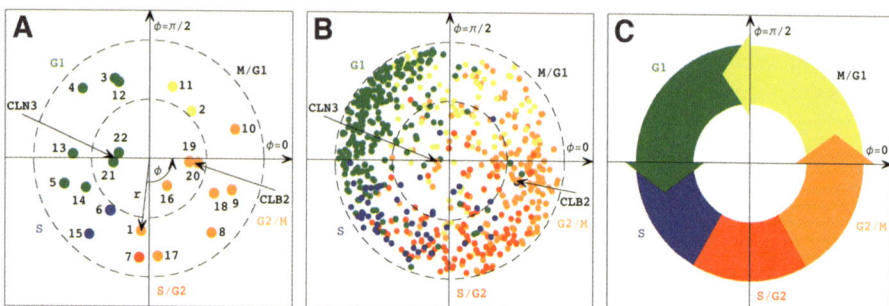

Fig. 6. The normalized yeast RNA expression in the SVD cell cycle subspace. (**A**) Correlations of the normalized expression of each of the 22 arrays with the first and second rotated eigenarrays along the x- and y-axes, color-coded according to the classification of the arrays into the five cell cycle stages: M/G_1 (yellow), G_1 (green), S (blue), S/G_2 (red), and G_2/M (orange). The dashed unit and half-unit circles out-line 100% and 25% of overall normalized array expression in this subspace. (**B**) Correlations of the normalized expression of each of the 646 cell cycle-regulated genes with the first and second rotated eigengenes along the x- and y-axes, color-coded according to either the traditional or microarray classifications. (**C**) The SVD picture of the yeast cell cycle.

Classification of the yeast arrays and genes in the subspaces spanned by these two rotated eigenarrays and corresponding eigengenes gives a picture that resembles the traditional understanding of yeast cell cycle regulation *(32)*: G_1 cyclins, such as *CLN3*, and G_2/M cyclins, such as *CLB2*, drive the cell cycle past either one of two antipodal checkpoints, from G_1 to S and from G_2/M to M/G_1, respectively (**Fig. 6C**).

2.3. SVD Model for Genome-Wide RNA Expression During the Cell Cycle Parallels the Analog Harmonic Oscillator

With all 4579 genes sorted, the normalized cell cycle expression approximately fits a traveling wave, varying sinusoidally across both genes and arrays (**Fig. 7A**). The normalized expression in the *CLB2*- and *CLN3*-overactive arrays approximately fits standing waves, constant across the arrays and varying sinusoidally across the genes only, which appear anticorrelated and correlated with the first eigenarray, respectively. The gene variations of the first and second rotated eigenarrays fit normalized cosine and sine functions of one period across all genes, respectively (**Fig. 7B,C**). In this picture, all 4579 genes, about three-quarters of the yeast genome, appear to exhibit periodic expression during the cell cycle. This picture is in agreement with the recent observation by Klevecz et al. *(33; see also* Li and Klevecz, **ref. 34***)* that DNA replication is gated by genome-wide RNA expression oscillations, which suggests that the whole yeast genome might exhibit expression oscillations during the cell cycle.

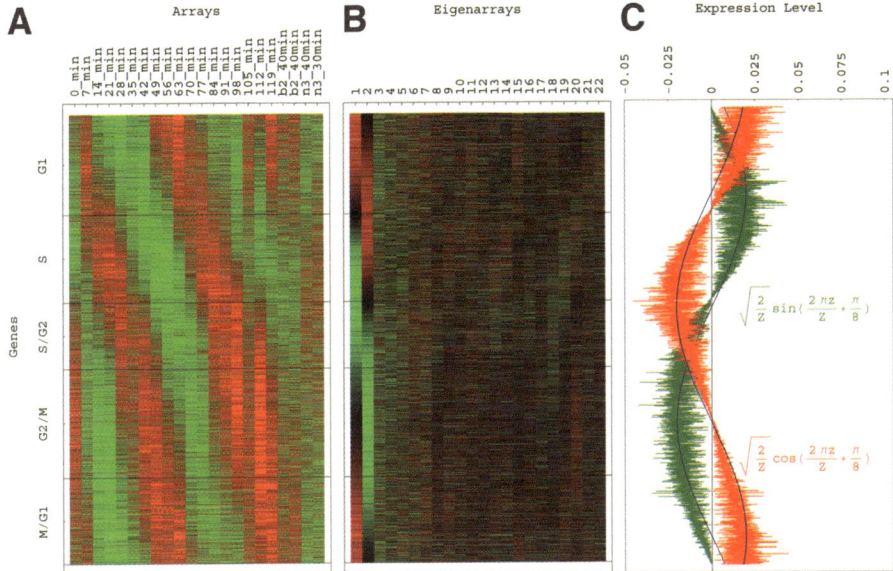

Fig. 7. The sorted and normalized yeast cell cycle RNA expression dataset and its sorted and rotated eigenarrays. (**A**) Raster display of the normalized expression of the 4579 genes across the 22 arrays. The genes are sorted by relative correlation of their normalized expression patterns with the first and second rotated eigengenes. This raster display shows a traveling wave of expression during the cell cycle and standing waves of expression in the *CLB2*- and *CLN3*-overactive arrays. (**B**) Raster display of the rotated eigenarrays, where the expression patterns of the first and second eigenarrays, which correspond to the first and second eigengenes, respectively, display the sorting. (**C**) Line-joined graphs of the first (red) and second (green) rotated eigenarrays, fit normalized cosine and sine functions of one period across all genes.

It is still an open question whether all yeast genes or only a subset of the yeast genes, and if so, which subset, show periodic expression during the cell cycle.

This SVD model describes, to first order, the RNA expression of most of the yeast genome during the cell cycle program as being driven by the activities of two periodically oscillating cellular elements or modules, which are orthogonal, i.e., $\pi/2$ out of phase relative to one another. The underlying genetic network or circuit suggested by this model might be parallel in its design to the analog harmonic oscillator. This well-known oscillator design principle is at the foundations of numerous physical oscillators, including (1) the mechanical pendulum, the position and momentum of which oscillate periodically in time with a phase difference of $\pi/2$; (2) the electronic LC circuit, where the charge on the capacitor and the current flowing through the inductor oscillate periodically in time with a phase difference of $\pi/2$; and (3) the chemical Lotka-Volterra irreversible autocatalytic reaction model, where, far from thermodynamic equilibirum, the concentrations of two intermediate reactants exhibit periodic oscillations in time that are $\pi/2$ out of phase relative to one another *(35–37)*.

3. GSVD for Comparative Modeling of DNA Microarray Datasets

This section reviews the GSVD comparative model for DNA microarray datasets *(19)*. GSVD is a comparative BSS algorithm that simultaneously decomposes two measured signals, i.e., the measured gene and array patterns of, e.g., RNA expression in two organisms, into mathematically decoupled "genelets" and two sets of "arraylets." The correspondence between these mathematical patterns uncovered in the measured signals and the similar and dissimilar among the biological programs that compose each of the two signals is illustrated with a comparative analysis of genome-scale RNA expression data from yeast *(6)* and human *(7)* during their cell cycle programs. One common picture of RNA expression oscillations during both the yeast and human cell cycles emerges from this analysis, which suggests an underlying eukaryotic genetic network or circuit that parallels the digital ring oscillator.

Comparisons of DNA sequence of entire genomes already give new insights into evolutionary, biochemical, and genetic pathways. Recent studies showed that the addition of RNA expression data to DNA sequence comparisons improves functional gene annotation and might expand the understanding of how gene expression and diversity evolved. For example, Stuart et al. *(38)* and independently also Bergmann, Ihmels, and Barkai *(39)* identified pairs of genes for which RNA coexpression is conserved, in addition to their DNA sequences, across several organisms. The evolutionary conservation of the coexpression of these gene pairs confers a selective advantage to the functional relations of these genes. The GSVD comparative model is not limited to genes of conserved DNA sequences, and as such it elucidates universality as well as specialization of molecular biological mechanisms that are truly on genomic scales. For example, the GSVD comparative model might be used to identify genes of common function across different organisms independently of the DNA sequence similarity among these genes, and therefore also to study nonorthologous gene displacement *(40)*.

3.1. Mathematical Framework of GSVD

Let the matrix $\hat{e}_1$ of size N_1-genes $\times$ M_1-arrays tabulate the genome-scale signal, e.g., RNA expression levels, measured in a set of M_1 samples using M_1 DNA microarrays. As before, the mth column vector in the matrix $\hat{e}_1$, $|a_{1,m}\rangle$, lists the expression signal measured in the mth sample by the mth array across all N_1 genes simultaneously. The nth row vector in the matrix $\hat{e}_1$, $\langle g_{1,n}|$, lists the signal measured for the nth gene across the different arrays, which correspond to the different samples. Let the matrix $\hat{e}_2$ of size N_2-genes $\times$ M_2-arrays tabulate the genome-scale signal, e.g., RNA expression levels, measured in a set of M_2 samples under M_2 experimental conditions that correspond one-to-one to the M_1 conditions

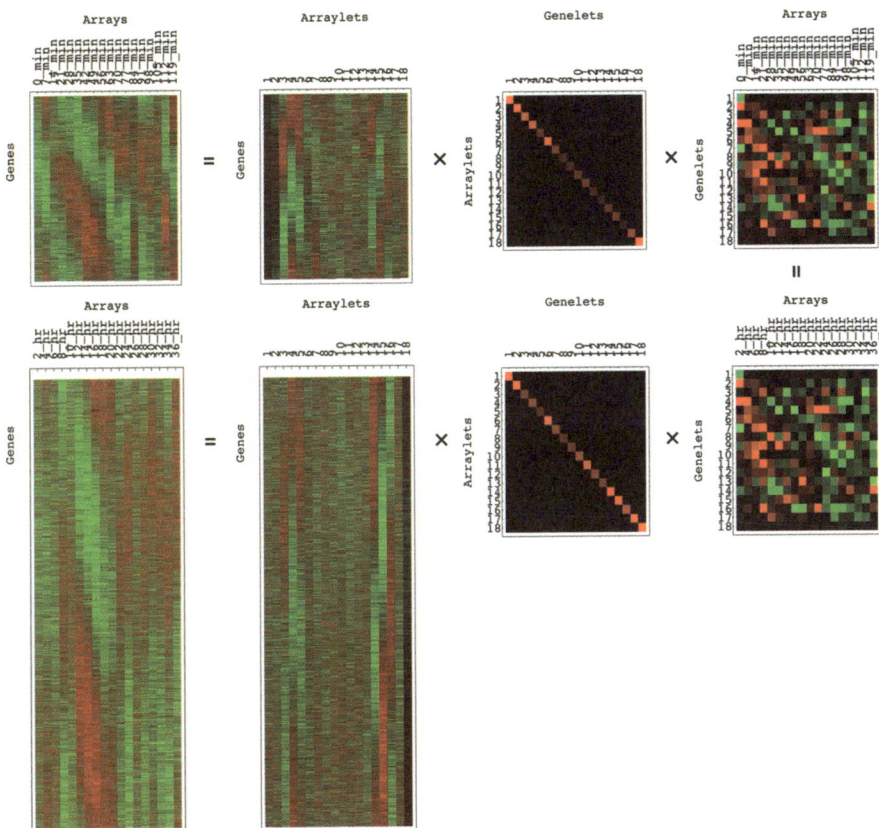

Fig. 8. Raster display of the GSVD of the yeast and human cell cycle RNA expression datasets, with overexpression (red), no change in expression (black), and underexpression (green) centered at the gene- and array-invariant expression of the 4523 yeast and 12,056 human genes. GSVD is a linear transformation of the yeast and human data from the 4523-yeast and 12,056-human genes × 18-arrays spaces to the reduced diagonalized 18-arraylets × 18-genelets spaces, which are spanned by the 4523- and 12,056-genes × 18-arraylets bases, respectively, and by the 18-genelets × 18-arrays shared basis.

underlying $\hat{e}_1$, such that $M_2 = M_1 \equiv M < \max\{N_1, N_2\}$. This one-to-one correspondence between the two sets of conditions is at the foundation of the GSVD comparative analysis of the two datasets, and should be mapped out carefully.

GSVD is a simultaneous linear transformation of the two expression datasets $\hat{e}_1$ and $\hat{e}_2$ from the two N_1-genes × M-arrays and N_2-genes × M-arrays spaces to the two reduced M-arraylets × M-genelets spaces (**Fig. 8**),

$$\hat{e}_1 = \hat{u}_1 \hat{\varepsilon}_1 \hat{x}^{-1},$$
$$\hat{e}_2 = \hat{u}_2 \hat{\varepsilon}_2 \hat{x}^{-1}. \tag{5}$$

In these spaces the data are represented by the diagonal nonnegative matrices ε_1 and ε_2. Their diagonality means that each genelet is decoupled of all other genelets in both datasets simultaneously, such that each genelet is expressed only in the two corresponding arraylets, each of which is associated with one of the two datasets.

The antisymmetric "angular distances" between the datasets $\{\theta_m\}$ are calculated from the "generalized eigenexpression levels" $\{\varepsilon_{1,l}\}$ and $\{\varepsilon_{2,l}\}$, which are listed in the diagonals of ε_1 and ε_2,

$$0 \le \theta_m = \arctan(\varepsilon_{1,m}/\varepsilon_{2,m}) - \pi/4 \le \pi/4 \tag{6}$$

These angular distances indicate the relative significance of each genelet, i.e., its significance in the first dataset relative to that in the second dataset, in terms of the ratio of expression information captured by this genelet in the first dataset to that in the second. An angular distance of 0 indicates a genelet of equal significance in both datasets, with $\varepsilon_{1,m} = \varepsilon_{2,m}$. An angular distance of $\pm\pi/4$ indicates no significance in the second dataset relative to the first, with $\varepsilon_{1,m} \gg \varepsilon_{2,m}$, or in the first dataset relative to the second, with $\varepsilon_{1,m} \ll \varepsilon_{2,m}$, respectively.

The transformation matrix $\hat{x}^{-1}$ defines the M-genelets $\times$ M-arrays basis set, which is shared by both datasets. The transformation matrices $\hat{u}_1$ and $\hat{u}_2$ define the N_1-genes $\times$ M-arraylets and N_2-genes $\times$ M-arraylets basis sets, that correspond to the first and second datasets, respectively. The mth row vector in $\hat{x}^{-1}$, $\langle \gamma_m | \equiv \langle m | \hat{x}^{-1}$, lists the expression signal of the mth genelet across the different arrays in both datasets simultaneously. The mth column vector in $\hat{u}_1$ or $\hat{u}_2$, $|\alpha_{1,m}\rangle \equiv \hat{u}_1 | m\rangle$ or $|\alpha_{2,m}\rangle \equiv \hat{u}_2 | m\rangle$, lists the genome-scale signal of the mth arraylet of either the first or the second dataset, respectively. The genelets are normalized, but not necessarily orthogonal, superpositions of the genes of the first dataset and, at the same time, also the second dataset. The arraylets of the first or the second datasets are orthonormal superpositions of the arrays of the first and second datasets, respectively. In general, $\hat{x}^{-1}$ is nonorthogonal, while $\hat{u}_1$ and $\hat{u}_2$ are both orthogonal,

$$\hat{x}^{-1}\hat{x} \ne \hat{u}_1^T\hat{u}_1 = \hat{u}_2^T\hat{u}_2 = \hat{I}, \tag{7}$$

where $\hat{I}$ is the identity matrix. The expression of each arraylet of either dataset is, therefore, not only decoupled but also decorrelated from that of all other arraylets of this dataset. The genelets and arraylets are unique up to phase factors of ± 1 for real data matrices $\hat{e}_1$ and $\hat{e}_2$, such that each genelet and arraylet capture both parallel and antiparallel gene and array expression patterns, except in degenerate subspaces, defined by subsets of equal angular distances. GSVD is, therefore, data driven, except in degenerate subspaces.

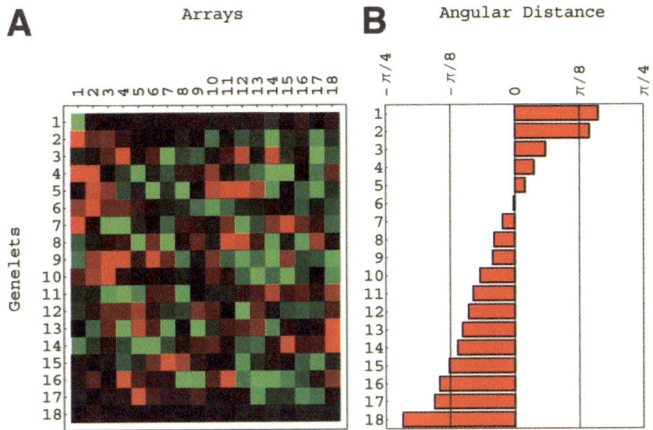

Fig. 9. The genelets of the yeast and human cell cycles RNA expression datasets. (**A**) Raster display of the expression of 18 genelets in the 18 yeast and 18 human arrays, simultaneously, centered at their array-invariant levels. (**B**) Bar chart of the angular distances, showing the first and second genelets highly significant in the yeast data relative to the human data, the third through the sixth and the 14th through the 16th almost equally significant in both datasets, and the 17th and 18th genelets highly significant in the human data relative to the yeast data. All other genelets are neither significant in the yeast data nor in the human data *(19)*.

3.2. GSVD Comparative Analysis of Yeast and Human Cell Cycle RNA Expression Data

In this example, GSVD is applied to two datasets, which tabulate RNA expression of 4523 yeast genes and 12,056 human genes in 18 samples each of time courses of α factor-synchronized yeast culture *(6)* and double thymidine block-synchronized HeLa cell line culture *(7)*, respectively. The yeast and human time courses span more than two and less than two and a half periods in the yeast and human cell cycles, respectively. Both yeast and human time courses are sampled at equal time intervals.

3.2.1. Common Genelets and Corresponding Arraylets Span the Common Yeast and Human Cell Cycle Subspace

Consider the 18 genelets of the yeast and human cell cycle datasets (**Fig. 9A**). Six genelets are almost equally significant in the yeast and human datasets (**Fig. 9B**): The third, fourth, and fifth genelets are slightly more significant in the yeast dataset than in the human dataset, with $0 < \theta_3 < \theta_4 < \theta_5 < \pi/16$. The 14th, 15th and 16th genelets are slightly more significant in the human dataset, with $-\pi/6 < \theta_{14} < \theta_{15} < \theta_{16} < 0$. The time-, i.e., array variations of the third, fourth

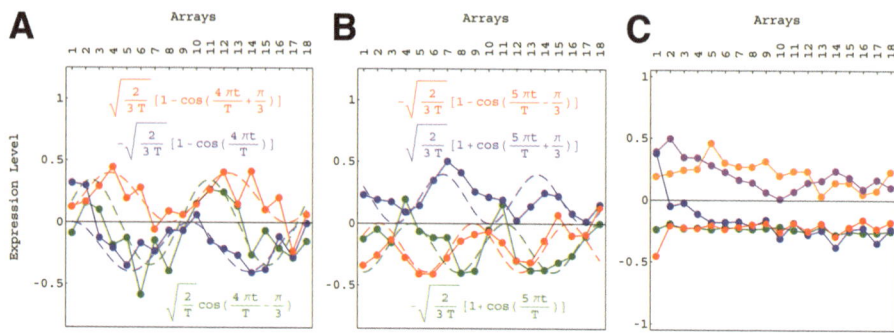

Fig. 10. Line-joined graphs of the expression levels of the significant genelets. (**A**) The third (red), fourth (blue), and fifth (green) genelets, which are associated with the common yeast and human cell cycle gene expression oscillations, fit dashed graphs of normalized cosines of two periods and initial phases of π/3 (red), 0 (blue) and –π/3 (green), respectively. (**B**) The 14th (red), 15th (blue) and 16th (green) genelets, which are also associated with cell cycle gene expression oscillations, fit dashed graphs of normalized cosines of two and a half periods and initial phases of –π/3 (red), π/3 (blue) and 0 (green), respectively. (**C**) The first (red) and second (blue) genelets are associated with the exclusive yeast response to the pheromone α factor, the 17th (orange) and 18th (green) are associated with the exclusive human stress response, and the sixth (violet) is associated with both the yeast and human transitions from synchronization responses into the cell cycle.

and fifth genelets fit normalized cosine functions of two periods and initial phases of π/3, 0 and –π/3, respectively, superimposed on time-invariant expression (**Fig. 10A**). The 14th, 15th and 16th genelets fit normalized cosines of two and a half periods and initial phases of –π/3, π/3, and 0, respectively (**Fig. 10B**). The time variations of the six common genelets suggest that they span the cell cycle subspace, which is common to both the yeast and human genomes, and is manifested in both datasets.

The corresponding six yeast and six human arraylets are associated by annotation with the corresponding yeast and human cell cycle cellular states, following the p-values for the distribution of the 604 yeast genes and 750 human genes, that were microarray-classified, and the 77 yeast genes and 73 human genes that were traditionally classified as cell cycle regulated, among all 4523 yeast and 12,056 human genes and among each of the subsets of 100 genes with largest and smallest levels of expression in each of the arraylets. The associations of the yeast and human arraylets are in agreement with the expression patterns of the genelets, taking into account the initial synchronization of the yeast culture in the cell cycle stage M/G_1 and that of the human culture in S. For example, the expression pattern of the fourth genelet is of 0 initial phase, suggesting that this genelet is correlated with the yeast cell cycle expression oscillations that peak at the stage M/G_1 and the human cell cycle expression

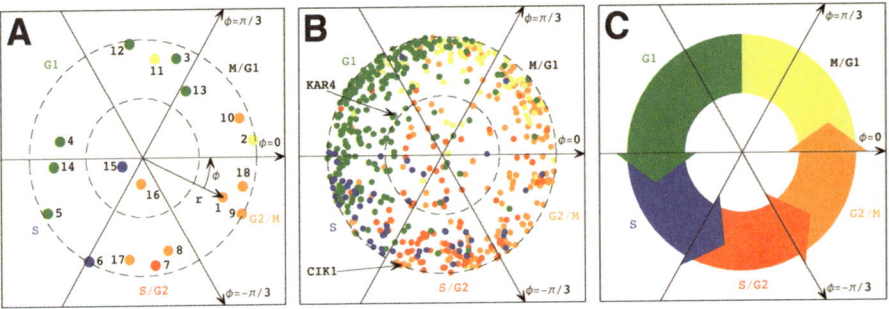

Fig. 11. Reconstructed yeast RNA expression in the GSVD common cell cycle subspace. **(A)** Projections of the expression of each of the 18 arrays, after reconstruction in the six-dimensional GSVD cell cycle subspace, onto the two-dimensional subspace that least-squares approximates it. The arrays are color coded according to their classification into the five cell cycle stages: M/G$_1$ (yellow), G$_1$ (green), S (blue), S/G$_2$ (red), and G$_2$/M (orange). The dashed unit and half-unit circles outline 100% and 50% of added up, rather than cancelled out, contributions of the six arraylets to the overall projected expression. The arrows describe the projections of the $-\pi/3$-, 0-, and $\pi/3$-phase arraylets. **(B)** Projections of the expression of each of the 612 cell cycle-regulated genes, reconstructed in the six-dimensional GSVD subspace, onto the two-dimensional subspace that approximates it. The genes are color coded according to either the traditional or microarray classifications. The expression patterns of *KAR4* and *CIK1* are anticorrelated. **(C)** The GSVD picture of the yeast cell cycle.

oscillations that peak at S. Following the traditional classifications, the corresponding yeast arraylet, i.e., the fourth yeast arraylet, is associated in parallel with the yeast cell cycle stage M/G$_1$, while the fourth human arraylet is associated in parallel with the human cell cycle stage S.

3.2.2. Simultaneous Reconstruction and Classification of the Yeast and Human Data in the Common Subspace Outlines the Biological Similarity in the Regulation of the Yeast and Human Cell Cycle Programs

The six-dimensional genelets subspace that represents the common yeast and human cell cycle expression oscillations is least squares-approximated with a two-dimensional subspace that is spanned by two orthonormal vectors $\langle x|$ and $\langle y|$. Projecting the expression of the 18 yeast arrays from the corresponding six-dimensional yeast arraylets subspace onto the corresponding approximate two-dimensional subspace **(Fig. 11A)** reveals that 50% or more of the contributions of the six arraylets add up, rather than cancel out, in the overall expression of 16 of the arrays. Sorting the arrays in this subspace gives an array order similar to that of the cell cycle time-points measured by the arrays. This order of the arrays describes the yeast cell cycle progression from the M/G$_1$ stage through G$_1$, S,

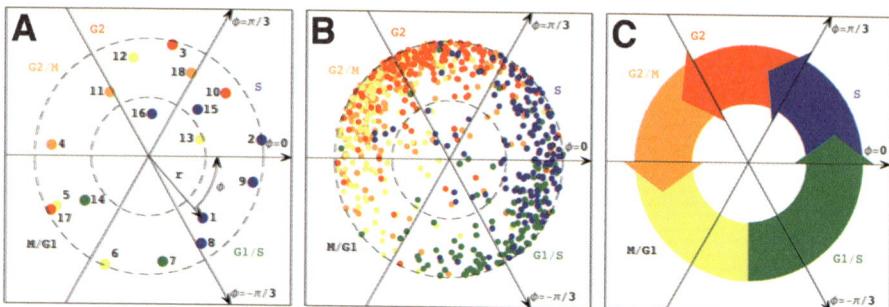

Fig. 12. Reconstructed human RNA expression in the GSVD common cell cycle subspace. (**A**) Projections of the expression of each of the 18 arrays, after reconstruction in the six-dimensional GSVD cell cycle subspace, onto the two-dimensional subspace that approximates it. The arrays are color coded according to their classification into the five cell cycle stages. The dashed unit and half-unit circles outline 100% and 50% of added up, rather than cancelled out, contributions of the six arraylets to the overall projected expression. The arrows describe the projections of $-\pi/3$-, 0- and $\pi/3$-phase arraylets. (**B**) Projections of the expression of each of the 774 cell cycle-regulated genes, reconstructed in the six-dimensional GSVD subspace, onto the two-dimensional subspace that approximates it. The genes are color coded according to either the traditional or microarray classifications. (**C**) The GSVD picture of the human cell cycle.

S/G_2, G_2/M back to M/G_1 twice. Projecting the expression of the 18 human arrays from the six-dimensional human arraylets subspace onto the approximate two-dimensional subspace reveals that 50% or more of the contributions of the six arraylets add up in the expression of 16 of the arrays (**Fig. 12A**). Sorting the arrays describes the human cell cycle progression from S through G_2, G_2/M, M/G_1, G_1/S back to S two and a half times. Note that, the fourth and 16th yeast arraylets, which correspond to the two 0-phase genelets, correlate with the cell cycle transition from G_2/M to M/G_1, in which the yeast culture is synchronized initially, and anticorrelate with that from G_1 to S. Consistently, the fourth and 16th human arraylets anticorrelate with the transition from G_2/M to M/G_1, and correlate with that from G_1 to S, in which the human culture is synchronized initially.

Projecting the expression of the yeast and human genes from the six-dimensional genelets subspace onto the two-dimensional subspace that least squares-approximates it reveals that 50% or more of the contributions of the six genelets add up in the overall expression of 552 of the 612 yeast and 731 of the 774 human genes that were traditionally or microarray-classified as cell cycle-regulated (**Figs. 11B** and **12B**). These genes include, for example, 14 of 16 human histones, which were not microarray-classified as cell cycle-regulated based on their overall expression (*19*). Simultaneous classification of the yeast and human genes into the five cell cycle stages describes the progression of yeast and human cell cycles along the yeast and human genes, respectively, and is in good agreement with both yeast and human microarray and traditional classifications. Note that, the two 0-phase genelets, the fourth and 16th genelets, correlate with cell cycle expression oscillations, which peak at the initial stages of synchronization of both yeast and human genes.

Simultaneous reconstruction and classification of the yeast and human arrays and genes in the subspaces spanned by the six yeast and six human arraylets, and six shared genelets, respectively, gives a picture that resembles the traditional understanding of the biological similarity in the regulation of the yeast and human, and perhaps all eukaryotic, cell cycles *(32)* of two antipodal checkpoints, at the transition from G_1 to S and at that from G_2/M to M/G_1, that are regulated independently of other cell cycle events (**Figs. 11C** and **12C**).

3.2.3. Exclusive Genelets and Corresponding Arraylets Span the Exclusive Yeast and Human Synchronization Responses Subspaces

The first and second genelets, which capture most of the expression information in the yeast dataset, yet very little of the expression information in the human dataset, with $\theta_1,\theta_2 > \pi/7$ (**Fig. 9B**), describe initial transient increase and decrease in expression, respectively (**Fig. 10C**). A theme of yeast response to pheromone synchronization emerges from the annotations of the genes with the largest and smallest levels of expression in the first and second yeast arraylets. The sixth genelet, equally significant in both datasets, with $\theta \sim 0$, describes an initial transient increase in expression superimposed on cosinusidial variation. A theme of transition from the response to the pheromone α factor into cell cycle progression emerges from the annotations of the yeast genes with the largest and smallest expression levels in the sixth yeast arraylet. These three genelets and corresponding three yeast arraylets are associated with the pheromone response program, which is exclusive to the yeast genome. Classification of the yeast genes and arrays into stages in the pheromone response in the subspaces spanned by these genelets and arraylets, respectively (**Fig. 13**), is in good agreement with the traditional understanding of this program *(41)*.

The 17th and 18th genelets are insignificant in the yeast dataset relative to that of the human, with q17,q18 < –p/4. A theme of human synchronization stress response emerges from the annotations of the genes with the largest and smallest expression levels in the 17th and 18th genelets. Also, from the annotations of the human genes with the largest and smallest expression levels in the sixth human arraylet emerges a theme of transition from stress response into cell cycle progression. These three genelets and corresponding three human arraylets are associated with this human exclusive stress response. Classification of the human genes and arrays into stress response

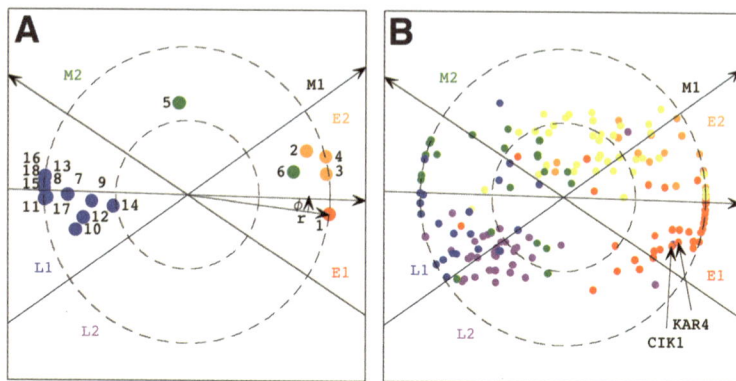

Fig. 13. Reconstructed yeast RNA expression in the GSVD yeast exclusive synchronization response subspace. (**A**) Projections of the expression of each of the 18 arrays, reconstructed in the three-dimensional GSVD synchronization response subspace, onto the two-dimensional subspace that least-squares approximates it. The arrays are color coded according to their classification into six stages in this response to synchronization program, which outlines the response to the pheromone α factor and the transition into cell cycle progression: early E_1 (red) and E_2 (orange), middle M_1 (yellow) and M_2 (green), and late L_1 (blue) and L_2 (violet). The dashed unit and half-unit circles outline 100% and 50% of added up, rather than cancelled out, contributions of the three arraylets to the overall projected expression. The arrows describe the projections of the three arraylets. (**B**) Projections of the expressions of 172 genes, reconstructed in the three-dimensional GSVD subspace, onto the two-dimensional subspace that approximates it. The genes are color coded according to the traditional understanding of the α factor synchronization response program. Genes that peak in E_1 are known to be involved in α factor response, mating, adaptation-to-mating signal, and cell cycle arrest; E_2 – filamentous and pseudohyphal growths and cell polarity; M_1 – ATP synthesis; M_2 – chromatin modeling; L_1 – chromatin binding and architecture; and L_2 – phosphate and iron transport. The expression patterns of *KAR4* and *CIK1* are correlated.

stages in the subspaces spanned by these genelets and arraylets, respectively *(19)*, is in agreement with the current, somewhat limited, understanding of this program *(7)*.

3.2.4. Data Reconstruction and Classification in the Common and Exclusive Subspaces Simulate Observation of Differential Expression in the Cell Cycle and Synchronization Response Programs

According to their expression in the yeast exclusive pheromone response subspace, the RNA expression patterns of the yeast genes *KAR4* and *CIK1* are correlated: The expression of both genes peaks early in the time course together with the expression of other genes known to be involved in the response to the α factor (**Fig. 13B**). In the common cell cycle subspace *KAR4* and *CIK1* are anti-correlated: *KAR4* peaks at the G_1 cell cycle stage, whereas *CIK1* peaks almost half a cell cycle period later (and also earlier) at S/G_2 (**Fig. 13B**). This difference in the relation of the expression patterns of *CIK1* and *KAR4* in the response to pheromone program as compared with that of the cell cycle is in agreement with the experimental observation of Kurihara et al. *(42)* that induction of *CIK1* depends on that of *KAR4* during mating, which is mediated by the α factor pheromone, and is independent of *KAR4* during the mitotic cell cycle.

In the human exclusive stress response subspace, most human histones reach their expression minima early. In the common cell cycle subspace, most histones peak early, together with other genes known to peak in the cell cycle stage S. This differential expression of most histones may explain why these histones do not appear to be cell cycle regulated based on their overall expression *(7)*: The superposition of the expression of the histones during the cell cycle and that in response to the synchronization leads to an overall steady-state expression early in the time course *(19)*.

GSVD uncovers the program-dependent variation in the expression patterns of the human histones, as well as the program-dependent variation in the relations between the expression patterns of the yeast genes *KAR4* and *CIK1*.

3.3.1. GSVD Comparative Model for Genome-Scale RNA Expression During the Yeast and Human Cell Cycles Parallels the Digital Ring Oscillator

With all 4523 yeast and 12,056 human genes sorted according to their phases in the GSVD common cell cycle subspace, the reconstructed yeast and human expressions approximately fit traveling waves of one period cosinusoidal variation across the genes, and of two or two and a half periods across the arrays, respectively (**Fig. 14A**). The gene variations of the six yeast and six human arraylets approximately fit one period cosines of π/3, 0, and –π/3 initial phases, such that the initial phase of each arraylet is similar to that of its corresponding genelet (**Fig. 14B,C**). In this picture, all 4523 yeast genes, about three-quarters of the yeast genome, as well as all 12,056 human genes, about two-thirds of the human genome according to current estimates *(35)*, appear to exhibit periodic expression during the cell cycle.

This GSVD model describes, to first order, the RNA expression of most of the yeast and human genomes during their common cell cycle programs as being driven by the activities of three periodically oscillating cellular elements or modules, which are π/3 out of phase relative to one another. The underlying eukaryotic genetic network or circuit suggested by this model might be parallel in its design to the digital three-inverter ring oscillator. Elowitz and Leibler *(44)*

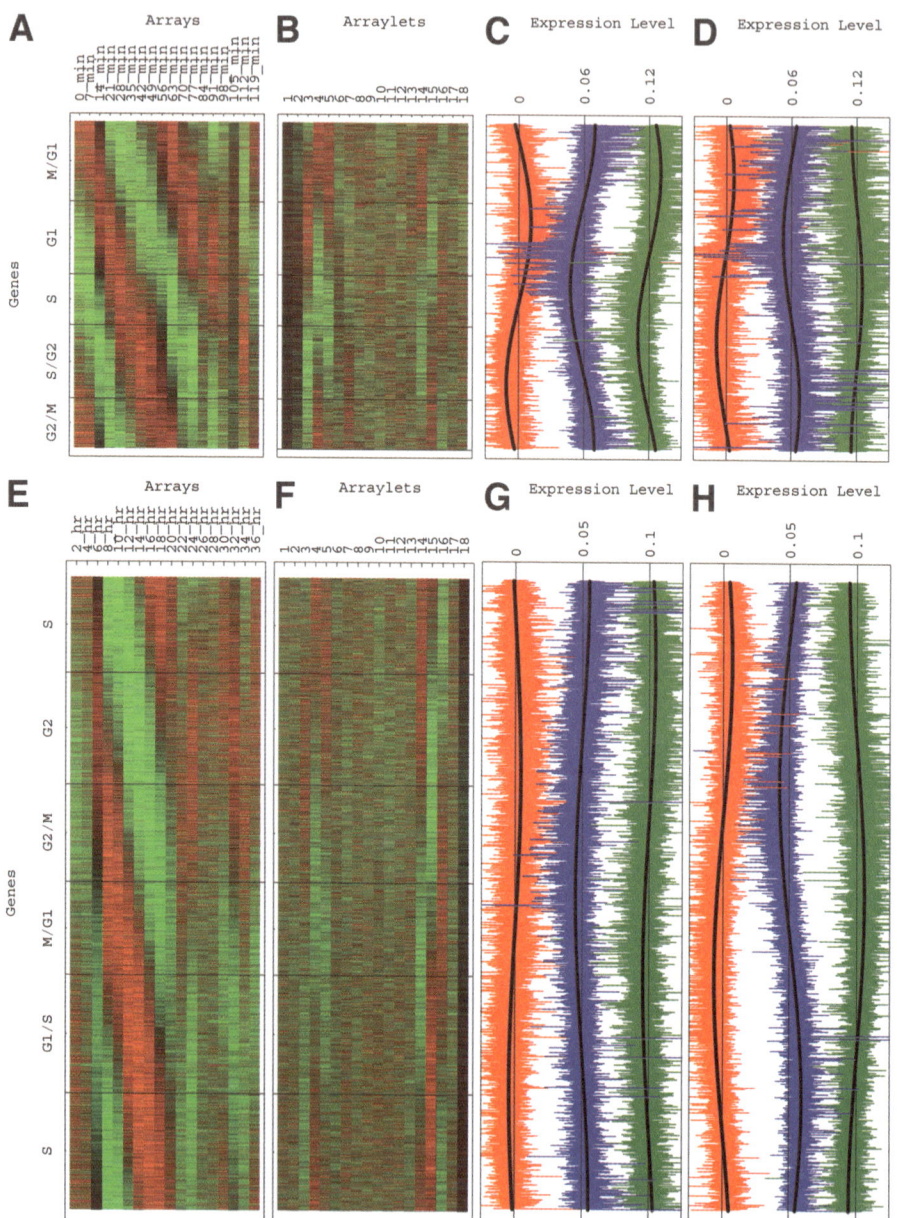

Fig. 14. Yeast and human cell cycles' RNA expression, reconstructed in the six-dimensional GSVD common subspace, with genes sorted according to their phases in the two-dimensional subspace that approximates it. (A) Yeast expression of the sorted 4523 genes in the 18 arrays, centered at their gene- and array-invariant levels, showing a traveling wave of expression. (B) Yeast expression of thesorted 4523 genes

recently demonstrated a synthetic genetic circuit analogous to this digital ring oscillator (*see also* Fung et al., **ref. 45**).

4. Pseudoinverse Projection for Integrative Modeling of DNA Microarray Datasets

Integrative analysis of different types of global signals, such as these measured by DNA microarrays from the same organism, promises to reveal global causal co-ordination of cellular activities. For example, Bussemaker, Li, and Siggia **(46)** predicted new regulatory motifs by linear regression of profiles of genome-scale RNA expression in yeast vs profiles of the abundance levels, or counts of DNA oligomer motifs in the promoter regions of the same yeast genes. Lu, Nakorchevskiy, and Marcotte **(47)** associated the knockout phenotype of individual yeast genes with cell cycle arrest by deconvolution of the RNA expression profiles measured in the corresponding yeast mutants into the RNA expression profiles measured during the cell cycle for all yeast genes that were microarray-classified as cell cycle regulated.

This section reviews the pseudoinverse projection integrative model for DNA microarray datasets and other large-scale molecular biological signals **(20,21)**. Pseudoinverse projection is an integrative BSS algorithm that decomposes the measured gene patterns of any given "data" signal of, e.g., proteins' DNA-binding into mathematically least squares-optimal pseudoinverse correlations with the measured gene patterns of a chosen "basis" signal of, e.g., RNA expression, in a different set of samples from the same organism. The measured array patterns of the data signal are least squares-approximated with a decomposition into the measured array patterns of the basis. The correspondence between these mathematical patterns that are uncovered in the measured signals and the independent

---

Fig. 14. *(Continued)* in the 18 arraylets, centered at their array-invariant levels. The expression patterns of the third through fifth and 14th through 16th arraylets display the sorting. (**C**) The third (red), fourth (blue), and fifth (green) yeast arraylets fit one period cosines of $\pi/3$ (red), 0 (blue) and $-\pi/3$ (green) initial phases. (**D**) The 14th (red), 15th (blue), and 16th (green) yeast arraylets fit one period cosines of $-\pi/3$- (red), $\pi/3$- (blue), and 0- (green) phases. (**E**) Human expression of the sorted 12,056 genes in the 18 arrays, centered at their gene- and array-invariant levels, showing a traveling wave of expression. (**F**) Human expression of the sorted 12,056 genes in the 18 arraylets, centered at their array-invariant levels. The expression patterns of the third through fifth and 14th through 16th arraylets display the sorting. (**G**) The third (red), fourth (blue), and fifth (green) human arraylets fit one period cosines of $\pi/3$- (red), 0- (blue), and $-\pi/3$- (green) phases. (**H**) The 14th (red), 15th (blue) and 16th (green) human arraylets fit one period cosines of $-\pi/3$- (red), $\pi/3$- (blue) and 0- (green) phases.

activities of cellular elements that compose the signals is illustrated with an integration of yeast genome-scale DNA-binding occupancy of cell cycle transcription factors *(8)* and DNA replication initiation proteins *(9)* with RNA expression during the cell cycle, using as basis sets the eigenarrays and arraylets determined by SVD and GSVD, respectively. One consistent picture emerges that predicts novel correlation between DNA replication initiation and RNA transcription during the yeast cell cycle. This novel correlation, which might be due to a previously unknown mechanism of regulation, demonstrates the power of the SVD, GSVD, and pseudoinverse projection models to predict previously unknown biological principles.

4.1. Mathematical Framework of Pseudoinverse Projection

Let the basis matrix $\hat{b}$ of size N-genomic sites or open reading frames (ORFs) $\times$ M-basis profiles tabulate M genome-scale molecular biological profiles of, e.g., RNA expression, measured from a set of M samples or extracted mathematically from a set of M or more measured samples. As before, the mth column vector in the matrix $\hat{b}$, $|b_m\rangle \equiv \hat{b}|m\rangle$, lists the signal measured in the mth sample by the mth array across all N ORFs simultaneously. The nth row vector in the matrix $\hat{b}$, $\langle n|\hat{b}$, lists the signal measured in the nth ORF across the different arrays, which correspond to the different samples. Let the data matrix $\hat{d}$ of size N-ORFs $\times$ L-data samples tabulate L genome-scale molecular biological profiles of, e.g., proteins' DNA binding, measured for the same ORFs in L samples from the same organism. The lth column vector in the matrix $\hat{d}$, $|d_l\rangle \equiv \hat{d}|l\rangle$, lists the signal measured in the lth sample across all N ORFs simultaneously.

Moore–Penrose pseudoinverse projection of the data matrix $\hat{d}$ onto the basis matrix $\hat{b}$ is a linear transformation of the data $\hat{d}$ from the N-ORFs $\times$ L-data samples space to the M-basis profiles $\times$ L-data samples space **(Fig. 15)**,

$$\hat{d} \to \hat{b}\hat{c},$$
$$\hat{b}^{\dagger}\hat{d} \equiv \hat{c}, \tag{8}$$

where the matrix $\hat{b}^{\dagger}$, that is, the pseudoinverse of $\hat{b}$, satisfies

$$\hat{b}\hat{b}^{\dagger}\hat{b} = \hat{b},$$
$$\hat{b}^{\dagger}\hat{b}\hat{b}^{\dagger} = \hat{b}^{\dagger},$$
$$(\hat{b}\hat{b}^{\dagger})^{T} = \hat{b}\hat{b}^{\dagger}, \tag{9}$$
$$(\hat{b}^{\dagger}\hat{b})^{T} = \hat{b}^{\dagger}\hat{b},$$

such that the transformation matrices $\hat{b}\hat{b}^{\dagger}$ and $\hat{b}^{\dagger}\hat{b}$ are orthogonal projection matrices for a real basis matrix $\hat{b}$.

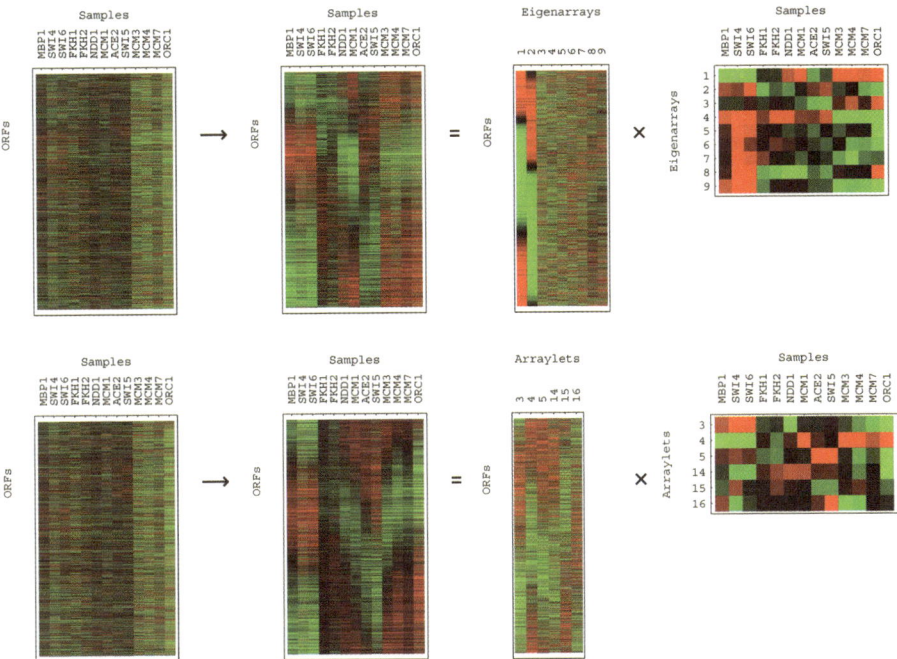

Fig. 15. Raster display of the pseudoinverse projection of the yeast cell cycle transcription factors and replication initiation proteins' DNA-binding data onto the SVD and GSVD cell cycle RNA expression bases, with overexpression (red), no change in expression (black) and underexpression (green) centered at ORF- and sample-invariant expression, and with the ORFs sorted according to their SVD and GSVD phases, respectively. Pseudoinverse projection is a linear transformation of the proteins' binding data from the 2227 ORFs × 13-data samples space to the nine eigenarrays of the SVD basis × 13-data samples space *(upper)*, and also of the proteins' binding data from the 2139 ORFs × 13-data samples space to the six arraylets of the GSVD basis × 13-data samples space *(lower)*.

In this space the data matrix $\hat{d}$ is represented by the pseudoinverse correlations matrix $\hat{c}$. The vector in the mth row of the matrix $\hat{c}$, $\langle c_m | \equiv \langle m | \hat{c}$, lists the pseudoinverse correlations of the L data profiles with the mth basis profile. The pseudoinverse correlations matrix $\hat{c}$ is unique, i.e., data driven.

4.2. Pseudoinverse Projection Integrative Analysis of Yeast Cell Cycle RNA Expression and Proteins' DNA-Binding Data

In this example, a data matrix that tabulates DNA-binding occupancy levels of nine yeast cell cycle transcription factors *(8)* and four yeast replication initiation proteins *(9)* across 2928 yeast ORFs is pseudoinverse projected onto (1) the SVD cell cycle RNA expression basis matrix, which tabulates the expression of the nine most significant eigenarrays of the α factor, *CLB2*, and *CLN3* dataset, including the two eigenarrays that span the SVD cell cycle subspace, across 4579 ORFs, 2227 of which are present in the data matrix; and (2) the GSVD cell cycle RNA expression basis matrix, which tabulates the expression of the six arraylets that span the GSVD cell cycle subspace across 4523 ORFs, 2139 of which are present in the data matrix.

4.2.1. Pseudoinverse Correlations Uncovered in the Data Correspond to Reported Functions of Transcription Factors

The nine transcription factors are ordered, following Simon et al. *(8)*, from these that have been reported to function in the cell cycle stage G_1, through these that have been reported to function in S, S/G_2, G_2/M, and M/G_1: Mbp1, Swi4, Swi6, Fkh1, Fkh2, Ndd1, Mcm1, Ace2, and Swi5. With this order, the SVD- and GSVD-pseudoinverse correlations approximately fit cosine functions of one period and of varying initial phases across the nine transcription factors' samples and are approximately invariant across the four samples of the replication initiation proteins, Mcm3, Mcm4, Mcm7, and Orc1 **(Fig. 16)**. Transcription factors that have been reported to function in antipodal cell cycle stages, such as Mbp1, Swi4, and Swi6 that are known to function in G_1 and Mcm1 that is known to function in G_2/M, consistently exhibit anticorrelated levels of DNA-binding in all patterns of pseudoinverse correlations. Each pattern of pseudoinverse correlations $\langle c_m |$ represents the activity of the transcripition factors during the cell cycle stage that the corresponding basis profile $\langle b_m |$ correlates with. For example, the first SVD basis profile, i.e., the first eigenarray, correlates with RNA expression oscillations at the transition from the cell cycle stage G_2/M to M/G_1 and anticorrelates with oscillations at the transition from G_1 to S **(Fig. 6C)**. Correspondingly, the first pattern of SVD-pseudoinverse correlations describes enhanced activity of the transcription factor Mcm1 and reduced activity of Mbp1, Swi4, and Swi6 **(Fig. 16B)**.

4.2.2. Pseudoinverse Reconstruction of the Data in the Basis Simulates Experimental Observation of Only the Cellular States Manifest in the Data that Correspond to Those in the Basis

The proteins' DNA-binding data is SVD- and independently also GSVD-reconstructed using pseudoinverse projections in the intersections of the SVD and GSVD bases matrices with the data matrix **(Fig. 17)**. With the 2227 and 2139 ORFs sorted according to their SVD and GSVD cell cycle phases, respectively,

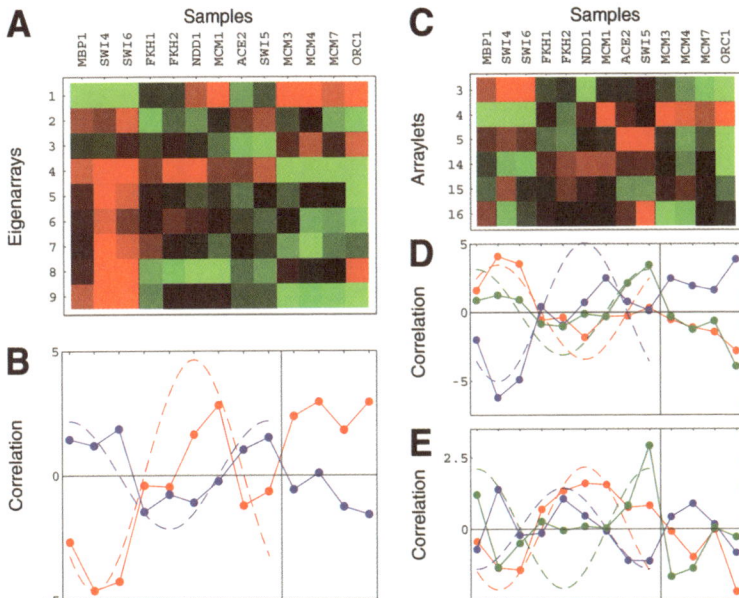

Fig. 16. Pseudoinverse correlations of the proteins' DNA-binding data with the SVD and GSVD cell cycle RNA expression. (**A**) Raster display of the correlations with the nine eigenarrays that span the SVD basis. (**B**) Line-joined graphs of the correlations with the first (red) and second (blue) most significant eigenarrays that span the SVD subspace. (**C**) Raster display of the correlations with the six arraylets that span the GSVD basis and the GSVD subspace. (**D**) Line-joined graphs of the correlations with third (red), fourth (blue), and fifth (green) arraylets, and (**E**) the 14th (red), 15th (blue), and 16th (green) arraylets.

the variations of the SVD- and GSVD-reconstructed binding profiles across the ORFs approximately fit cosine functions of one period and of varying initial phases.

The SVD- and GSVD-reconstructed transcription factors' data approximately fit traveling waves, cosinusoidally varying across the ORFs as well as the nine samples. Simon et al. (*8*) observed a similar traveling wave in the binding data from the nine transcription factors, ordered as in **Subheading 4.2.1.** above, across only 213 ORFs. These traveling waves are in agreement with current understanding of the progression of cell cycle transcription along the genes and in time as it is regulated by DNA binding of the transcription factors at the promoter regions of the transcribed genes. Pseudoinverse reconstruction of the data in both the SVD and GSVD bases, therefore, simulates experimental observation of only the proteins' DNA-binding cellular states that correspond to those of RNA expression during the cell cycle.

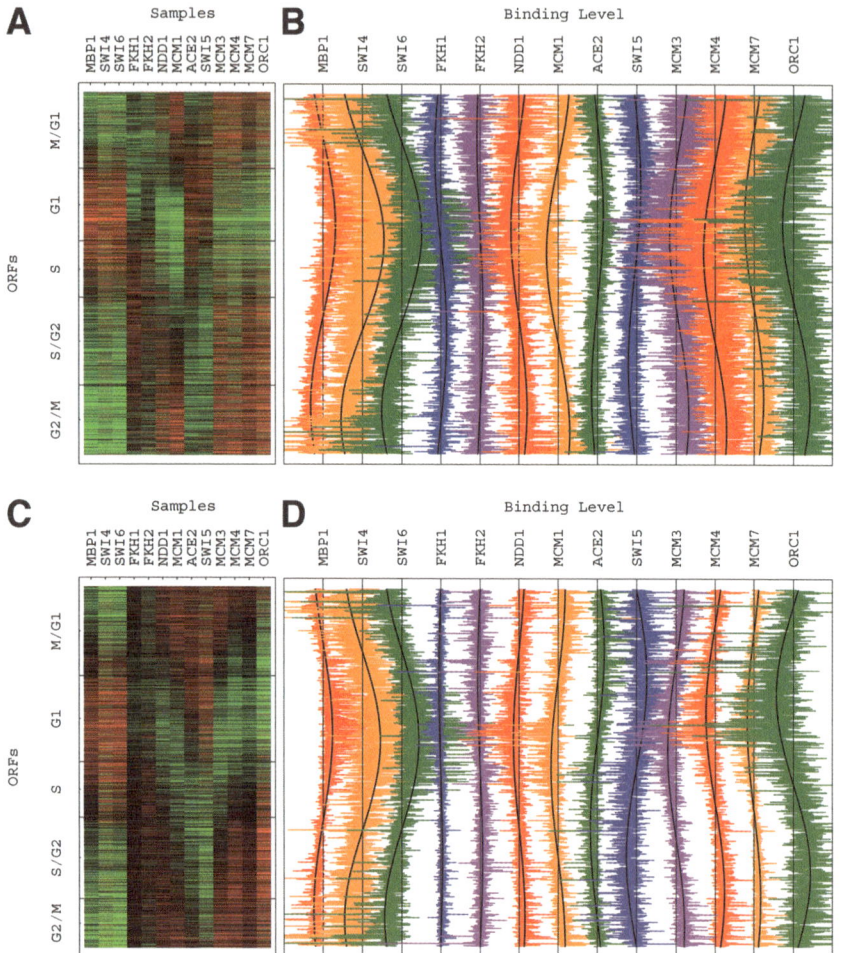

Fig. 17. Pseudoinverse reconstructions of the proteins' DNA-binding data in the SVD and GSVD cell cycle RNA expression bases, with the open reading frames sorted according to their SVD and GSVD phases, respectively, showing a traveling wave in the nine transcription factors and a standing wave in the four replication initiation proteins. **(A)** Raster display of the SVD-reconstructed data. **(B)** Line-joined graphs of the SVD-reconstructed data profiles. **(C)** Raster display of the GSVD-reconstructed data. **(D)** Line-joined graphs of the GSVD-reconstructed data profiles.

The SVD- and GSVD-reconstructed replication initiation proteins' data approximately fit a standing wave, cosinusoidally varying across the ORFs and constant across the four samples. These replication initiation proteins' reconstructed profiles are antiparallel to the reconstructed profiles of Mbp1, Swi4, and Swi6, and parallel to that of Mcm1.

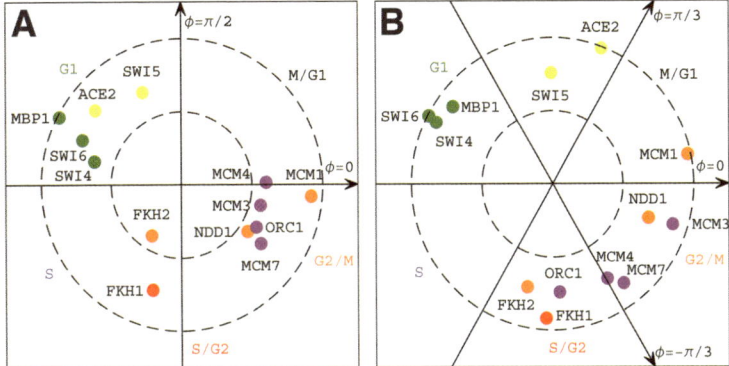

Fig. 18. Reconstructed yeast proteins' DNA-binding data in the RNA expression bases. (**A**) Correlations of the reconstructed binding of each of the 13 proteins with the first and second rotated eigenarrays along the *x*- and *y*-axes. The transcription factors are color coded according to their classification into the five cell cycle stages: M/G$_1$ (yellow), G$_1$ (green), S (blue), S/G$_2$ (red), and G$_2$/M (orange). The replication initiation proteins are colored violet. The dashed unit and half-unit circles outline 100% and 25% of overall normalized array expression in this subspace. (**B**) Projections of the binding of each of the nine transcription factors and four replication initiation proteins, after reconstruction in the six-dimensional GSVD cell cycle subspace, onto the two-dimensional subspace that least-squares approximates it. The dashed unit and half-unit circles outline 100% and 50% of added up, rather than cancelled out, contributions of the six arraylets to the overall projected reconstructed binding. The arrows describe the projections of the $-\pi/3$-, 0-, and $\pi/3$-phase arraylets.

4.2.3. Classification of the Basis-Reconstructed Data Samples Maps the Cellular States of the Data Onto Those of the Basis and Gives a Global Picture of Possible Causal Coordination of These States

Projected from the SVD basis, that is spanned by nine eigenarrays, onto the SVD cell cycle subspace, that is spanned by the two most significant of these eigenarrays, all SVD-reconstructed samples have at least 25% of their binding profiles in this subspace, except for Fkh2 (**Fig. 18A**). Projected from the six-dimensional GSVD cell cycle subspace, that is spanned by six arraylets, onto the two-dimensional subspace that approximates it, 50% or more of the contributions of the six arraylets to each GSVD-reconstructed sample add up, rather than cancel out (**Fig. 18B**).

Sorting the samples according to their SVD or GSVD phases gives an array order that is similar to that of Simon et al. (*8*), and describes the yeast cell cycle progression from the cellular state of Mbp1's binding through that of Swi5's. The SVD and GSVD mappings of the transcription factors' binding profiles onto the expression subspaces are also in agreement with the current understanding of the cell cycle program. Mapping the binding of Mbp1, Swi4, and Swi6 onto the cell cycle expression stage G_1 corresponds to the biological coordination between the binding of these factors to the promoter regions of ORFs and the subsequent peak in transcription of these ORFs during G_1. The mapping of Mbp1, Swi4, and Swi6 onto G_1, which is antipodal to G_2/M, also corresponds to their binding to promoter regions of ORFs that exhibit transcription minima or shutdown during G_2/M, and to their minimal or lack of binding at promoter regions of ORFs which transcription peaks in G_2/M. Similarly, the mapping of Mcm1 onto G_2/M corresponds to its binding to the promoter regions of ORFs that are subsequently transcribed during the transition from G_2/M to M/G_1. The binding profiles of the replication initiation proteins are SVD- and GSVD-mapped onto the cell cycle stage that is antipodal to G_1. These SVD and GSVD mappings are consistent with the reconstructed profiles of Mcm3, Mcm4, Mcm7, and Orc1 being antiparallel to those of Mbp1, Swi4, and Swi6 and parallel to that of Mcm1.

The parallel and antiparallel associations by annotation of the proteins' DNA-binding profiles with the cellular states of RNA expression during the cell cycle are also consistent with the SVD and GSVD mappings. These associations follow the p-values for the distribution of the 400 and 377 ORFs that were microarray-classified and the 58 and 60 ORFs that were traditionally classified as cell cycle regulated among all 2227 and 2139 ORFs that are mapped onto the SVD and GSVD subspaces, respectively, and among each of the subsets of 200 ORFs with largest and smallest levels of binding occupancy in each of the profiles. Again, the binding profiles of all four DNA replication initiation proteins, Mcm3, Mcm4, Mcm7, and Orc1 are anticorrelated with RNA expression in the cell cycle stage G_1, together with the profile of the transcription factor Mcm1, whereas the profiles of the transcription factors Mbp1, Swi4, and Swi6 that are known to drive the cell cycle stage G_1, are correlated with RNA expression in this stage *(20,21)*.

Thus, DNA-binding of Mcm3, Mcm4, Mcm7, and Orc1 adjacent to ORFs is pseudoinverse-correlated with minima or even shutdown of the transcription of these ORFs during the cell cycle stage G_1. This novel correlation suggests a previously unknown genome-scale coordination between DNA replication initiation and RNA transcription during the cell cycle in yeast.

The correlation between Mcm3, Mcm4, Mcm7, and Orc1 and the transcription factor Mcm1 suggests a genome-scale, or maybe even a genome-wide coordination in the activities of the DNA replication initiation proteins and Mcm1. One possible explanation of this correlation may be provided by the recent suggestion by Chang et al. *(48; see also* Donato, Chang and Tye, **ref. *49)*** that Mcm1 binds origins of replication, and thus functions as a replication initiation protein in addition to its function as a transcription factor. However, this correlation does not necessarily mean that Mcm1 colocalizes with origins. It is the tendency of ORFs adjacent to Mcm1's binding sites to exhibit transcription minima during the cell cycle stage G_1, which correlates with a similar tendency of those ORFs that are adjacent to binding sites of the replication initiation proteins.

4.3. Pseudoinverse Projection Integrative Model for Genome-Scale RNA Transcription and DNA-Binding of Cell Cycle Transcription Factors and Replication Initiation Proteins in Yeast

One consistent picture emerges upon integrating the genome-scale proteins' DNA-binding data with the SVD and GSVD cell cycle RNA expression bases, which is in agreement with the current understanding of the yeast cell cycle program *(50–53)*, and is supported by recent experimental results *(49)*. This picture correlates for the first time the binding of replication initiation proteins with minima or shutdown of the transcription of adjacent ORFs during the cell cycle stage G_1, under the assumption that the measured cell cycle RNA expression levels are approximately proportional to cell cycle RNA transcription activity. It was shown by Diffley et al. *(50)* that replication initiation requires binding of Mcm3, Mcm4, Mcm7, and Orc1 at origins of replication across the yeast genome during G_1 (*see also* **ref. 51**). And, it was shown by Micklem et al. *(52)* that these replication initiation proteins are involved with transcriptional silencing at the yeast mating loci (*see also* **ref. 53**). Either one of at least two mechanisms of regulation may be underlying this novel genome-scale correlation between DNA replication initiation and RNA transcription during the yeast cell cycle: the transcription of genes may reduce the binding efficiency of adjacent origins. Or, the binding of replication initiation proteins to origins of replication may repress, or even shut down, the transcription of adjacent genes.

This is the first time that a data-driven mathematical model, where the mathematical variables and operations represent biological or experimental reality, has been used to predict a biological principle that is truly on a genome scale. The ORFs in either one of the basis or data matrices were selected based on data quality alone, and were not limited to ORFs that are traditionally or microarray-classified as cell cycle regulated, suggesting that the RNA transcription signatures of yeast cell cycle cellular states may span the whole yeast genome.

5. Are Genetic Networks Linear and Orthogonal?

The SVD model, the GSVD comparative model, and the pseudoinverse projection integrative model are all mathematically linear and orthogonal. These models formulate genome-scale molecular biological signals as linear superpositions of mathematical patterns, which correlate with activities of cellular elements, such as regulators or transcription factors, that drive the measured signal and cellular states where these elements are active. These models associate the independent cellular states with orthogonal, i.e., decorrelated, mathematical profiles suggesting that the overlap or crosstalk between the genome-scale effects of the corresponding cellular elements or modules is negligible.

Recently, Ihmels, Levy, and Barkai *(54)* found evidence for linearity as well as orthogonality in the metabolic network in yeast. Integrating genome-scale RNA expression data with the structural description of this network, they showed that at the network's branchpoints, only distinct branches are coexpressed, and concluded that transcriptional regulation biases the metabolic flow toward linearity. They also showed that individual isozymes, i.e., chemically distinct but functionally similar enzymes, tend to be corregulated separately with distinct processes. They concluded that transcriptional regulation uses isozymes as means for reducing crosstalk between pathways that use a common chemical reaction.

Orthogonality of the cellular states that compose a genetic network suggests an efficient network design. With no redundant functionality in the activities of the independent cellular elements, the number of such elements needed to carry out a given set of biological processes is minimized. An efficient network, however, is fragile. The robustness of biological systems to diverse perturbations, e.g., phenotypic stability despite environmental changes and genetic variation, suggests functional redundancy in the activities of the cellular elements, and therefore also correlations among the corresponding cellular states. Carslon and Doyle *(55)* introduced the framework of "highly optimized tolerance" to study fundamental aspects of complexity in, among others, biological systems that appear to be naturally selected for efficiency as well as robustness. They showed that trade-offs between efficiency and robustness might explain the behavior of such complex systems, including occurrences of catastrophic failure events.

Linearity of a genetic network may seem counterintuitive in light of the nonlinearity of the chemical processes, which underlie the network. Arkin and Ross *(56)* showed that enzymatic reaction mechanisms can be thought to compute the mathematically nonlinear functions of logic gates on the molecular level. They also showed that the qualitative logic gate behavior of such a reaction mechanism may not change when situated within a model of the cellular program that uses the reaction. This program functions as a biological switch from one pathway to another in response to chemical signals, and thus computes a nonlinear logic gate function on the cellular scale. Another cellular program that can be thought to compute nonlinear functions is the well-known genetic switch in the bacteriophage λ, the program of decision between lysis and lysogeny *(57)*. McAdams and Shapiro *(58)* modeled this program with a circuit of integrated logic components. However, even if the kinetics of biochemical reactions are nonlinear, the mass balance constraints that govern these reactions are linear. Schilling and Palsson *(59)* showed that the underlying pathway structure of a biochemical network, and therefore also its functional capabilities, can be extracted from the linear set of mass balance constraints corresponding to the set of reactions that compose this network.

That genetic networks might be modeled with linear and orthogonal mathematical frameworks does not necessarily imply that these networks are linear and orthogonal(e.g., **refs. *60–62***). Dynamical systems, linear and nonlinear, are regularly studied with linear orthogonal transforms *(63)*. For example, SVD might be used to reconstruct the phase-space description of a dynamical system from a series of observations of the time evolution of the coordinates of the system. In such a reconstruction, the experimental data are mapped onto a subspace spanned by selected patterns that are uncovered in the data by SVD. The phase-space description of linear systems, for which the time evolution, or "motion," of the coordinates is periodic, such as the analog harmonic oscillator, is the "limit cycle." The phase-space description of nonlinear systems, for which the coordinates' motion is chaotic, such as the chemical Lotka-Volterra irreversible autocatalytic reaction *(35–37)*, is the "strange attractor." Broomhead and King *(64)* were the first to use SVD to reconstruct the strange attractor.

Although it is still an open question whether genetic networks are linear and orthogonal, linear and orthogonal mathematical frameworks have already proven successful in describing the physical world, in such diverse areas as mechanics and perception. It may not be surprising, therefore, that linear and orthogonal mathematical models for genome-scale molecular biological signals (1) provide mathematical descriptions of the genetic networks that generate and sense the measured data, where the mathematical variables and operations represent biological or experimental reality; (2) elucidate the design principles of cellular systems as well as guide the design of synthetic ones; and (3) predict previously unknown biological principles.

These models may become the foundation of a future in which biological systems are modeled as physical systems are today.

Acknowledgments

The author thanks D. Botstein and P. O. Brown for introducing her to genomics, G. H. Golub for introducing her to matrix and tensor computation and M. van de Rijn for introducing her to translational cancer research. The author also thanks T. M. Baer, G. M. Church, J. F. X. Diffley, J. Doyle, S. R. Eddy, P. Green, R. R. Klevecz, E. Rivas, and J. J. Wyrick for thoughtful and thorough reviews of parts of the work presented in this chapter. This work was supported by a National Human Genome Research Institute Individual Mentored Research Scientist Development Award in Genomic Research and Analysis (K01 HG00038-05) and by a Sloan Foundation and Department of Energy Postdoctoral Fellowship in Computational Molecular Biology (DE-FG03-99ER62836).

References

1. Fodor, S. P., Rava, R. P., Huang, X. C., Pease, A. C., Holmes, C. P., and Adams, C. L. (1993) Multiplexed biochemical assays with biological chips. *Nature* **364,** 555–556.
2. Schena, M., Shalon, D., Davis, R. W., and Brown, P. O. (1995) Quantitative monitoring of gene expression patterns with a complementary DNA microarray. *Science* **270,** 467–470.
3. Brown, P. O., and Botstein, D. (1999) Exploring the new world of the genome with DNA microarrays. *Nat. Genet.* **21,** 31–37.
4. Pollack, J. R., and Iyer, V. R. (2002) Characterizing the physical genome. *Nat. Genet.* **32,** 515–521.
5. Sherlock, G., Hernandez-Boussard, T., Kasarskis, A., et al. (2001) The Stanford microarray database. *Nucleic Acids Res.* **29,** 152–155.
6. Spellman, P. T., Sherlock, G., Zhang, M. Q., et al. (1998) Comprehensive identification of cell cycle-regulated genes of the yeast Saccharomyces cerevisiae by microarray hybridization. *Mol. Biol. Cell* **9,** 3273–3297.
7. Whitfield, M. L., Sherlock, G., Saldanha, A., et al. (2002) Identification of genes periodically expressed in the human cell cycle and their expression in tumors. *Mol. Biol. Cell* **13,** 1977–2000.
8. Simon, I., Barnett, J., Hannett, N., et al. (2001) Serial regulation of transcriptional regulators in the yeast cell cycle. *Cell* **106,** 697–708.
9. Wyrick, J. J., Aparicio, J. G., Chen, T., et al. (2001) Genome-wide distribution of ORC and MCM proteins in S. cerevisiae: high-resolution mapping of replication origins. *Science* **294,** 2301–2304.
10. Newton, I. (1999) *The Principia: Mathematical Principles of Natural Philosophy.* (Cohen, I. B., and Whitman, A., trans.) University of California Press, Berkeley, CA.
11. Hubel, D. H., and Wiesel, T. N. (1968) Receptive fields and functional architecture of monkey striate cortex. *J. Physiol.* **195,** 215–243.
12. Barlow, H. B. (1972) Single units and sensation: a neuron doctrine for perceptual psychology? *Perception* **1,** 371–394.
13. Olshausen, B. A., and Field, D. J. (1996) Emergence of simple-cell receptive field properties by learning a sparse code for natural images. *Nature* **381,** 607–609.
14. Bell, A. J., and Sejnowski, T. J. (1997) The "independent components" of natural scenes are edge filters. *Vision Res.* **37,** 3327–3338.
15. Golub, G. H., and Van Loan, C. F. (1996) *Matrix Computation, 3rd ed.*, Johns Hopkins University, Press, Baltimore, MD.
16. Alter, O., Brown, P. O., and Botstein, D. (2000) Singular value decomposition for genome-wide expression data processing and modeling. *Proc. Natl. Acad. Sci. USA* **97,** 10,101–10,106.

17. Alter, O., Brown, P. O., and Botstein, D. (2001) Processing and modeling genome-wide expression data using singular value decomposition. In: *Microarrays: Optical Technologies and Informatics, vol. 4266* (Bittner, M. L., Chen, Y., Dorsel, A. N., and Dougherty, E. R., eds.), Int. Soc. Optical Eng., Bellingham, WA, pp. 171–186.

18. Nielsen, T. O., West, R. B., Linn, S. C., et al. (2002) Molecular characterisation of soft tissue tumours: a gene expression study. *Lancet* **359**, 1301–1307.

19. Alter, O., Brown, P. O., and Botstein, D. (2003) Generalized singular value decomposition for comparative analysis of genome-scale expression data sets of two different organisms. *Proc. Natl. Acad. Sci. USA* **100**, 3351–3356.

20. Alter, O., Golub, G. H., Brown, P. O., and Botstein, D. (2004) Novel genome-scale correlation between DNA replication and RNA transcription during the cell cycle in yeast is predicted by data-driven models. In: *Proc. Miami Nat. Biotechnol. Winter Symp. on the Cell Cycle, Chromosomes and Cancer, vol. 15* (Deutscher, M. P., Black, S., Boehmer, P. E., et al., eds.), Univ. Miami Sch. Med., Miami, FL, www.med.miami.edu/mnbws/Alter-.pdf.

21. Alter, O. and Golub, G. H. (2004) Integrative analysis of genome-scale data by using pseudoinverse projection predicts novel correlation between DNA replication and RNA transcription. *Proc. Natl. Acad. Sci. USA* **101**, 16,577–16,582.

22. Alter, O., and Golub, G. H. (2005) Reconstructing the pathways of a cellular system from genome-scale signals using matrix and tensor computations. *Proc. Natl. Acad. Sci. USA* **102**, 17,559–17,564.

23. Alter, O., and Golub, G. H. (2006) Singular value decomposition of genome-scale mRNA lengths distribution reveals asymmetry in RNA gel electrophoresis band broadening. *Proc. Natl. Acad. Sci. USA* **103**, 11,828–11,833.

24. Alter, O. (2006) Discovery of principles of nature from mathematical modeling of DNA microarray data. *Proc. Natl. Acad. Sci. USA* **103**, 16,063–16,064.

25. Wigner, E. P. (1960) The unreasonable effectiveness of mathematics in the natural sciences. *Commun. Pure Appl. Math.* **13**, 1–14.

26. Hopfield, J. J. (1999) Odor space and olfactory processing: collective algorithms and neural implementation. *Proc. Natl. Acad. Sci. USA* **96**, 12,506–12,511.

27. Sirovich, L., and Kirby, M. (1987) Low-dimensional procedure for the characterization of human faces. *J. Opt. Soc. Am. A* **4**, 519–524.

28. Turk, M., and Pentland, A. (1991) Eigenfaces for recognition. *J. Cogn. Neurosci.* **3**, 71–86.

29. Landau, L. D., and Lifshitz, E. M. (1976) *Mechanics, 3rd ed.* (Sykes, J. B., and Bell, J. S., trans.), Butterworth-Heinemann, Oxford, UK.

30. Tavazoie, S., Hughes, J. D., Campbell, M. J., Cho, R. J., and Church, G. M. (1999) Systematic determination of genetic network architecture. *Nat. Genet.* **22**, 281–285.

31. Roberts, C. J., Nelson, B., Marton, M. J., et al. (2000) Signaling and circuitry of multiple MAPK pathways revealed by a matrix of global gene expression profiles. *Science* **287**, 873–880.

32. Alberts, B., Bray, D., Lewis, J., Raff, M., Roberts, K., and Watson, J. D. (1994) *Molecular Biology of the Cell, 3rd ed.*, Garland Pub., New York, NY.

33. Klevecz, R. R., Bolen, J., Forrest, G., and Murray, D. B. (2004) A genomewide oscillation in transcription gates DNA replication and cell cycle. *Proc. Natl. Acad. Sci. USA* **101,** 1200–1205.

34. Li, C. M., and Klevecz, R. R. (2006) A rapid genome-scale response of the transcriptional oscillator to perturbation reveals a period-doubling path to phenotypic change. *Proc. Natl. Acad. Sci. USA* **103,** 16,254–16,259.

35. Nicolis, G. and Prigogine, I. (1971) Fluctuations in nonequilibrium systems. *Proc. Natl. Acad. Sci. USA* **68,** 2102–2107.

36. Rössler O. E. (1976) An equation for continuous chaos. *Phys. Lett. A* **35,** 397–398.

37. Roux, J. -C., Simoyi, R. H., and Swinney, H. L. (1983) Observation of a strange attractor. *Physica D* **8,** 257–266.

38. Stuart, J. M., Segal, E., Koller, D., and Kim, S. K. (2003) A gene-coexpression network for global discovery of conserved genetic modules. *Science* **302,** 249–255.

39. Bergmann, S., Ihmels, J., and Barkai, N. (2004) Similarities and differences in genome-wide expression data of six organisms. *PLoS Biol* **2,** E9.

40. Mushegian, A. R., and Koonin, E. V. (1996) A minimal gene set for cellular life derived by comparison of complete bacterial genomes. *Proc. Natl. Acad. Sci. USA* **93,** 10,268–10,273.

41. Dwight, S. S., Harris, M. A., Dolinski, K., et al. (2002) Saccharomyces Genome Database (SGD) provides secondary gene annotation using the Gene Ontology (GO). *Nucleic Acids Res.* **30,** 69–72.

42. Kurihara, L. J., Stewart, B. G., Gammie, A. E., and Rose, M. D. (1996) Kar4p, a karyogamy-specific component of the yeast pheromone response pathway. *Mol. Cell. Biol.* **16,** 3990–4002.

43. Ewing, B. and Green, P. (2000) Analysis of expressed sequence tags indicates 35,000 human genes. *Nat. Genet.* **25,** 232–234.

44. Elowitz, M. B., and Leibler, S. (2000) A synthetic oscillatory network of transcriptional regulators. *Nature* **403,** 335–338.

45. Fung, E., Wong, W. W., Suen, J. K., Butler, T., Lee, S. G., and Liao, J. C. (2005) A synthetic gene-metabolic oscillator. *Nature* **435,** 118–122.

46. Bussemaker, H. J., Li, H., and Siggia, E. D. (2001) Regulatory element detection using correlation with expression. *Nat. Genet.* **27,** 167–171.

47. Lu, P., Nakorchevskiy, A., and Marcotte, E. M. (2003) Expression deconvolution: a reinterpretation of DNA microarray data reveals dynamic changes in cell populations. *Proc. Natl. Acad. Sci. USA* **100,** 10,370–10,375.

48. Chang, V. K., Fitch, M. J., Donato, J. J., Christensen, T. W., Merchant, A. M., and Tye, B. K. (2003) Mcm1 binds replication origins. *J. Biol. Chem.* **278,** 6093–6100.

49. Donato, J. J., Chung, S. C., and Tye, B. K. (2006) Genome-wide hierarchy of replication origin usage in *Saccharomyces cerevisiae*. *PLoS Genet.* **2,** E9.

50. Diffley, J. F. X., Cocker, J. H., Dowell, S. J., and Rowley, A. (1994) Two steps in the assembly of complexes at yeast replication origins in vivo. *Cell* **78,** 303–316.

51. Kelly, T. J. and Brown, G. W. (2000) Regulation of chromosome replication. *Annu. Rev. Biochem.* **69,** 829–880.

52. Micklem, G., Rowley, A., Harwood, J., Nasmyth, K., and Diffley, J. F. X. (1993) Yeast origin recognition complex is involved in DNA replication and transcriptional silencing. *Nature* **366,** 87–89.

53. Fox, C. A. and Rine, J. (1996) Influences of the cell cycle on silencing. *Curr. Opin. Cell Biol.* **8,** 354–357.

54. Ihmels, J., Levy, R., and Barkai, N. (2004) Principles of transcriptional control in the metabolic network of *Saccharomyces cerevisiae*. *Nat. Biotechnol.* **60,** 86–92.

55. Carlson, J. M. and Doyle, J. (1999) Highly optimized tolerance: a mechanism for power laws in designed systems. *Phys. Rev. E* **60,** 1412–1427.

56. Arkin, A. P. and Ross, J. (1994) Computational functions in biochemical reaction networks. *Biophys. J.* **67,** 560–578.

57. Ptashne, M. (1992) *Genetic Switch: Phage Lambda and Higher Organisms*, 2nd ed., Blackwell Publishers, Oxford, UK.

58. McAdams, H. H. and Shapiro, L. (1995) Circuit simulation of genetic networks. *Science* **269,** 650–656.

59. Schilling, C. H. and Palsson, B. O. (1998) The underlying pathway structure of biochemical reaction networks. *Proc. Natl. Acad. Sci. USA* **95,** 4193–4198.

60. Yeung, M. K., Tegner, J., and Collins, J. J. (2002) Reverse engineering gene networks using singular value decomposition and robust regression. *Proc. Natl. Acad. Sci. USA* **99,** 6163–6168.

61. Price, N. D., Reed, J. L., Papin, J. A., Famili, I., and Palsson, B. O. (2003) Analysis of metabolic capabilities using singular value decomposition of extreme pathway matrices. *Biophys. J.* **84,** 794–804.

62. Vlad, M. O., Arkin, A. P., and Ross, J. (2004) Response experiments for nonlinear systems with application to reaction kinetics and genetics. *Proc. Natl. Acad. Sci. USA* **101,** 7223–7228.

63. Doyle, J. and Stein, G. (1981) Multivariable feedback design: Concepts for a classical/modern synthesis. *IEEE Trans. Automat. Contr.* **26,** 4–16.

64. Broomhead, D. S. and King, G. P. (1986) Extracting qualitative dynamics from experimental-data. *Physica D* **20,** 217–236.

3

Online Analysis of Microarray Data Using Artificial Neural Networks

Braden Greer and Javed Khan

Summary

Herein we have set forth a detailed method to analyze microarray data using artificial neural networks (ANN) for the purpose of classification, diagnosis, or prognosis. All aspects of this analysis can be carried out online via a website. The reader is guided through each step of the analysis including data partitioning, preprocessing, ANN architecture, and learning parameter selection, gene selection, and interpretation of the results. This is one possible method of many but we have found it suitable to microarray data and attempted to discuss universal guidelines for this type of analysis along the way.

Key Words: Microarray; gene expression; artificial neural networks; neural networks; machine learning; artificial intelligence; cancer; ANN; disease classification; disease diagnosis; disease prognosis.

1. Introduction

Artificial neural networks (ANNs) are computer learning algorithms that are patterned after the ability of the human neuron to learn by example. When a human neuron is presented with a similar signal repeatedly it can rewire its synapses to more efficiently recognize and transmit a signal. Similarly, when an artificial neuron is presented with a repeated signal (the training data), it can adjust its weighting factors through a process of error minimization according to the pertinent features of the input data and efficiently recognizes subsequent examples (the testing data). For a more detailed background of the theory of ANNs and their use, the reader is directed to several reviews and books *(1–6)*. ANNs are being increasingly developed and applied to classify, diagnose, and predict prognosis of diseases according to their gene expression signatures as measured by microarrays *(7–24)*. The wealth and complexity of microarray data lends itself well to the application of ANNs, and the ultimate promise of

From: *Methods in Molecular Biology, vol. 377, Microarray Data Analysis: Methods and Applications*
Edited by: M. J. Korenberg © Humana Press Inc., Totowa, NJ

the combination of these two technologies is accurate, inexpensive, and rapid diagnosis and prognosis in the clinic. To date, cancer research has nearly monopolized this powerful combination *(7–24)* with the exception of a study predicting the risk of coronary artery disease *(22)*. Although their diverse genetic mutations and misregulations make cancers excellent candidates for microarray and ANN, cancer is certainly not the only context that stands to benefit—the treatment and understanding of nearly every genetic disease could be advanced. In this chapter, the reader is guided through each step of the analysis process, from data partitioning, preprocessing, ANN architecture and learning parameter selection, gene selection, and interpretation of the results. It is our hope that the clear step-by-step instructions in this chapter and the user-friendly website we have developed will further the use of this powerful combination and benefit the greater research and medical communities.

2. Materials

1. Microarray data in tab-delimited .TXT format from samples with some known differential phenotype.
2. A computer with internet access.

3. Methods
3.1. Partition Data Into Training and Testing Sets

Care needs to be taken in this first very crucial step. An ample number of samples should be selected for training the networks lest they be naïve, and an ample number of samples should be selected for testing to give credence to the training. A rule of thumb we have used is to have at the very least 10 examples from each class for training (the heterogeneity of your data may require additional samples, but it is not recommended to use fewer) (*see* **Note 1**). In addition, the samples should be randomly distributed between training and testing such that no known distinctions delineate the two groups. One must avoid the temptation of putting the trouble samples into the training set and thereby artificially enhance the testing results. Finally, replicate samples are acceptable in the training set but should be not be split between training and testing sets.

3.2. Preparing the Input Files

There are two input files necessary to perform the ANN analysis via our website: a class file (*see* **Table 1**) and a data file (*see* **Table 2**). The data file should be in tab-delimited text format with the genes in rows and the samples in columns. The first column must be gene identifiers that must contain at least one non-numerical character in each gene name (i.e., '12345' is not acceptable, but 'Gene12345' is acceptable). The data file should have exactly one header row with the names of the samples in the exact row order of the samples in the class file (*see* **Table 1**).

Table 1
Sample Class File

Sample	Classname	Color	Train1;Test0
Sample1	Class1	4	1
Sample2	Class1	4	1
Sample3	Class1	4	1
Sample4	Class2	2	1
Sample5	Class2	2	1
Sample6	Class2	2	1
Sample7	TEST	5	0
…	…	…	…
SampleM	TEST	5	0

This file must be in tab-delimited text (.TXT) format. The header should be included but the exact column titles do not matter, only the order of the columns. The samples in rows should be in the exact column-order of the samples in the data file (*see* **Table 2**). Class name must not be exclusively numeric but must contain some text. In a leave-one-out analysis and a gene minimization analysis, the samples designated as "test" (0 in the Train/Test Column) will be discarded.

The class file should also be in tab-delimited text format and its purpose is to convey the *a priori* class information, as well as to designate samples for training or testing (*see* **Table 1**). The first column is a list of sample names that should each contain at least one character, similar to the gene identifier in the data file. It is imperative that the rows of samples in the class file be in the exact column order of the samples in the data file. The second column is the class name used for display purposes in the results, which should also include at least one character. For test samples, it is sufficient to put "Test" as the class name. The third column tells the program which color you want each class to be associated with. The colors and their numbers are listed on the website (**Fig. 1**). There should a one-to-one correspondence between the "Class" and "Color" columns. The fourth and last column tells the program which samples are to be used for training and which are to be used for testing. Assign a '1' to all the training samples and a '0' to all the testing samples.

3.3. Preprocess the Data

There are two major steps for data preprocessing available at our website: normalization and dimension reduction via principal components analysis (PCA).

3.3.1. Normalize the Data

Normalization is an important step in any data analysis. If the data is not normalized appropriately the rest of the analysis suffers. If you are analyzing ratio data, it is recommended that you always log the data prior to any analysis.

Table 2
Sample Data File

GeneID	Sample1	Sample2	Sample3	...	SampleM
Gene1	0.46	0.41	0.86	...	0.47
Gene2	0.16	0.80	0.29	...	0.55
Gene3	0.36	0.71	0.64	...	0.71
Gene4	0.23	0.24	0.80	...	0.92
Gene5	0.02	0.01	0.88	...	0.58
Gene6	0.28	0.71	0.05	...	0.33
Gene7	0.21	0.37	0.47	...	0.46
Gene8	0.31	0.71	0.59	...	0.98
Gene9	0.72	0.03	0.25	...	0.58
Gene10	0.51	0.26	0.04	...	0.59
...	...	...	...	...	...
GeneN	0.98	0.19	0.47	...	0.75

This file must be in tab-delimited text (.TXT) format. The sample columns should be in the exact row order of the samples in the class file (*see* **Table 1**). Class name must not be exclusively numeric but must contain some text. Gene name must not be exclusively numeric but must contain some text. In a leave-one-out analysis and a gene minimization analysis, the samples designated as "test" (0 in the train/test column) will be discarded.

This gives equal weighting to ratios between 0 and 1 and ratios greater than 1. The option on the website is only given so that those whose data is already logged can skip this step.

Next is the option of centering or Z-scoring the data by the mean or median (*see* **Note 2**). Centering the data subtracts the mean or the median of each gene (row) from each data-point in that row. Z-scoring the data centers the data first and then divides each data-point by the standard deviation of all the data-points of its row (*see* **Note 3**). The default settings are to log the data but not to Z-score the data. The option is given to the user, however, for those who would like to explore other normalization options. For Affymetrix data we recommend that intensities *not* be logged, but if the input is a ratio of intensities (based on a reference or a sample median), these should be logged.

3.3.2. Reduce the Dimensionality of the Data

ANN analysis with microarray data if not carefully performed will suffer from the "curse of dimensionality," in which the number of variables (genes) is much greater than the number of observations (samples). In a typical microarray dataset of 40,000 genes with 100 samples from two populations, an ANN will very likely find genes that will follow the desired pattern of differential expression between the two populations just by the sheer numbers of experiments (i.e., genes measured) performed. Because we are searching for biological

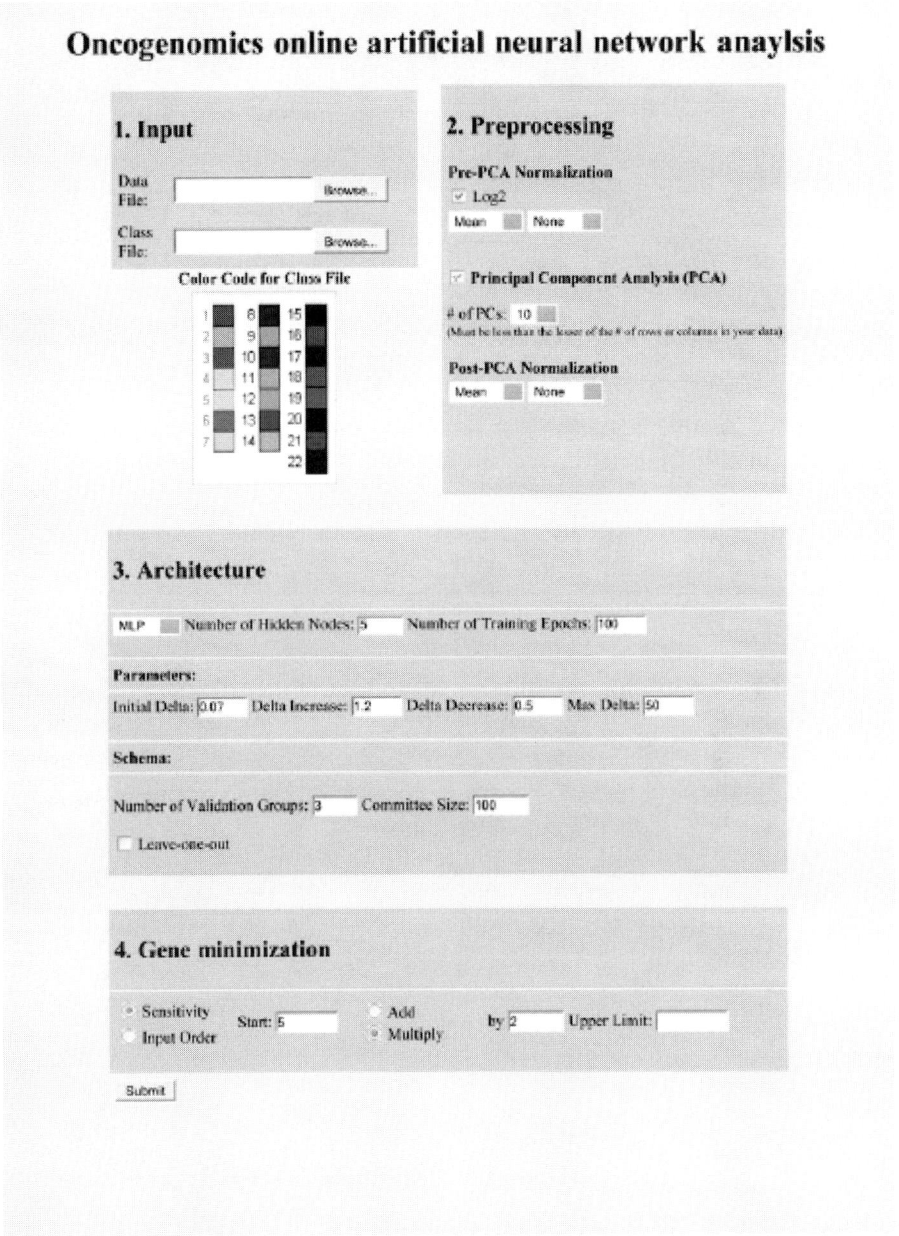

Oncogenomics online artificial neural network anaylsis

1. Input

Data File: [   ] Browse...

Class File: [   ] Browse...

Color Code for Class File

1		8		15	
2		9		16	
3		10		17	
4		11		18	
5		12		19	
6		13		20	
7		14		21	
				22	

2. Preprocessing

Pre-PCA Normalization

☑ Log2

Mean ▢ None ▢

☑ Principal Component Analysis (PCA)

# of PCs: 10 ▢
(Must be less than the lesser of the # of rows or columns in your data)

Post-PCA Normalization

Mean ▢ None ▢

3. Architecture

MLP ▢ Number of Hidden Nodes: [5] Number of Training Epochs: [100]

Parameters:

Initial Delta: [0.07] Delta Increase: [1.2] Delta Decrease: [0.5] Max Delta: [50]

Schema:

Number of Validation Groups: [3] Committee Size: [100]

▢ Leave-one-out

4. Gene minimization

○ Sensitivity
○ Input Order   Start: [5]   ○ Add   ● Multiply   by [2]   Upper Limit: [   ]

[Submit]

Fig. 1. Screenshot of the Oncogenomics online ANN user interface.

differences and not random noise, we must reduce the dimensionality of the dataset. This can be done by at least two common methods. The first is to select a subset of genes using a statistical filter (e.g., *t*-test, variance filter) where the number of genes is less than or equal to the number of samples. A second method, PCA (*see* **Note 4**), is available in the preprocessing stage. In brief, PCA transforms the data by first identifying the direction of greatest variance in the high-dimensional dataset and then creating new axes such that the first dimension is along the direction of greatest variance and subsequent axes capture less and less of the original variance. The result is that one can use the first 2 to 10 dimensions (components) of the transformed data for example, and not lose much information. This generates an input dataset that does not suffer from the "curse of dimensionality" because the number of variables (i.e., components in rows) is now much smaller than the number of observations (i.e., samples in columns). Next, the number of components used for input to the network must be selected. This decision depends on the complexity of the data. Somewhere in the range of 5 to 10 components should suffice for most microarray datasets on the order of 50k genes. Beyond 10 components the data will likely capture very little of the original variance in the data. As the number of genes in an experiment increases dramatically, the number of principal components necessary to capture the variance of the data may also increase. The default is to perform PCA and use the top 10 components as input.

3.3.3. Normalize the Reduced Data

The final step in preprocessing is to normalize the dimensional-reduced dataset. Some believe it is good practice to Z-score the reduced dataset prior to training to give equal variance to each of the components to aid training. Similar normalization options are available as described in **Subheading 3.3.1**. The default is to Z-score the principal components.

3.4. Architecture

In this section we will discuss the methods and parameters for learning. The first decision is the choice between a linear network and a multilayer perceptron (MLP) network. The linear network has only two layers: an input and an output layer; whereas the MLP network inserts one hidden layer between the input and output layers (in principle many hidden layers can be used, but we have implemented only one hidden layer, which should be sufficient for most microarray studies). The hidden layer in the MLP allows the network to learn more complex nonlinear signals from the data (*see* **Note 5**). If MLP is selected the number of hidden nodes needs to be chosen. There are a wide variety of rules of thumb for selecting the appropriate number of hidden nodes and some are listed next. We are not in favor of any of these because they do not take into account several factors including number of training cases, noise, and so on. We have included them, however, to give the user a starting point to work from.

1. Size of this (hidden) layer to be somewhere between the input layer size and the output layer size *(25)*.
2. Number of inputs + outputs * (2/3).
3. Never require more than twice the number of hidden units as you have inputs in an MLP with one hidden layer *(26,27)*.
4. As many hidden nodes as dimensions (principal components) needed to capture 70–90% of the variance of the input dataset *(28)*.

Trial-and-error starting from one or more of these rules of thumb is our suggested method. Remember, though, that the greater number of hidden nodes, the more complicated signal the networks can learn. We have found only minimal benefit, to more than three to five hidden nodes for our datasets. Do your own experimenting however, and determine how many nodes will suit your particular situation. Finally, the number of training epochs or cycles needs to be set. The default value of 100 epochs should be sufficient for most applications. Often the error has reached its lower limit well before 100 epochs, but it is better to perform too many epochs rather than too few. The risk of overtraining through too many epochs is minimal if one has taken care to reduce the dimensionality of the data appropriately and incorporate an appropriate cross-validation scheme (*see* **Subheading 3.4.2.**).

3.4.1. Learning Parameters

We chose to use the resilient back-propagation algorithm to train the neural networks for our website for its speed and ease of use. This algorithm has the desirable property that it is relatively insensitive to changes in the learning parameters *(29)*. This is an excellent property for someone who wants to use ANNs but is not get bogged down endlessly tuning a host of learning parameters. Nonetheless, the pertinent learning parameters for this algorithm are adjustable from the user interface. Resilient back-propagation employs a tuning parameter, referred to as "delta," which controls the degree to which the weights of the network will be penalized for error. "Initial delta" is the penalty for the first error, after which the penalty will increase and decrease according to "delta increase" and "delta decrease," respectively. "Max delta" sets the upper limit for the delta penalty factor. For most applications it will be sufficient to leave these parameters at their defaults. The defaults are as follows: initial delta, 0.07; max delta, 50; delta increase, 1.2; delta decrease, 0.5.

3.4.2. Cross-Validation

Cross-validation is an important procedure to ensure properly trained networks. In this context, validation is a technique whereby a subset of training samples are set aside during the learning process and used to validate the trained networks. The classification error of the validation samples is monitored as the learning process cycles through the specified number of epochs. The classification error of the validation samples should decrease rapidly and remain low. If the validation error increases with increasing epochs, then the network is learning features of the training set that are not generalizable, but are sample specific, and training is stopped. The validation samples act as a kind of warning for the network to stop learning to prevent what is known as "over-training." Our software allows you to partition the training data into a specified number of randomly selected validation groups. This works as follows: if the user chooses m validation groups, and there are N training samples, then $N(m-1)/m$ samples will be used to train and N/m samples will be used to validate the network. The program will iterate through each of the m groups such that each one will be employed as a validation group exactly one time, for a total of m training iterations. A general rule of thumb for choosing the number of validation groups is to ensure at least $\frac{1}{2}$ of your training samples from the category with the fewest samples will always be in the $N(m-1)/m$ group. For example, if you have 30 (N) samples from 2 populations and the least-represented population has 10 samples, then 6 (m) validation groups would be a good choice because the validation groups would have 5 samples and there would never be a situation where there were very few training samples from either population (*see* **Note 6**). Another consideration is to ensure that all populations will be represented in the validation group. If you split your 30 samples into 15 validation groups, it's very likely that many of your randomly selected groups of two will only have one population represented. If the number of training samples, N, is not divisible by the number of selected groups, m, the program will compensate and form validation groups with slightly different sample sizes.

3.4.3. Committee Voting

When randomly selecting groups for cross-validation (**Subheading 3.4.2.**) it is possible that one could introduce a bias by grouping all of a certain sample type, or problem samples together in a validation group. To avoid this possible error, it is important to repeat the process of randomly selecting groups, training, and validating many times over and report results based on averages of these analyses. In addition, repeating the training process many times allows us to calculate an empirical confidence interval from the training data by which we can accept or reject the output votes for the testing set. The default value of 100 should be sufficient for most applications, but it is a good idea to verify this by monitoring the results with several increments of votes (*see* **Note 7**).

3.4.4. Leave-One-Out Analysis

The leave-one-out option (*see* **Note 8**) is useful to see what would happen if each of your samples was presented to the fully-trained network as a blind test sample. This is a separate consideration from cross-validation discussed in **Subheading 3.4.2.** In this case, 1 sample from the N total samples is set aside and is not used in the learning process at all. After each network is trained, the 1 sample is presented as a test and the resulting vote is stored. After all of the networks have completed training (the number of which is decided by the number of committees) and the test sample is tested each time, the average vote for the test sample is calculated. Next, the test sample is replaced into the dataset and a new test sample is selected and the process is repeated until each of the N samples has been presented to the network as a blind test sample exactly one time. The results are as if all of your samples were in the testing set. It is a very conservative way to analyze your data. As you could imagine it can take a long time to run—sometimes several days of computing are required (*see* **Note 9**).

3.5. Gene Minimization

In a typical microarray experiment the expression of tens of thousands of genes is measured, and in a typical study the number of genes that are significantly differentially expressed is on the order of tens or hundreds, occasionally thousands. It is therefore advantageous to remove the uninteresting genes and thereby reduce the noise in the dataset, as well as discover meaningful biology through the identification of genes implicated in a disease or process. To achieve these ends we have implemented a gene minimization algorithm that will rank the genes based on their importance to the classification and then retrain the networks using increasing numbers of the top-ranking genes while monitoring the classification error (*see* **Note 10**). One can then select the subset of top-ranking genes that produces the minimum error to train and then test blinded samples. The option is given to you also to perform the minimization using the "Input Order" if your data file is already sorted according to your favorite gene ranking statistic (e.g., *t*-test, rank-sum test) (*see* **Note 11**). The order should be from most to least important (e.g., the first gene should have the highest *t*-value or lowest *p*-value).

The "Start" parameter allows you to choose the number of top-ranking genes to train with in the first run. You can then choose to increase the number of top-ranking genes to use in successive training by adding or multiplying the current number by a user-defined factor. For example, if you start with 5 and multiply by 2, you will train with the top 5, 10, 20, 40, 80, and so on genes. You can also limit the number of additional trainings by defining the upper limit. For example, if you started with 100 genes and added 100 genes and defined the upper limit as 500, you would train with the top 100, 200, 300, 400, and 500 genes. The default is to start with the top 5 genes and multiply by 2 while the number of selected genes is less than or equal to the total number of genes.

3.6. Results

When the program has completed analyzing your data, you will be notified via email with a link you can download your results from. The files will be as follows:

1. A .TXT file with "Votes" in the file name and columns with the sample names, train/test value, class number, ANN prediction, confidence intervals (when the number of classes > 2), average committee vote (i.e., validation votes for training samples and test votes for testing samples), and standard error of the committee votes.
2. A .JPG file with "Votes" in the file name visually representing the voting data in the .TXT file described in **item 1** only if there are two classes, if there are more it is difficult to visualize this.
3. A .TXT file with "GeneRank" in the file name with the columns GeneID, rank, total sensitivity, sensitivity, and sign. (Sensitivity and sign will be repeated for each class in analyses with three or more classes. In the case of two classes, there is only one output and therefore, one sensitivity measure.)
4. A .JPG file with the "Legend" in the name which contains the class names and colors for each of the output figures described in **items 2** and **5**.
5. A .JPG file of the first three principal components of the data (if PCA was performed).
6. A .JPG file with "GeneMinimization" in the name, which is a barplot of the average number of misclassifications of training samples (y-axis) including standard error with training based on increasing numbers of the top-ranking genes (x-axis).

4. Notes

1. The number of training samples necessary to perform a valid analysis is also proportional to the complexity of the question being asked. In the case of diagnosis between different tumor types, for example, 10 samples might be sufficient. On the other hand, a prognosis study might require many more samples because the difference between the classes is likely to be much more subtle and the expression profile within a class more heterogeneous.
2. The choice between mean or median for centering purposes will not usually alter the results too drastically. In fact with increasing number of samples from a normal distribution, the median should approximate the mean. The median is helpful to reduce the influence of an extreme outlier that could affect the mean of a dataset with a small number of samples. With increasing sample size, however, the influence of an outlier on the mean is diminished.
3. All of the normalization options on our website perform normalization in the gene direction (in our case, row-wise). If you have systematic sample-specific bias owing to different microarray print lots or who performed the experiment, you should remove these via normalization in the sample direction (in our case, column-wise). *See* **ref. 30** for a review of normalization techniques.
4. For a more in-depth description of the theory of principal components analysis, *see* **ref. 31**.

5. The choice between linear and MLP is dependent on the complexity of the input signal. From our experience, an MLP will yield somewhat better results. With many datasets, though, a linear network will yield sufficient results. The reader is encouraged to explore both options with their dataset.

6. This rule of thumb is very conservative. In reality, when choosing m samples at random from a dataset with N samples across several populations, the expected number of randomly selected samples, p, from the least represented population with r samples is, of course, proportional: $p = mr/N$. So, one would expect the least-represented population to have the least samples in the validation group. Where sample size is relatively equal across populations, the rule of thumb from **Subheading 3.4.2.** should be followed. If sample size is very unequal across populations, then one may use the above expectation value as a guide to selecting the validation group size. Remember that the fewer the validation groups, the faster the run time.

7. In particular, watch that the confidence interval and gene ranking stabilize. The voting results should stabilize with relatively few votes, but the confidence interval and gene ranking require more votes to stabilize.

8. There is sometimes some misunderstanding regarding the leave-one-out analysis. It is important not to confuse this with the cross-validation step. The leave-one-out analysis is outside of the cross-validation step, in that the cross-validation has no knowledge of the left-out sample. Indeed, it would not be prudent to perform a leave-one-out cross-validation as the one validation sample would not be representative of the entire training population and the result would be a training process tailored to the one validation sample. In the leave-one-out analysis in **Subheading 3.4.4.**, the left-out sample has no affect on the training of the networks whatsoever. It is as if you performed as many analyses as you had samples each time designating one sample for testing (marked with a '0' in the train/test column in the class file) and concatenated the testing results into one spreadsheet or one visualization. Therefore, the training of the networks is not tailored to the one left-out sample in this analysis.

9. One important caveat is that you should not perform any supervised gene selection prior to the leave-one-out analysis. If you do, the blind test sample is no longer blind because it has influenced the selection of the genes. This is why the test is usually a more conservative estimate of the ability of your data to predict blind test samples. If you do an analysis with separate training and testing datasets, you will be able to minimize the genes (and thereby reduce noise), and an increase in the prediction accuracy should be realized. Leave-one-out analysis results should be interpreted with this in mind.

10. The sensitivity of a gene is calculated by taking the derivative of the output divided by the derivative of the input. For complete details *see* the Supplementary Methods in **ref. *18***.

11. It was noted before in **Note 9**, but it is worth repeating that any supervised gene selection should not have included the test samples. If you select your genes taking the test samples into consideration, they are no longer blind test samples.

References

1. Peterson, C. and Ringner, M. (2003) Analyzing tumor gene expression profiles. *Artif. Intell. Med.* **28**, 59–74.
2. Ringner, M. and Peterson, C. (2003) Microarray-based cancer diagnosis with artificial neural networks. *Biotechniques* **Suppl**, 30–35.
3. Ringner, M., Peterson, C., and Khan, J. (2002) Analyzing array data using supervised methods. *Pharmacogenomics* **3**, 403–415.
4. Greer, B. T. and Khan, J. (2004) Diagnostic classification of cancer using DNA microarrays and artificial intelligence. *Ann. NY Acad. Sci.* **1020**, 49–66.
5. Dayhoff, J. E. and Deleo, J. M. (2001) Artificial neural networks: opening the black box. *Cancer* **91**, 1615–1635.
6. Haykin, S. (1999) *Neural Networks: A Comprehensive Foundation* Prentice-Hall, Upper Saddle River, NJ.
7. Ando, T., Suguro, M., Hanai, T., Kobayashi, T., Honda, H., and Seto, M. (2002) Fuzzy neural network applied to gene expression profiling for predicting the prognosis of diffuse large B-cell lymphoma. *Jpn. J. Cancer Res.* **93**, 1207–1212.
8. Berrar, D. P., Downes, C. S., and Dubitzky, W. (2003) Multiclass cancer classification using gene expression profiling and probabilistic neural networks. *Pac. Symp. Biocomput.* 5–16.
9. Bicciato, S., Pandin, M., Didone, G., and Di Bello, C. (2003) Pattern identification and classification in gene expression data using an autoassociative neural network model. *Biotechnol. Bioeng.* **81**, 594–606.
10. Bloom, G., Yang, I. V., Boulware, D., et al. (2004) Multi-platform, multi-site, microarray-based human tumor classification. *Am. J. Pathol.* **164**, 9–16.
11. Ellis, M., Davis, N., Coop, A., et al. (2002) Development and validation of a method for using breast core needle biopsies for gene expression microarray analyses. *Clin. Cancer Res.* **8**, 1155–1166.
12. Futschik, M. E., Reeve, A., and Kasabov, N. (2003) Evolving connectionist systems for knowledge discovery from gene expression data of cancer tissue. *Artif. Intell. Med.* **28**, 165–189.
13. Gruvberger, S., Ringner, M., Chen, Y., et al. (2001) Estrogen receptor status in breast cancer is associated with remarkably distinct gene expression patterns. *Cancer Res.* **61**, 5979–5984.
14. Gruvberger, S. K., Ringner, M., Eden, P., et al. (2003) Expression profiling to predict outcome in breast cancer: the influence of sample selection. *Breast Cancer Res.* **5**, 23–26.
15. Gruvberger-Saal, S. K., Eden, P., Ringner, M., et al. (2004) Predicting continuous values of prognostic markers in breast cancer from microarray gene expression profiles. *Mol. Cancer Ther.* **3**, 161–168.
16. Kan, T., Shimada, Y., Sato, F., et al. (2004) Prediction of lymph node metastasis with use of artificial neural networks based on gene expression profiles in esophageal squamous cell carcinoma. *Ann. Surg. Oncol.* **11**, 1070–1078.

17. Linder, R., Dew, D., Sudhoff, H., et al. (2004) The 'subsequent artificial neural network' (SANN) approach might bring more classificatory power to ANN-based DNA microarray analyses. *Bioinformatics* **20**, 3544–3552.
18. Khan, J., Wei, J. S., Ringner, M., et al. (2001) Classification and diagnostic prediction of cancers using gene expression profiling and artificial neural networks. *Nat. Med.* **7**, 673–679.
19. Liu, B., Cui, Q., Jiang, T., and Ma, S. (2004) A combinational feature selection and ensemble neural network method for classification of gene expression data. *BMC Bioinformatics* **5**, 136.
20. O'Neill, M. C. and Song, L. (2003) Neural network analysis of lymphoma microarray data: prognosis and diagnosis near-perfect. *BMC Bioinformatics* **4**, 13.
21. Selaru, F. M., Xu, Y., Yin, J., et al. (2002) Artificial neural networks distinguish among subtypes of neoplastic colorectal lesions. *Gastroenterology* **122**, 606–613.
22. Tham, C. K., Heng, C. K., and Chin, W. C. (2003) Predicting risk of coronary artery disease from DNA microarray-based genotyping using neural networks and other statistical analysis tool. *J. Bioinform. Comput. Biol.* **1**, 521–539.
23. Wei, J. S., Greer, B. T., Westermann, F., et al. (2004) Prediction of clinical outcome using gene expression profiling and artificial neural networks for patients with neuroblastoma. *Cancer Res.* **64**, 6883–6891.
24. Xu, Y., Selaru, F. M., Yin, J., et al. (2002) Artificial neural networks and gene filtering distinguish between global gene expression profiles of Barrett's esophagus and esophageal cancer. *Cancer Res.* **62**, 3493–3497.
25. Blum, A. (1992) *Neural Networks in C++.* Wiley, New York, NY.
26. Swingler, K. (1996) *Applying Neural Networks: A Practical Guide* Academic Press, London, UK.
27. Berry, M. A. L., G (1997) *Data Mining Techniques* John Wiley and Sons, New York, NY.
28. Boger, Z. and Guterman, H. (1997) Knowledge extraction from artificial neural network models. IEEE Systems, Man, and Cybernetics Conference, Orlando, FL.
29. Demuth, H. B., Mark (2001) *Neural Network Toolbox* 4th. The Mathworks, Natick, MA.
30. Yang, Y. H., Dudoit, S., Luu, P., et al. (2002) Normalization for cDNA microarray data: a robust composite method addressing single and multiple slide systematic variation. *Nucleic Acids Res.* **30**, e15.
31. Joliffe, I. T. (2002) *Principal Component Analysis, 2nd ed.,* Springer-Verlag, New York, NY.

4

Signal Processing and the Design of Microarray Time-Series Experiments

Robert R. Klevecz, Caroline M. Li, and James L. Bolen

Summary

Recent findings of a genome-wide oscillation involving the transcriptome of the budding yeast *Saccharomyces cerevisiae* suggest that the most promising path to an understanding of the cell as a dynamic system will proceed from carefully designed time-series sampling followed by the development of signal-processing methods suited to molecular biological datasets. When everything oscillates, conventional biostatistical approaches fall short in identifying functional relationships among genes and their transcripts. Worse, based as they are on steady-state assumptions, such approaches may be misleading. In this chapter, we describe the continuous gated synchrony system and the experiments leading to the concept of genome-wide oscillations, and suggest methods of analysis better suited to dissection of oscillating systems. Using a yeast continuous-culture system, the most precise and stable biological system extant, we explore analytical tools such as wavelet multiresolution decomposition, Fourier analysis, and singular value decomposition to uncover the dynamic architecture of phenotype.

Key Words: Genome-wide; transcription; oscillation; attractor; microarray; singular value decomposition; SVD; replicates.

1. Introduction

The idea that the cell is an oscillator, an attractor, and that time is a variable of the system, though well supported by both theory and experimental findings, is still something of a novelty in genomics *(1–4)*. Prior to the development of genome-wide assays, experimental support for viewing the cell as an attractor was limited to measurement of single constituents or to analysis of the response of cells to intentional perturbations to the cell cycle *(5)*.

Now, for the first time, we have the capacity to make precise measurements of all of the transcripts of a cell, most of the metabolites and, soon, one might project, all of the proteins in a quantitative manner. Recently, we took advantage of microarray technology to measure all of the transcripts of yeast cells

From: *Methods in Molecular Biology, vol. 377, Microarray Data Analysis: Methods and Applications*
Edited by: M. J. Korenberg © Humana Press Inc., Totowa, NJ

growing synchronously with respect to their respiratory/reductive cycle (6). This cycle, which switches its redox state from respiration to reduction with great precision, gives us the first glimpse into the evolutionary early molecular organization of cells as they dealt with the transition from a reductive to an oxidizing environment. The metabolic state of these cultures appears to be an excellent benchmark and manifestation of the temporal organization of transcription. As a practical matter, the precision and stability of the cycle allows the ready development of techniques for time-series analysis of microarray data that can be used in mammalian systems.

Feasibility forces the consideration of when genome-wide oscillations can be exploited to give a clearer insight into cellular regulatory mechanisms and when, because of limited control over the biological system, they can, at best, only be accounted for and not exploited. In either case, it is no longer sufficient to assume because no particular effort has been put into synchronizing a cellular system, that it is necessarily random or exponential. If cell-to-cell signaling in a single-celled organism such as yeast gives rise to spontaneous oscillations and gated synchrony in the culture as a whole, then mammalian cell cultures and tissues, where cell-to-cell connectivity and signaling are well recognized, partial synchronization is a near certainty, and the deviation from randomness that this represents, becomes a problem for microarray analysis.

Most important for the microarray field at the present moment is the realization that it may be much more informative to take a careful sampling of a system through time than to take multiple samples without regard to time. We will show evidence in this work that once the uncertainty from time variation in gene expression is removed, the Affymetrix system is capable of remarkable precision with signal to noise of 60 decibels in respiratory-phase transcripts. In these studies, only a few of the samples were done in duplicate or triplicate in the conventional statistical sense. Rather, close time sampling through multiple cycles were taken giving the option of phase aligning and averaging the data into a single cycle, and by this act, generating a combined biological and oscillator-phase replicate, or displaying the dataset as an oscillation and analyzing it using signal-processing methods.

All of the data presented and analyzed here is derived from expression-array analysis using the Affymetrix yeast S98 chip and the new Yeast2 chip. In order to optimize new analysis methods, we felt it would be best to use the most accurate biological and measurement systems. Spotted-array analyses were not included because of their greater inherent noise and platform-to-platform variability.

2. Materials

1. Fermenters (B. Braun Biotech, Aylesbury, Buckinghamshire, UK; model: Biolab CP; working volume of 650 mL).
2. KH_2PO_4 monobasic, $CaCl_2 \cdot 2H_2O$, $(NH_4)_2SO_4$, $MgSO_4 \cdot 7H_2O$, $CuSO_4 \cdot 5H_2O$, and $MnCl_2 \cdot 4H_2O$ (J. T. Baker, Philipsburg, NJ); H_2SO_4, acid-washed glass beads,

2- mercaptoethanol, antifoam A and D(+)-glucose monohydrate (Sigma, St. Louis, MO); $FeSO_4 \cdot 7H_2O$ (Mallinckrodt, Paris, KY); $ZnSO_4 \cdot 7H_2O$ (EM Science, Darmstadt, Germany); yeast extract (Difco, Sparks, MD); RNA later, GeneChip Expression Kit, and poly(A) standards (Ambion, Ambion, TX); RLT buffer, RNA easy mini kit, and DNase (Qiagen, Valencia, CA).
3. The Mini Bead beater (BioSpec Products, Inc., Bartlesille, OK.).
4. RNA was examined for quality using capillary electrophoresis with the Agilent 2100 Biosizer (Agilent Technologies, Palo Alto, CA).
5. RNA Lab-On-A-Chip (Caliper Technologies Corp., Mountain View, CA).
6. Yeast arrays, GeneChip hybridization oven 640, Fluidics Station 450, and GeneArray scanner (Affymetrix, Santa Clara, CA).
7. Mathcad is from Mathsoft Inc. (Cambridge, MA); Mathematica is from Wolfram Research (Champaign, IL); SigmaPlot is from Systat Software Inc. (Point Richmond, CA); and MatLab is from The Mathworks Inc. (Natick, MA).

3. Methods

3.1. Culture Conditions and Monitoring of the Oscillation

1. The basic medium: $(NH_4)_2SO_4$ (5 g/L), KH_2PO_4 (2 g/L), $MgSO_4$ (0.5 g/L), $CaCl_2$ (0.1 g/L), $FeSO_4$ (0.02 g/L), $ZnSO_4$ (0.01 g/L), $CuSO_4$ (0.005 g/L), $MnCl_2$ (0.001 g/L), 70% H_2SO_4 (1 mL/L), and yeast extract (1 g/L).
2. Glucose medium is supplemented with 22 g/L glucose monohydrate and 0.2 mL/L antifoam A.
3. The fermenters are operated at an agitation rate of 750 rpm, an aeration rate of 150 mL/min, a temperature of 30°C, and a pH of 3.4 or 4.0. Cultures are not nutrient limited and glucose levels oscillate between 50 and 200 μM in each cycle.
4. The oscillations reported are not unique to this strain, IFO 0233, and are achieved under culture conditions suited to an acidophile, such as *Saccharomyces cerevisiae*. The system for establishing and continuously monitoring synchrony has been carefully engineered to make it possible to perform molecular, biological, and cell biological sampling as frequently as required without perturbation. The strains have been analyzed by flow cytometry together with a number of commonly used haploid and diploid strains to show that it is a diploid. The diploid strains IFO 0224, NCYC 87, NCYC 240, and PC 3087 have also been tested and show oscillatory dynamics under different conditions (unpublished). Along with IFO 0233, these are all wild-type brewing, distilling, bread and/or spoilage strains of *S. cerevisiae*.
5. Continuous synchrony cultures of yeast are typically maintained and monitored for many weeks after their initial establishment (**Fig. 1**). Measurement of the dissolved oxygen (DO) concentration, O_2, CO_2, and H_2S levels are made every 10 s and determination of the period of the oscillation and its variability is made each day. Periods typically are in the range of 40–45 ± 0.5 min (*7–10*). As part of the standard procedure in the lab, the oscillation in dissolved O_2 is monitored before, during, and following sampling for RNA isolation. In this way, it is possible to reduce concerns regarding the degree of synchrony, the absence of perturbation, and the stability of the oscillation.

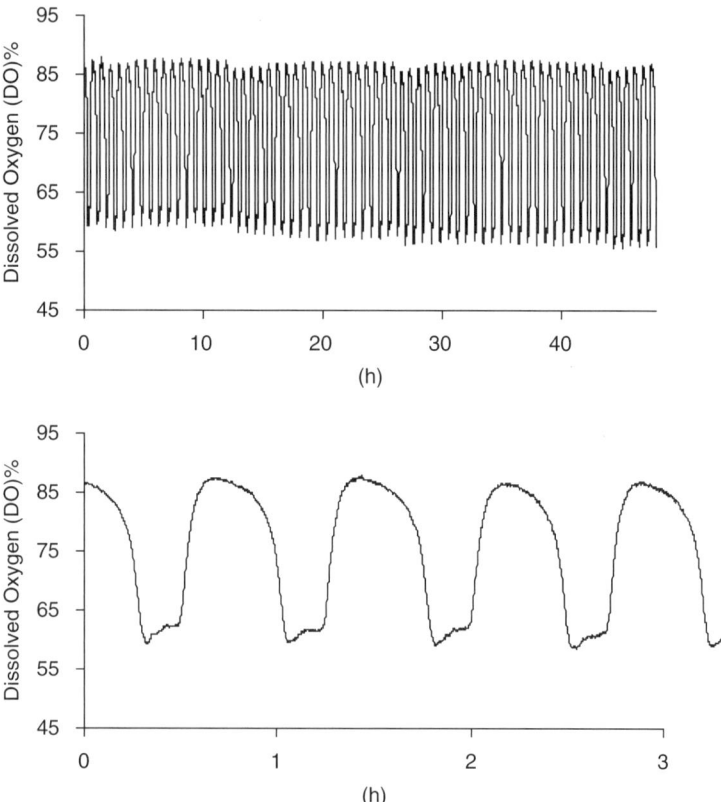

Fig. 1. Respiratory oscillations in continuous cultures. Respiratory oscillations begin soon after inoculation and continue with little change in period or amplitude. Dissolved oxygen level is shown for 48 h. The shape of the oscillation can be seen more clearly in the lower panel, where a segment of the curve of the upper panel has been expanded.

3.2. Oscillations in Batch Cultures

DO levels or other measures of the respiratory oscillation are not routinely monitored in most laboratories, and yet, synchronization of the respiratory–reductive cycle appears to be a widespread occurrence in batch cultures. Monitoring of DO levels in batch cultures shows that 18–24 h after inoculation, at a point where glucose levels have fallen below 200 μM and cell number is greater than ~5 × 10^7 cells/mL, the oscillation begins and typically endures for 6–10 cycles (**Fig. 2**). Autonomous oscillations in yeast have been known for many years, and appear to involve a mutual synchronization or entrainment between member cells in the population (*11,12*). The emergence of oscillations following synchronization is a reflection of the fact that single

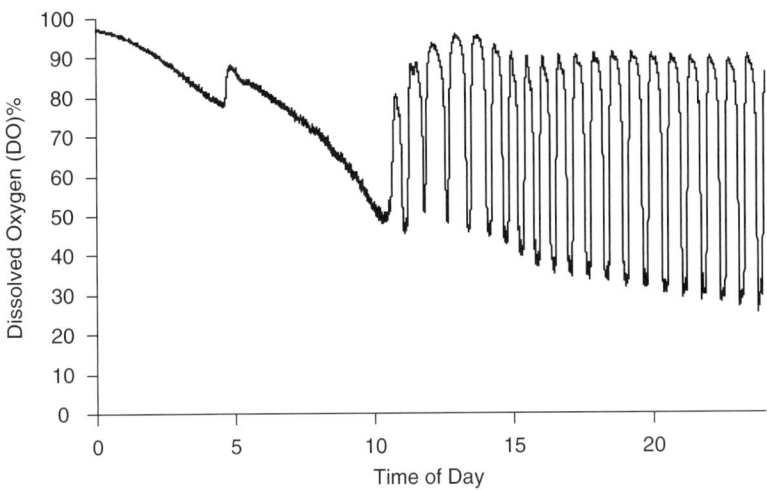

Fig. 2. Dissolved oxygen oscillations in *S. cerevisiae* grown on glucose medium. The fermenter was inoculated with 2×10^7 cells in 650 mL and grown as described in **Subheading 3**. Fermentative growth on glucose was observed during the first 12 h following inoculation. Oscillatory dynamics typically appear beginning 16 to 24 h after inoculation and 6 or more short period cycles are then observed. Once all the available carbon sources are catabolized, the culture enters stationary phase where oxygen consumption ceases. To initiate oscillations in plateau phase, culture medium is added and removed at a rate of 0.086/min. Once established, oscillatory dynamics remain largely unchanged for weeks to months. Normally, periodicity remains between 40–45 min. Dissolved oxygen levels and carbon dioxide release are the most accessible output from the oscillator and are characterized by a phase of high respiration followed by a shift to a low respiration phase. No difference in oscillation was seen in light or darkness. The oscillation is dependent on pH, aeration, and carbon dioxide. Oscillation also occurs when glucose, ethanol, or acetaldehyde is used as a carbon source.

cells are autonomous oscillators. In an effort to define the underlying mechanism, culture conditions favoring stable, continuous oscillatory behavior have been worked out. However, as **Fig. 2** shows, these oscillations can occur spontaneously in "overnight" batch cultures where no particular effort has been made to facilitate their appearance by manipulation of culture conditions. The occurrence of oscillations in these "overnight" cultures is one of the most repeatable behaviors seen in this culture system. For reasons that are not clear, these batch-culture oscillations are almost invariant, whereas setting up conditions to achieve the optimal amplitude and stability and long-term oscillations in a continuous cultures system is more uncertain. One concern should be that these oscillations, if undetected in other laboratories, could contribute to a seemingly intractable biological variability in many experimental designs.

3.3. Total RNA Preparation

1. Cells from the fermenter (0.5 mL) were collected every 4 min (*see* **Note 1** for sampling interval). The cells are pelleted, the supernatant decanted, and the pellet is placed in a dry ice acetone bath or in liquid nitrogen. Samples are stored at –80°C. The time from removal of the sample to freezing is less than 1 min. Cell numbers are kept between $0.5–1 \times 10^9$/mL.

2. For RNA isolation, the pellet is resuspended in 0.5 mL of RNA later containing 10 µL 2-mercaptoethanol/mL RNA later. Cells are lysed by beating in a Mini Bead beater for 3 min with 0.5 mL acid-washed glass beads. After the cell lysate is removed, the beads are washed three times with 0.5 mL Qiagen RLT buffer containing 10 µL 2-mercaptoethanol/mL RLT buffer by bead beater (1 min each wash). The cell lysate and washes are pooled. An equal volume of 70% ethanol is added, and RNA is purified with RNA easy columns according to the manufacturer. DNA is digested on the columns according to the instructions. RNA is eluted two times in RNase-free water with a volume of 50 µL each time so that the total volume is 0.1 mL. The final RNA samples are analyzed by capillary electrophoresis. Typical total RNA yields are 20–40 µg with absorbance 260/280 ratios of 1.8–2.2.

3. In a synchronous cell system, where there is reason to think that the level of mRNA is not constant through the cycle, a method for adjusting for differences in recovery, for amplification, and for hybridization is essential (*see* **Note 2**). In order to normalize RNA yields between different samples, a fixed amount of polyadenylated *B. subtilis lys, phe, thr,* and *dap* poly(A) standards are added to cells before lysis. Fourteen microliters of 1:500 premixed poly(A) standards are added to every 0.5 mL pellet of cells resuspended in 0.5 mL of RNA later before cell lysis and RNA purification in order to achieve a reasonable signal on the microarray.

4. The new yeast S2 chip contains the complete probe set for both *S. cerevisiae* and *S. pombe,* and this combination offers a second and potentially more robust method of normalization. A constant number of *S. pombe* cells (about 5% of the *S. cerevisiae* cells) is added to each experimental sample, and the two RNAs were isolated together. Control experiments have shown that less than 20 of the 5000 pombe transcripts bind at greater than background levels to the *S. cerevisiae* probes. By setting the total hybridization or a selected subset of the hybridized transcripts to a constant value, variations in mRNA yields between samples can be normalized. More details are described in **Subheading 3.7.**

3.4. Target Preparation/Processing for Affymetrix GeneChip Analysis

1. Purified total RNA samples are processed as recommended in the Affymetrix GeneChip Expression Analysis Technical Manual. RNA samples are adjusted to a final concentration of 1 µg/µL. Typically, 25–250 ng are loaded onto an RNA Lab-On-A-Chip and analyzed in an Agilent Bioanalyzer 2100.

2. Double-stranded cDNA is synthesized from 5 µg of total RNA using GeneChip Expression 3′-Amplification Reagents One-Cycle cDNA Synthesis Kit and oligo-dT primers containing a T7 RNA polymerase promoter.

3. Double-stranded cDNA is used as a template to generate biotinylated cRNA using the GeneChip Expression 3'-Amplification Reagents for IVT Labeling (*see* **Notes 3** and **4**). The biotin-labeled cRNA is fragmented to 35–200 bases following the Affymetrix protocol.

4. Five micrograms of fragmented cRNA is hybridized to Yeast 2.0 Affymetrix arrays at 45°C for 16 h in a hybridization oven.

5. The GeneChip arrays were washed and then stained with streptavidin-phycoerythrin on an Affymetrix Fluidics Station 450, followed by scanning on an Affymetrix GeneArray scanner.

3.5. Data Analysis

In the Notes section, we describe the standard path for analysis of microarray experiments. Raw results are collected first into Excel where the P,M,A, (present, marginal, or absent) discrimination is made. Adjustments are then made for hybridization and RNA-recovery differences and the intensity values were scaled accordingly. These adjustments could also be done using the Affymetrix GCOS software. In some instances, the Excel files are converted back to .txt or .csv to permit further processing. These files are then put into Mathcad, Mathematica, SigmaPlot, or MatLab. Intensity values for each of the verified open reading frames (ORFs) in the S98 chip and the yeast S2 chip are linked to the SGD (Saccharomyces Genome Database) site and both their genetic and physical map locations can be associated with the intensity values for each gene. The results for all ORFs scored as present using the default Affymetrix settings are identified according to the original sample number and the phase in the DO oscillation to which are mapped for presentation. Further analysis was performed for all ORFs present in all samples in each of the three cycles. In a recent experiment, of the ORFs scored as present by these criteria, all 5443 had average p-values less than 0.035 and 5254 had p-values less than 0.01.

3.6. Normalization With Constitutive or Maintenance Genes

One important issue that must be considered relates to the general applicability of the proposed time-series analyses. The findings reported here indicate that the choice of controls must involve more than the assumption that if a culture has not been intentionally synchronized or perturbed, it is necessarily random or stable. In several microarray-assay systems, housekeeping genes have been used as internal standards or as a means of estimating noise in the assay.

The use of actin and other constitutive, maintenance, or housekeeping genes as normalizing standards is a time-honored practice in PCR and other amplification assays. Both the singular value decomposition (SVD) and wavelet decomposition studies rely in different ways on the global behavior of transcription to make their case. It is now clear from our earlier study that the constitutive gene transcripts are not constant through the transcriptional cycle. Earlier, Warrington et al. *(18)* addressed this question in an analysis of human adult and fetal tissues. Of the 535 genes identified as highly expressed in all tissues examined, all but 47 varied by greater than 1.9-fold. They caution that further analysis might find regular variations in these transcripts as well. A gene may be constitutive even though its transcript is not maintained at a constant level through a cycle. Constitutive expression is not constant expression.

3.7. Normalizing for RNA Recovery, Copying, Amplification, and Hybridization

At each stage in the process of measuring transcript levels in the Affymetrix system, the protocol calls for bringing the amount of material to the same concentration. Upon completion of the procedures, each chip is scaled to a target value. This raises a point of interest. How can one expect to quantify, or even qualitatively detect differences between samples using this approach? It assumes that the total message synthesis and the levels of specific messages will be very similar between samples. As we have seen, this appears not to be the case in the gated synchrony system. Because there is evidence in our system, as well as mammalian systems, that constitutive transcripts are not constant through the cycle, their use as a standard for normalization is not correct. However, because the amplitude of their oscillation is low with an average 1.25- fold peak-to-trough ratio, they can be used semiquantitatively to verify that there is a change in those transcripts showing high-amplitude oscillations. This is not an entirely satisfactory solution to the problem. We have sought other methods to normalize the data.

There is the potential for a phase obliteration artifact in the standard methods of expression-array analysis using Affymetrix chips or one-color-spotted arrays. Consider an extreme instance where 90% of the transcripts are made at one brief phase of the cycle with the remaining transcripts made uniformly through the remainder of the cycle. Adding equal amounts of message to the copying and amplification mix will reduce the contribution of the high transcript phase significantly. If we further normalize by requiring equal total hybridization in all samples, then we have pretty much insured that all phases of the cycle will have equal numbers of transcripts maximally expressed. The only sure way to avoid this is to spike into the samples at the time of RNA isolation a set of standards not expressed by the cells of interest and normalize each microarray to constant expression in these standards.

Our approach using the S2 chip and early experiments with the S98 chip is to use the *B. subtilis* poly(A) standards spiked into the cell pellet at the beginning of the RNA isolation as a measure of both recovery and variations in amplification. This approach, although imperfect, gives at least some assurance that variations in total transcript levels for all transcripts in any one chip is not because of differences in recovery. It also overcomes the inherent bias in adjusting the input RNA to a constant level throughout the procedure.

What then should be the sequence of adjustments for a time-series experiment where samples have been prepared as described previously? The procedure we adopted works back from the chip results to the isolation. First, starting with the raw un-normalized data, adjust for differences in hybridization efficiency using the biotinylated *E. coli* transcript standards. Then adjust for amplification and recovery differences using the *B. subtilis* poly(A) standards and finally, if applicable, adjust for differences in mRNA recovery using the *S. pombe* spiked standard. In **Fig. 3**, two time-series expression profiles for a respiratory and a reductive phase transcript are shown to compare the raw data and the result using the poly(A) standards together with the hybridization standards. In this system, the adjustments for RNA recovery change the absolute level of expression but not the pattern of the oscillation.

Another solution to this problem using the yeast S2 chip, which contains both the *S. pombe* and *S. cerevisiae* probe sets, appears to be the use of an *S. pombe* cell spike. The correct amount of *S. pombe* to be used will depend on the isolation procedure. In contrast to the poly(A) spike, the cellular RNAs go through the same isolation procedures. Whether the recovery of RNA from pombe is different from *S. cerevisiae* is not a concern because the *S. pombe* spike is identical in all samples. Although this approach has the advantage that the *B. subtilis* standards can be used exactly as recommended by Affymetrix, allowing for independent evaluation of hybridization, copying, and recovery, it has not yet been fully evaluated by this laboratory.

In the original studies, transcripts were included in the analysis if at least three of the samples in each cycle were scored as present using the standard Affymetrix defaults. We find, using the new S2 chip, that the results can be improved by including only those transcripts present throughout the experimental series. The initial inclusion was done to avoid the possibility of eliminating samples whose oscillations were extreme. However, it appears that the algorithm used by Affymetrix does not eliminate any of the transcripts of interest even when levels fall to near zero. Among the 191 questionable genes, only a small fraction *(16)* show average expression levels greater than 100 and none show strong signal at the 40-min cycle time and all of these have p-values less than 0.05. Although we might choose to include this group into our analysis for some purposes, they can probably be eliminated from consideration in a study in which the global properties of the system are being examined. All of the genes with the most dramatic cyclic behavior were present in all 32 samples.

One question we wished to resolve was the lower limits of signal in a time-series analysis. The Affymetrix S2 chip has both the *S. cerevisiae* and *S. pombe* probe sets together and interspersed. This seemed to offer an opportunity to find

A   YGL184C

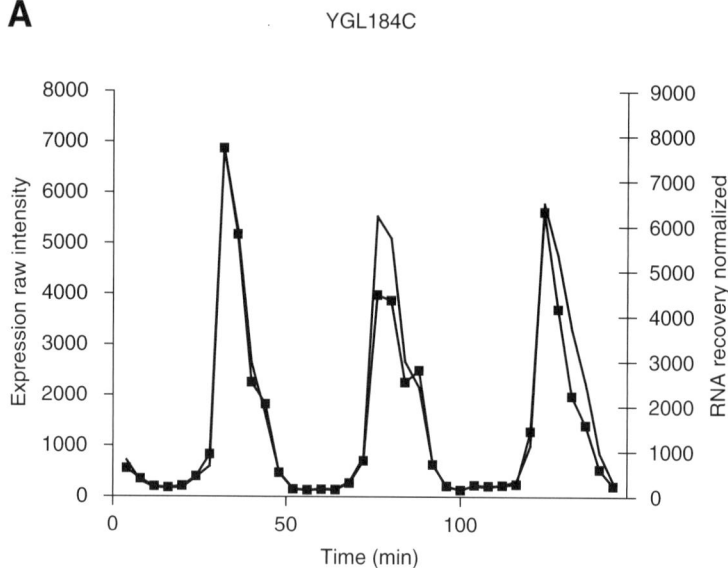

B   YOR186W

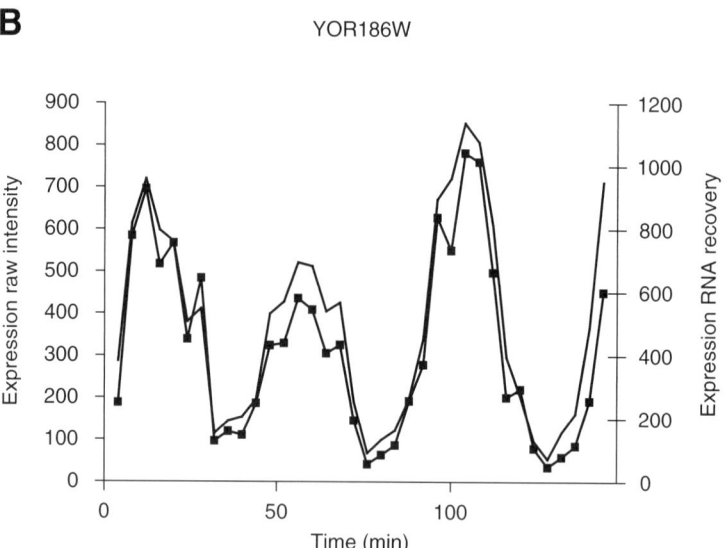

Fig. 3. Controlling for RNA recovery, copying, amplification, and hybridization. Levels of expression in two probe sets, YGL184C and YOR186W, are shown (represented by the line). Addition of the *B. subtilis* poly(A) RNA was made prior to disruption of the cells in the Mini-Bead-Beater. For each chip, the values of the two poly(A) standards, AFFX-r2-Bs-thr-3_s_at and AFFX-r2-Bs-phe-3_at, were determined and averaged with the entire series and then scaled by the average. The resulting ratio was

a true machine plus amplification plus a hybridization boundary below which we should find the system, as opposed to biological, noise. Of the 5000 *S. pombe* probe sets on the chip, all but 20 are entirely absent in all 32 samples for all genes. We used the values for the entire *S. pombe* scored as absent as a lower boundary for noise in our pair-wise comparisons. This lower boundary can be put under 16 intensity units in an experiment where the average intensity for all probe sets is greater than 2000 and the maximum intensity is greater than 16,000.

3.8. Being Misled by Scatterplots
and the Pair-Wise Comparison Paradigm

It has become commonplace to argue that many replicates are required to make a "change call" in expression. The numbers suggested are extraordinary, varying upward to 25. The time-averaged value of any oscillating constituent is a constant and one might expect that sampling done in ignorance of the dynamic state will tend to eliminate all of the most stable oscillatory components of the system leaving as "changed" the most unstable high-amplitude oscillations. We will argue that since the system is oscillatory, or in most cases, unknown, it makes more sense to take single samples through multiple cycles and use signal processing to characterize patterns of expression. The most important point to be taken from this work is the demonstration that biological variability is not intractable and that the notion that 25 biological replicates are necessary overlooks the obvious problem that the samples used to derive such a number are either not time resolved or resolved poorly.

As an example of how multiple samples done without knowledge of the underlying cellular dynamics might be misleading, we have taken two samples 40 min apart but taken from the same phase of the transcriptional cycle, and two samples taken 20 min apart from differing phases and compared them using the standard pair-wise comparison. Each gene scored as present in both samples is plotted vs itself. In **Fig. 4**, the raw data are shown. In doing the comparison in this way we are placing an additional burden on the biological system, the more so because it is difficult to impossible to sample at precisely the identical phase in two successive cycles. Nevertheless, the agreement is quite good as the left panel of **Fig. 4** shows. In contrast, the right panel shows the paired samples taken 20 min apart, but out of phase.

Consider the case in most yeast laboratories where no measurements of the respiratory state of the cell is taken. Even in the case where replicates are taken from

---

Fig. 3. *(Continued)* used to scale each transcript for all chips (represented by the line with squares). The disadvantage to this approach is that the poly(A) standards were intended to be used only to verify the quality of the copying and amplification, and not as a standard for recovery.

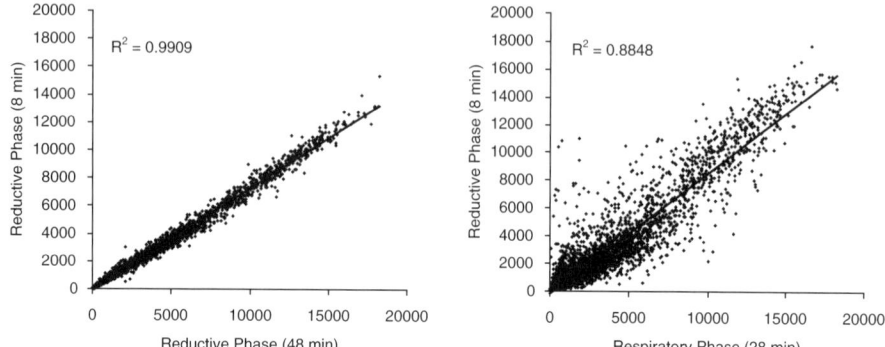

Fig. 4. Pair-wise comparison of samples resolved and purposely not resolved with respect to time of sampling. Each of the 5243 transcripts scored as present in all 32 of the Affymetrix chips through three cycles of the oscillation was included in this comparison. In the left panel, two samples taken approximately one cycle apart are compared. In the right, the two samples were taken at roughly one-half cycle apart.

the same culture, small differences in sampling time may be sufficient to yield quite different patterns of expression. In the respiratory phase of the cycle, half-lives of 2–4 min are common such that the time required to sample, centrifuge, and flash-freeze a sample before returning for a replicate would be sufficient to alter the pattern. This is perhaps an extreme example but consider a more realistic case where a treated and control series of samples are being taken from two overnight batch cultures, one treated and one control. Similar optical densities or cell counts are not adequate to insure an identical phase of the oscillation. What are, in fact, regular temporal patterns of expression would be incorrectly identified by conventional statistical treatments as outliers, part of the intractable noise—and the limit for making a change call would necessarily need to be increased; a lot more replicates would be recommended to no particular benefit.

3.9. Genome-Wide Oscillations in Transcription: Expression Microarray Analysis

Thus far the concern has been with the details of getting a reliable and quantitative measure from a time-series experiment. Far more crucial is the consequence of doing microarray experiments in the absence of any knowledge of the dynamics of the biological system being used.

Microarray analysis from a yeast continuous synchrony culture system shows a genome-wide oscillation in transcription. Maximums in transcript levels occur at three nearly equally spaced intervals in this approx 40-min cycle of respiration and reduction. **Fig. 2** in the published work (**6**) shows the time of maximum transcript level for all expressed genes as a color-contour plot. The time of maximum was determined by averaging the expression level in the three replicates from the same phase in three cycles of the oscillation. Note that these represent combined technical and biological replicates. Once the time of maximum was assigned it was fixed for all subsequent analyses. The results for all three cycles can be seen as a color "temperature map" in the supplemental data from the published work *(6)*.

The preferred representation for whole-genome data displays is the color "temperature" map in which high levels of expression are represented in reds and orange and low levels in blue *(6)*. Such maps can also be converted to a simpler contour map. Here we have taken the three cycles of expression data, averaged it, and ordered the genes according to when in the cycle they are maximally expressed (**Fig. 5**). Because every gene will have a maximum somewhere in the cycle, more quantitative measures may be needed if the claim of genome-wide periodicity is to be supported.

3.10. Fast Fourier Transform Filtering of Expression Microarray Data

The classical tool for investigating periodicity in sampled sequences is the discrete Fourier transform, realized almost exclusively as the fast Fourier transform (FFT) in the modern analytical toolbox. This tool is especially effective when the periodic nature of a sequence closely resembles a sine or cosine waveform. In this case the transformed sequence is singular or nearly so, indicating that perhaps the entire signal is represented, or matched, by a single function with a constant frequency. The FFT can be thought of as a high fidelity-matched filter producing an optimum representation.

Fourier analysis has the virtue of being the most mainstream of signal-processing methods, but has not been widely applied in molecular biological studies because the datasets usually available are short and sparsely sampled. This was the reason that our original reanalysis of the Stanford cell cycle data *(13,14)* employed wavelet multi resolution decomposition (WMD). In designing our own microarray experiments we sought to avoid some of these shortcomings by first optimizing sampling structures with signal processing or other nonlinear methods in mind. For techniques such as FFT, the data should encompass at least three cycles to permit detection of the period of interest. Equal sampling intervals throughout are essential and for some signal-processing treatments, such as FFT or wavelet decomposition, the total sample set should be dyadic (a power of two). Although this dyadic series limit can be overcome with selected wavelet families or the use of complex Fourier techniques, with some increase in computation time only the simple FFT is discussed here. A somewhat shorter series may be adequate for WMD and it appears that of the methods discussed here, SVD is the most forgiving in this regard *(15–17)*. Sampling frequencies of 8–10

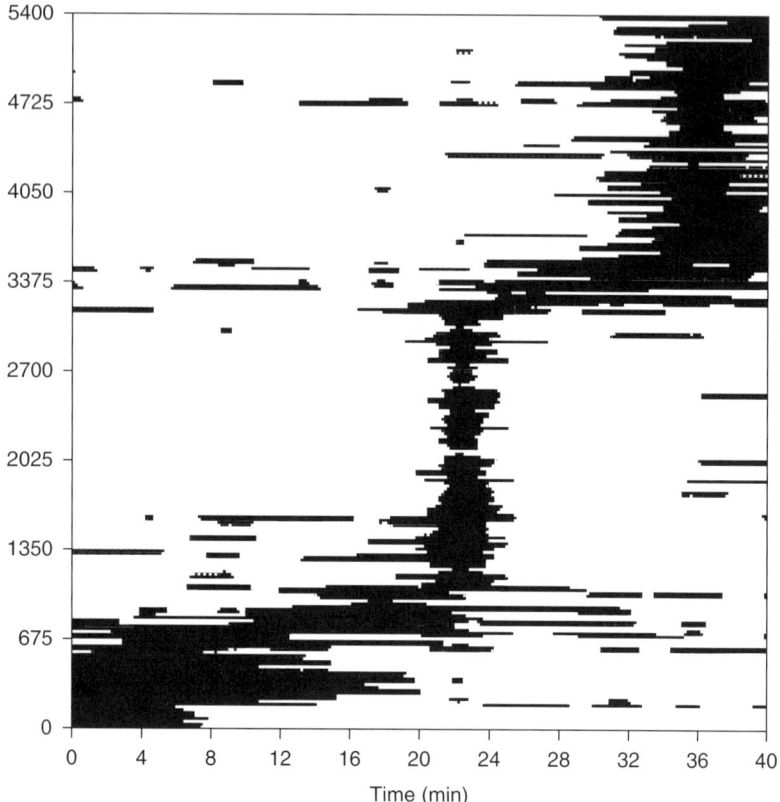

Fig. 5. Average expression levels from three cycles of the respiratory oscillation. A black-and-white contour (intensity) map of the expression levels of the 5329 expressed genes are shown for all 32 samples through 3 cycles of the dissolved oxygen oscillation. Genes were scored as present based on the Affymetrix default settings as discussed in **Subheading 3.5.** Values shown here were scaled by dividing the average expression level for each gene into each of the time-series samples for that gene. Transcripts were ordered according to their phase of maximum expression in the average of the three replicates.

samples/cycle would provide an adequate dataset for wavelet signal processing and would allow oscillations to be mapped into concentration space by means of lag plotting or other attractor reconstruction methods.

3.11. Analysis by FFT of the Genome-Wide Approx 40-Min Oscillations in Transcription

In **Fig. 6**, the FFTs, applied to each time-series expression pattern, were used as a filter, the power in the transform at frequencies near 40 min were sorted from greatest to least, and the original untransformed datasets ordered according to their

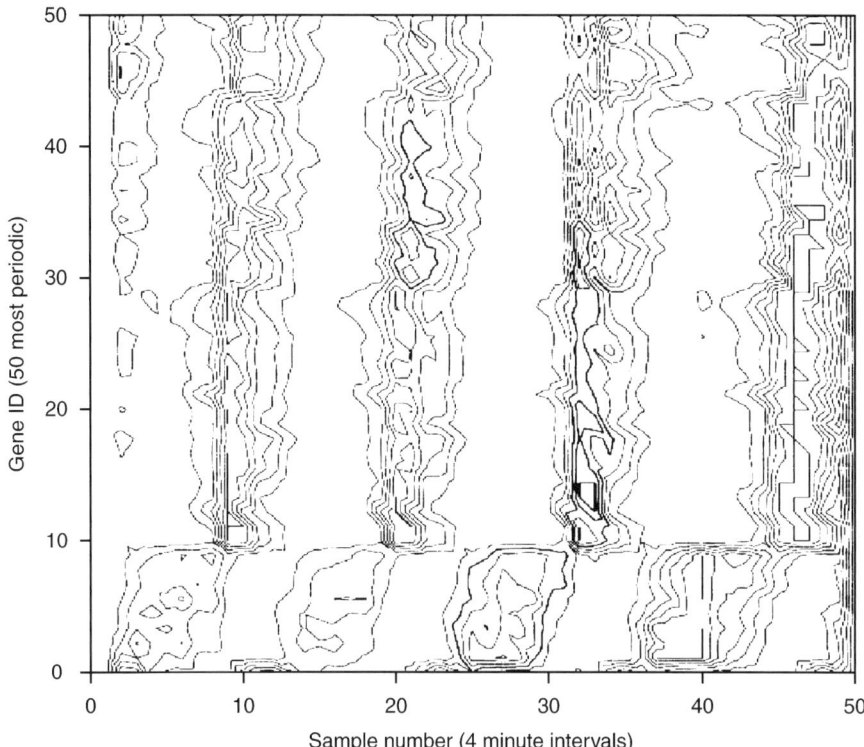

Fig. 6. Raw expression patterns sorted by fast Fourier transform (FFT) power at 40 min. All transcripts scored as present were analyzed individually using the default FFT function in Mathcad. The transformations were sorted according to their power and those with periods of approx 40 min were identified in the original untransformed data. The contour plot shown is for the 50 most periodic by this criterion taken from the raw Affymetrix dataset.

power at 40 min. Of the 5437 genes scored as present in a recent experiment, 4332 showed maximum power at 40 min. As an example of what might be seen using such a filter, compare **Fig. 5**, where all transcripts are organized according to their time of maximum, with **Fig. 6** in which the 50 most periodic (showing the strongest signal at 40 min) are plotted. In the transcriptome as a whole, respiratory-phase transcripts, those showing maximum expression in the respiratory phase, represent only about 16% of all transcripts, while in the Fourier filtered data, the relationship is reversed, with 85% being classed among the 50 most periodic.

3.12. Wavelet Match Filtering and Wavelet Decomposition

If the periodic sequence does not resemble a sine or cosine, or if the signal is nonstationary, then the effectiveness of the FFT for producing a matched filter representation may be very much reduced. In such cases, a different signal-processing approach should be sought despite the familiarity with FFT analysis. In earlier studies using data taken from spotted-array studies where the quality of the signal was poor, wavelet decomposition was used to uncover the 40- and 80-min oscillations *(16,17)*. This topic is beyond the scope of this analysis.

3.13. SVD

Some suggestion of a genome-wide cell cycle or half cell cycle quantized *(18)* oscillation in transcription appeared in a series of reanalyses of the Stanford cell cycle data where methods more suited to short, sparse, and noisy data were employed *(3–17)*. Alter et al. *(16,17)*, Rifkin and Kim *(15)* in their SVD-based analyses, Klevecz and Douse *(13)*, and Klevecz *(14)* using wavelet decomposition, all showed evidence for genome-wide oscillation in transcription. The amplitude of the oscillation was low, with about a twofold difference for the average of all non-cell cycle genes. There was not a consensus in these reports with respect to the period of the oscillation. SVD has proven to be an excellent method for developing a global representation of the expression profiles and seems as well to identify both biological perturbations and measurement variability. Perturbations because of serum or media additions were detected in the Alter et al. analysis *(17)*, and two major oscillatory components contributing to the global pattern of expression were seen, as well in the analysis of synchronized mammalian cell cultures. In our own study, SVD uncovered the discontinuity between the two experiments used based on small differences in phase and amplitude of the oscillation as shown in **Fig. 7**.

3.14. Analysis by SVD of the Genome-Wide Approx 40-Min Oscillations in Transcription

Application of SVD to the unscaled data in our recent results shown in **Fig. 5** led to the following interpretation: in the first four eigengene results (**Fig. 7**, left panel), eigengene 1 was directly related to the total intensity found in each expression profile whereas eigengene 2 found a discontinuity between the two independent experiments used in the original study *(6)* and suggested that the data was acquired from two independent experiments with slightly different period lengths and amplitudes. A plot of eigengene 3 vs eigengene 4 (**Fig. 7**, right panel) shows that the decomposition collected most of the oscillatory behavior into these two eigengenes. Assigning the same initial phase to the first time point in this graph then allows determination of phase assignments for the remaining time-points. This phase assignment was in good agreement with that used *(6)* based on their timing in the dissolved oxygen traces (**Fig. 4**).

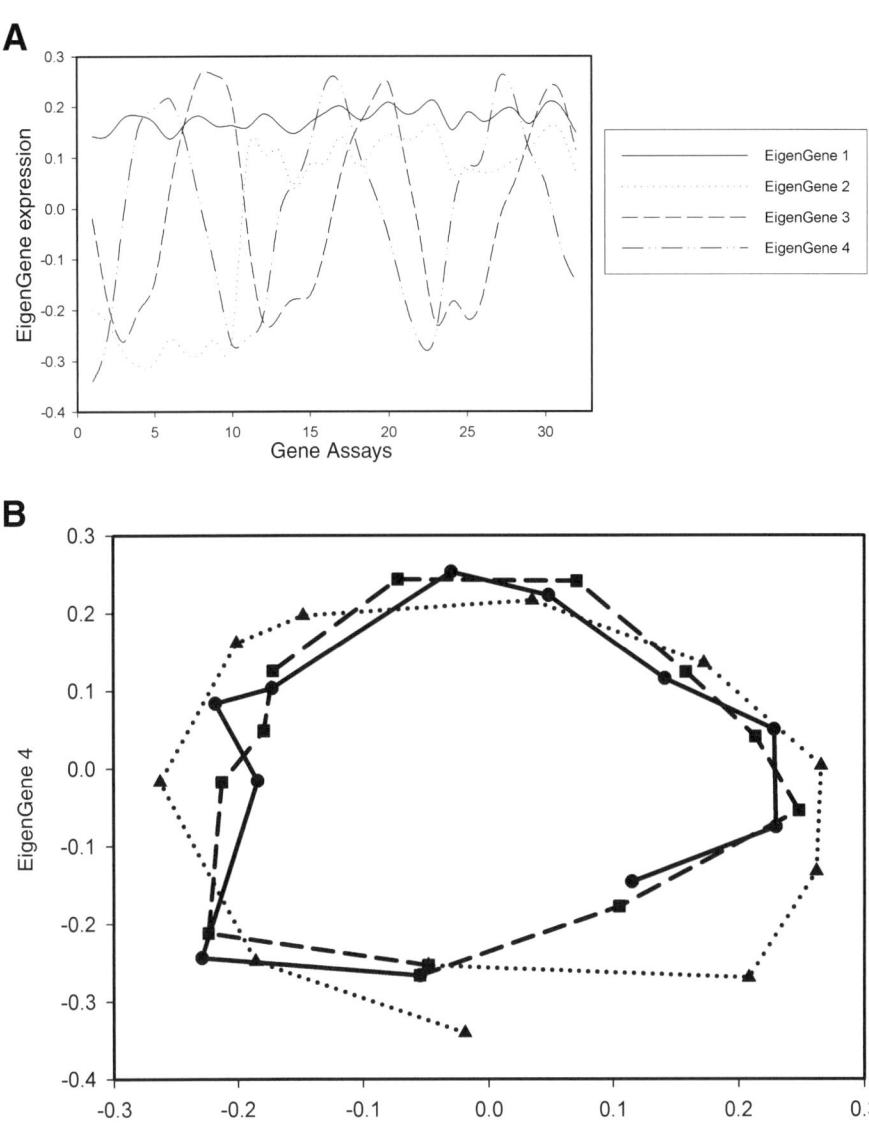

Fig. 7. Single value decomposition (SVD) principal eigengenes. On the left panel, the first four eigengenes are shown from the SVD of the intensity of expression profiles. The assay order is the same as that for the published data. On the right panel is the plot of eigengene 3 vs eigengene 4 from the SVD of the intensity of expression profiles. Three cycles are shown.

From a purely practical perspective, a significant effort should be put into resolving the question of genome-wide oscillations using the microarray technologies if for no other reason than to improve the signal-to-noise ratio. The application of analytical methods that are suited to nonlinearities in time-series data should also find a wider use. It seems clear that the most successful and widely applied method so far is SVD. In theory, wavelet analysis has some advantages over FFTs for the data length and densities likely to be encountered in expression-array studies. It will be much improved if optimized wavelet families are found that can represent the transcript or other biological signal of interest efficiently and accurately. Having said that, we were surprised to find that FFT filtering, that is, using the Fourier transform to sort those transcripts showing a particular frequency was very successful; though it must be added that this was a relatively long and densely sampled dataset.

3.15. Sampling in Clinical Studies

Exempted from the criticisms and conclusions developed here are clinical studies where the biology is unavoidably bad but where the solution—to do time-series analysis—is not feasible in most instances under prevailing protocols. It will be of interest to see whether using the limited information available regarding the time of day when a sample was taken can improve the diagnostic utility of expression microarrays and begin the process of uncovering the dynamics of expression in tumor cells.

The presence of genome-wide oscillations in yeast raises the possibility of similar dynamics in mammalian cells and tissues. Circadian and higher frequency oscillations have been known for more than 50 yr and are well characterized in extensive literature. In both dividing and nondividing mammalian tissues, oscillations with periods from a few hours to a day in length have been observed in essentially every constituent examined. For some genes important in chemotherapy, day-to-night variation can be as much as 10-fold. If samples are taken from differing tumor tissues without regard to time, with the idea that variation between samples may be exploitable for diagnostic clustering or treatment, the possibility that the variation may have more to do with circadian or regular higher frequency oscillations than with any exploitable intrinsic difference must be considered.

4. Notes

1. Optimize experimental design and sampling for time-series analysis. Take a minimum of 8 samples/cycle. Sampling interval should be such that 8 samples multiplied by the sample interval is exactly equal to the cycle time. For example, if the cycle time equals 43 min, then the sampling interval should be 5.38 min.

2. Total RNA content, and in particular mRNA content, may not be constant through the cycle. To control for biological vs recovery differences, all samples are spiked with a constant amount of a poly-A standard before beginning isolation. Other RNA standards can be used including *S. pombe* mRNA.

3. It should be possible to use the single-step amplification using the IVT kit.

4. If all samples cannot be done on the same day in the same batch, randomize the sample series. If time-series replicates are available run replicates separately in each batch.

5. Use raw data with all Affymetrix normalization and scaling factors set to 1.

6. Currently, no commercial software products have adequate time-series analysis algorithms. Paste Affymetrix txt files into Excel. Excel has the virtue that all data manipulation is open—there are no black boxes as there are in commercial packages.

7. Copy out *cerevisiae* and standards to separate worksheets.

8. To avoid missing interesting low expressers, retain all transcripts in which at least one sample in each cycle is called "P" (present).

9. For a cleaner less noisy result, remove all transcripts from the entire time series if any member of the time series contains an "A" (absent) calls.

10. Adjust all samples in the time series for differences in hybridization using the biotinylated standards and a polynomial fit. Calculate the mean of the hybridization standards. Fit a polynomial to these mean values. Correct each of the standards in the time-series data to the fitted result. Correct the signals for expressed transcripts by this same technique.

11. Test all samples for large differences in mRNA recovery using the *B. subtilis* poly(A) standards. Use the same routine as described in **Note 10**. If no large discrepancies are seen, use the result from **Note 10**.

12. A number of suitable Math packages are available including Bioconductor, an R-based collection, as well as the more standard Mathcad, Matlab, S-Plus, and JMP. Both Matlab and Mathcad have a very complete set of signal-processing routines including FFT, SVD, and WMD.

References

1. Kauffman, S. and Wille, J. J. (1975) The mitotic oscillator in Physarum polycephalum. *J. Theor. Biol.* **55,** 47–93.

2. Klevecz, R. R. and Shymko, R. M. (1985) Quasi-exponential generation time distributions from a limit cycle oscillator. *Cell Tissue Kinet.* **18,** 263–271.

3. Mackey, M. C. and Glass, L. (1977) Oscillation and chaos in physiological control systems. *Science* **197,** 287–289.

4. Klevecz, R. R. (1998) Phenotypic heterogeneity and genotypic instability in coupled cellular arrays. *Physica D* **124,** 1–10.

5. Klevecz, R. R., Kros, J., and Gross, S. D. (1978) Phase response versus positive and negative division delay in animal cells. *Exp. Cell Res.* **116,** 285–290.

6. Klevecz, R. R., Bolen, J., Forrest, G., and Murray, D. B. (2004) A genomewide oscillation in transcription gates DNA replication and cell cycle. *Proc. Natl. Acad. Sci. USA* **101,** 1200–1205.

7. Klevecz, R. R. and Murray, D. B. (2001) Genome wide oscillations in expression. *Mol. Biol. Reports* **28**, 73–82.

8. Murray, D. B., Klevecz, R. R., and Lloyd, D. (2003) Generation and maintenance of synchrony in Saccharomyces cerevisiae continuous culture. *Exp. Cell. Res.* **287**, 10–15.

9. Satroutdinov, A. D., Kuriyama, H., and Kobayashi, H. (1992) Oscillatory metabolism of Saccharomyces cerevisiae in continuous culture. *FEMS Microbiol Lett.* **77**, 261–267.

10. Murray, D. B., Engelen, F., Lloyd, D., and Kuriyama, H. (1999) Involvement of glutathione in the regulation of respiratory oscillation during a continuous culture of Saccharomyces cerevisiae. *Microbiol.* **145**, 2739–3747.

11. Mochan, E. and Pye, E. K. (1973) Respiratory oscillations in adapting yeast cultures. *Nat. New Biol.* **242**, 177–179.

12. Poole, R. K., and Lloyd, D. Oscillations of enzyme activities during the cell-cycle of a glucose-repressed fission-yeast, *Schizosaccharomyces pombe* 972h-. *Biochem. J.* **136**, 195–207.

13. Klevecz, R. R. and Dowse, H. B. (2000) Tuning in the transcriptome: basins of attraction in the yeast cell cycle. *Cell Proliferation* **33**, 209–218.

14. Klevecz, R. R. (2000) Dynamic architecture of the yeast cell cycle uncovered by wavelet decomposition of expression microarray data. *Funct. Integr. Genom.* **1**, 186–192.

15. Rifkin, S. A. and Kim, J. (2002) Geometry of gene expression dynamics. *Bioinformatics* **18**, 1176–1183.

16. Alter, O., Brown, P. O., and Botstein, D. (2000) Singular value decomposition for genome-wide expression data processing and modeling.) *Proc. Natl. Acad. Sci. USA* **100**, 3351–3356.

17. Alter, O., Brown, P. O., and Botstein, D. (2003) Generalized singular value decomposition for comparative analysis of genome-scale expression datasets of two different organisms. *Proc. Natl. Acad. Sci. USA* **97**, 10,101–10,106.

18. Warrington, J. A., Nair, A., Mahadevappa, M., and Tsyganskaya, M. (2000) Comparison of human adult and fetal expression and identification of 535 housekeeping/maintenance genes. *Physiol. Genomics* **2**, 143–147.

5

Predictive Models of Gene Regulation

Application of Regression Methods to Microarray Data

Debopriya Das and Michael Q. Zhang

Summary

Eukaryotic transcription is a complex process. A myriad of biochemical signals cause activators and repressors to bind specific *cis*-elements on the promoter DNA, which help to recruit the basal transcription machinery that ultimately initiates transcription. In this chapter, we discuss how regression techniques can be effectively used to infer the functional *cis*-regulatory elements and their cooperativity from microarray data. Examples from yeast cell cycle are drawn to demonstrate the power of these techniques. Periodic regulation of the cell cycle, connection with underlying energetics, and the inference of combinatorial logic are also discussed. An implementation based on regression splines is discussed in detail.

Key Words: Transcription regulation; regression; splines; cooperativity; correlation; yeast; cell cycle; *cis*-regulatory element; MARS.

1. Introduction

In the past decade, there have been tremendous advances in high-throughput molecular technologies for measuring mRNA levels genome wide. Such technologies not only provide information on which genes are over- or under-expressed, but along with genomic sequence data, also allow one to obtain a deeper insight into the *cis*-regulatory mechanisms that drive gene transcription. One problem that has been intensively studied in this context is to identify the *cis*-elements that control and regulate the transcription process. The traditional approach to solve this problem has been to cluster genes by their expression profiles across multiple conditions and to find over-represented motifs in promoters of genes in each cluster *(1)*. Clustering-based approaches gave researchers a starting tool kit to obtain a snapshot of key regulatory elements. However, it became increasingly clear that such approaches have several limitations. First, many genes often do not cluster tightly enough to allow for

From: *Methods in Molecular Biology, vol. 377, Microarray Data Analysis: Methods and Applications*
Edited by: M. J. Korenberg © Humana Press Inc., Totowa, NJ

identification of their regulatory elements with reasonable accuracy. Second, gene regulation is combinatorial with a significant amount of cooperativity, especially in mammals. Classifying genes into disjoint clusters can often lead to incomplete identification of functional motif combinations. Additionally, some genes in an expression cluster may exist because of secondary effects and may be regulated by elements different from those for the primary response genes. Most importantly, clustering methods require expression data from multiple conditions, which is not always available.

Over the past few years, a new paradigm has emerged involving methodologies that can efficiently extract information on functional *cis*-regulatory elements and their functional combinations from microarray data on just a few condition. We will review these interesting developments in this chapter. This is by no means an exhaustive survey. But, we hope to convey the essential points. We will primarily use yeast cell cycle expression data to compare the techniques.

2. Regression Approach to *Cis*-Regulatory Element Analysis
2.1. Basic Idea

In order to obtain functional regulatory motifs on promoter DNA from microarray expression data using regression, one correlates the motif occurrences with the logarithm of expression ratios *(2)*. The basic idea behind this can be explained as follows. For a given cell type, only a limited set of transcription factors (TFs) are active under any given condition. The extent to which genes are up/downregulated in these cells depends directly on the strength with which these TFs and their combinations bind to their promoter DNA, if they bind at all. For a low eukaryote like yeast, the motifs are largely nondegenerate and the strength of binding to a particular motif is directly related to its count in the promoter of each gene. Thus, the mRNA levels must directly correlate with the modulation of motif occurrences across the genes. A regulatory motif that is active would strongly correlate with the expression levels and vice versa. Regression analyses exploit these correlations to infer the functional *cis*-elements and their cooperativity.

Consider, for example, that we are interested in the effect of the MCB (*MLuI* cell-cycle box) element, ACGCGT, on yeast cell cycle at a particular time-point. To do this, one records the counts n_g of the MCB motif in the promoter of each gene g and also the logarithm of their expression ratios, $\log(E_g/E_{gC})$, where E_g is the mRNA level of gene g at the given time-point and E_{gC} is that for the control set C. The control can be, for example, a homogeneous mix of mRNAs across all the cell cycle phases. One then examines correlation between the $\log(E_g/E_{gC})$ values and the counts n_g by fitting a straight line:

$$y_g^p = a + bn_g \tag{1}$$

where $y_g = \log(E_g/E_{gC})$ and p indicates the predicted value of y. The coefficients a and b are obtained by minimizing the residual sum of squares, $\sum_g (y_g - y_g^p)^2$. The accuracy of the model is estimated by $\Delta\chi^2$, the percent reduction of variance (%RIV) present in the original expression data *(2,3)*:

$$\Delta\chi^2 = \left[\frac{1 - \sum_g (r_g - \bar{r})^2}{\sum_g (y_g - \bar{y})^2} \right] \times 100, \tag{2}$$

where $r_g = y_g - y_g^p$ is the residual, and $\bar{y}$ and $\bar{r}$ are the corresponding means. It is directly related to the residual sum of squares mentioned previously. If the MCB element is active under the given condition, its counts will correlate significantly with the expression data and $\Delta\chi^2$ will be large. If, on the other hand, it is inactive, there will not be any significant correlation and $\Delta\chi^2$ will be low. One can convert $\Delta\chi^2$ to p-values using an F-test *(3,4)* or an extreme value distribution *(2)*. In the above two situations, the p-values will be low and high, respectively. Some examples for the G_1/S element MCB are shown in **Fig. 1A,B**. %RIV for the MCB element is significantly higher in the G_1/S phase **(Fig. 1A)** than in the G_2/M phase **(Fig. 1B)**. Thus, $\Delta\chi^2$ quantifies the impact of each regulatory element on transcription and, hence, allows one to identify the active elements.

2.2. A Description Based on Energetics

In this subsection, we lay out some of the connections with energetics that underlie the regression approach. Let us consider the rate of change of mRNA level of a gene in a given system *(3)*:

$$\frac{dE_g}{dt} = K_A - K_D \cdot E_g, \tag{3}$$

where E_g denotes the number of mRNA molecules of gene g in the system, i.e., its expression level. Here, A stands for activation and D for decay. Under steady-state approximation, this rate ≈ 0, and hence,

$$\log(E_g) = \log(K_A) - \log(K_D) \tag{4}$$

Now, $K_A \propto p_{bind}$, the probability that the promoter DNA of the gene is bound by a TF. p_{bind} is given by *(5)*:

$$p_{bind} = \frac{1}{1 + e^{(\Delta G - \mu)/RT}} \approx e^{-(\Delta G - \mu)/RT}, \tag{5}$$

where ΔG is the change in free energy when a TF binds to the promoter. μ is related to the rate constant and corresponds to the gene activation threshold.

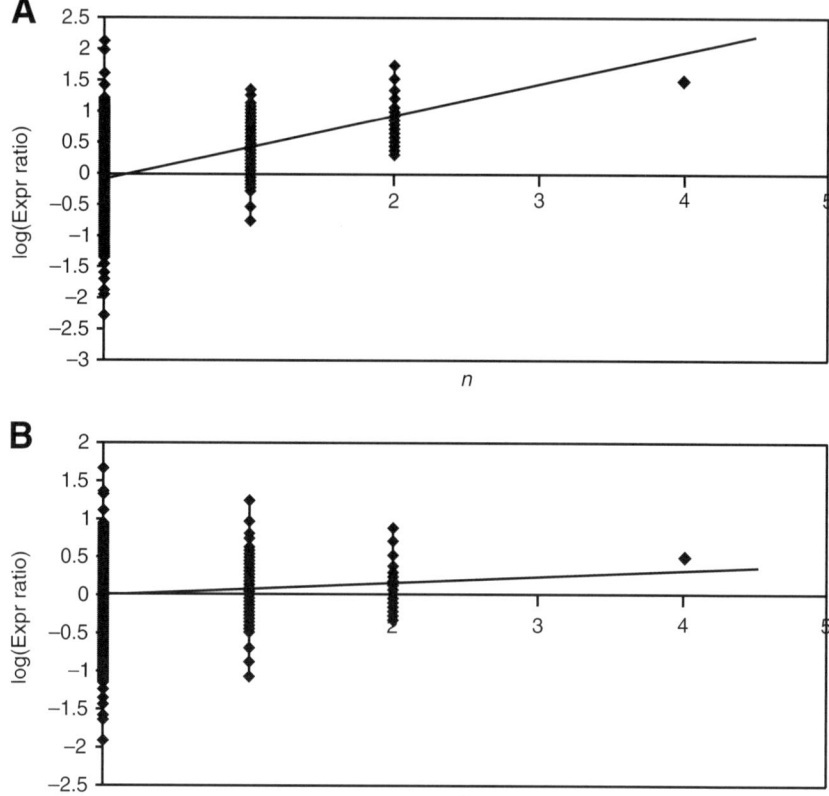

Fig. 1. Plots of logarithm of expression ratios vs the counts (*n*) of the MCB (*MLuI* cell-cycle box) element ACGCGT for the yeast cell cycle-specific genes. Expression ratios were obtained from the alpha-arrest experiments (*1*). (**A**) Linear fit for the 21-min time-point (G_1/S phase) yields a = –0.07 and b = 0.49 (**Eq. 1**), $\Delta\chi^2$ = 18.8% with *p*-value = 8.6e-32 (**Eq. 15**). (**B**) Linear fit for the 35-min time-point (G_2/M phase) yields a = –0.02, b = 0.09 (**Eq. 1**), $\Delta\chi^2$ = 1.1% with *p*-value = 0.004 (**Eq. 15**).

In the last part of **Eq. 5**, we have made the Boltzmann approximation, i.e., $\Delta G - \mu \gg RT$. Free energy contribution from a particular motif with *n* copies in a given promoter is:

$$\Delta G - \mu = \varepsilon_0 + n \cdot \varepsilon_1 \qquad (6)$$

where each copy leads to a free energy change ε_1, and ε_0 is the basal contribution. From **Eqs. 4–6**, we notice that following the Boltzmann approximation, log of the expression ratio is linear in *n* (*see* **Note 1**). Comparison with **Eq. 1** shows that

$$a = -\varepsilon_0, \ b = -\varepsilon_1. \qquad (7)$$

That is, the fit coefficients of regression models of expression ratios are a measure of binding free energy (*see* **Note 2**). This can be very nicely seen from the predicted time courses of MCB and SCB (Swi4/6 cell-cycle box) elements during the yeast cell cycle *(2)*. MCB and SCB elements are active during the G_1/S phase of the cell cycle. From **Fig. 1A** of **ref.** *2*, we notice that the fit coefficients are strongly positive near the G_1/S phase (time-points 21 and 77) and strongly negative near the G_2/M phase (time-point 56). Thus, according to the previous discussions, the binding energies are strongly negative at the G_1/S phase, i.e., it is favorable to bind the MCB and SCB elements in this phase. On the other hand, in the G_2/M phase, the binding energies are positive, and MCB and SCB elements are very unfavorable to be bound, i.e., they are inactive.

2.3. Combinatorial Regulation via Multivariate Linear Models

REDUCE (Regulatory Element Detection Using Correlation with Expression), proposed by Bussemaker et al. *(2)*, goes a step ahead and considers the effects of combinatorial regulation via multiple transcription factors. Here multiple motifs contribute additively to the log of expression ratio:

$$y_g^p = a + \sum_\mu b_\mu n_g^\mu \tag{8}$$

where the index μ indicates motif id and n_g^μ is the count of motif μ for gene g. The coefficient b_μ is the (free energy) contribution from the motif μ. The significant motifs are determined by a step-wise linear regression and the coefficients a and $\{b_\mu\}$ are obtained finally by a multivariate linear fit. Using the yeast cell cycle data as an example, Bussemaker et al. showed that REDUCE can verify many regulatory motifs important in the cell cycle obtained by the clustering approach *(1,6)*. MCB, SCB, SFF, Swi5, and stress response element STRE and Met31/32 are some such examples. Using Mcm1 as an example, they further showed that if a position weight matrix (PWM) score is used instead of word counts, the accuracy, as determined by %RIV, can go up by as much as 80% (*see* **Note 3**). A more comprehensive analysis using weight matrices was later done by Conlon et al. *(4)*. They designed the algorithm, MotifRegressor, which combines the *ab-initio* motif finder MDscan *(7)* with multivariate linear regression. Thus, MDscan was used to generate a large number of PWMs. A prioritized list of motifs was initially selected from this set by applying regression on individual motifs. The significant motifs were finally determined by step-wise linear regression on the prioritized set, leading to the model:

$$y_g^p = a + \sum_\mu b_\mu S_g^\mu \tag{9}$$

where S_g^μ is the PWM score *(4)* for the motif μ in the promoter of gene *g*. MotifRegressor, like REDUCE, could identify several key regulatory motifs in the yeast cell cycle and other experiments.

3. Cooperativity

The prior models do not account for cooperativity. Cooperativity among TFs is a salient aspect of eukaryotic transcription *(8,9)*. This is even more so in mammals, where transcription is considered to be almost promiscuous *(9)*. Hence, such synergistic effects must be incorporated in the computational models to get an accurate view of the underlying regulation process. Cooperativity among multiple motifs is reflected in more than additive contributions from such motifs, in contrast to what is captured by the linear models in the previous section.

3.1. Expression Coherence Score Approach

Models of cooperativity which did not rely on clustering were first proposed by Pilpel et al. *(10)* and later advanced by Banerjee et al. *(11,12)*. The method is based on the use of expression coherence scores. Here, one first finds motifs in the promoters of the genes and considers all possible pairs of motifs. For a given pair of motifs A and B, three sets of genes are considered: those that have both A and B, those that have A but not B, and those that have B but not A. For each set, an expression coherence (EC) score is calculated, which measures how tightly correlated the expression levels of an average pair of genes in the set (relative to a random pair) are based on a distance measure (Euclidean distance *[10]* or correlation coefficient *[11]*). For a synergistic motif pair, the gene set with both motifs A and B has a much higher EC score than those with either of them alone. Banerjee et al. *(11)* later quantified this difference in terms of a *p*-value based on a hypergeometric distribution. This method reproduced several well-known synergistic pairs in yeast *(10,11)*: Mcm1-SFF (cell cycle), Mcm1-Ste12 (sporulation), Bas1-Gcn4 (heat shock), Mbp1-Swi6 (cell cycle), Swi4-Swi6 (cell cycle), Ndd1-Stb1 (cell cycle). The last three pairs are cited from **ref.** *11*, where ChIP-chip data was used to identify the targets of a given TF, and then microarray data was used to obtain the cooperative TF pairs.

3.2. Toward a Synthesis: Regression Models of Cooperativity

The disadvantage of the EC score framework is that it is hard to quantify the relative impact of individual motifs and pairs of motifs on gene expression. Also, it needs expression data across multiple time-points to calculate the correlation measures. These limitations can be easily overcome if cooperativity is built directly into a regression model. This was implemented by Keles et al. *(13)* in the program SCVmotif, where cooperativity was introduced as product terms in the model. Thus, for example, for motifs 1 and 2, **Eq. 6** needs to be modified as:

$$\Delta G - \mu = \varepsilon_0 + \varepsilon_1 \cdot n_1 + \varepsilon_2 \cdot n_2 + \varepsilon_{11} \cdot (n_1 \cdot n_1) + \varepsilon_{22}(n_2 \cdot n_2) + \varepsilon_{12} \cdot (n_1 \cdot n_2) \quad (10)$$

Thus, two motifs make more than (or less than) additive contributions to the log expression ratio leading to synergistic effects. Here, relative distance, orientation, or other parameters related to the physical locations of the two motifs are not considered. Thus, the assumption here is that for a given number of motifs of type 1 and 2, each pair of these two motifs makes a similar free energy contribution on average upon TF binding, independent of their relative physical locations on the promoter DNA.

SCVmotif *(13)* considers interactions between all pairs of motifs. Thus, the model has the structure:

$$y_g^p = a + \sum_\mu b_\mu n_g^\mu + \sum_{\mu,v} c_{\mu v} n_g^\mu n_g^v, \quad (11)$$

where the Greek indices indicate motif ids. The authors used a variant of word counts that incorporated the probability distribution of the words in the promoter regions *(13)*. Interaction terms involving the same motif were ignored. Significant motifs and motif pairs were determined by a combination of forward and backward selection, and cross-validation. Yeast cell cycle was used to show that several motifs can be correctly predicted in the G_1/S phase by including interactions. MCB and SCB are two such examples. Interaction between them was also found to be significant.

4. Spline Models of Cooperative Gene Regulation

The previous methods provided a foundation for the regression approach to identification of functional motifs from gene-expression data. However, closer analysis revealed several limitations. For example, when applied to the yeast cell cycle data, we found that linear models learnt by REDUCE *(2)* lead to a %RIV of only 10% on average (noise level accounts for ~50% *[2]*). The models that include cooperativity, as discussed previously, are also limiting. With the feature selection approach proposed by Keles et al. *(13,14)*, we found that either the known pairs of motifs are not quite often correctly predicted or the accuracy of the regression model does not improve significantly (<5%) when interacting pairs are introduced in the model, which is inconsistent with the biological notion of synergistic gene regulation. Furthermore, gene transcription is strongly nonlinear *(8)*. None of these models captures the nonlinearities.

Many of these limitations can be avoided by using spline models *(3)*. We first note that the TF-binding probabilities have a sigmoidal dependence

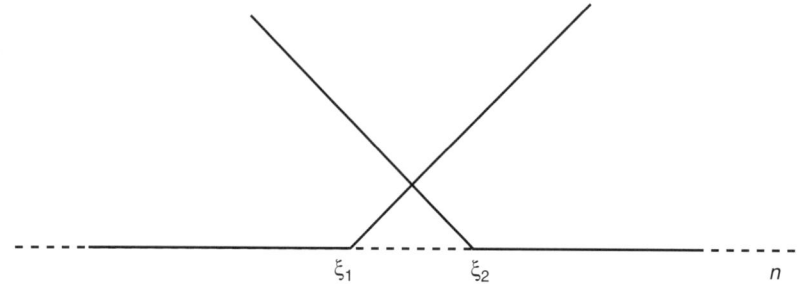

Fig. 2. Two types of linear splines.

(**Eq. 5**), the logarithm of which approximately has the shape of a linear spline. Furthermore, synergistic interactions among TFs that drive the transcriptional process lead to a switch-like behavior *(8)* as in a sigmoidal function. Thus, gene transcription is intrinsically nonlinear and spline models would provide a more faithful description of the underlying regulatory mechanism. The splines capture the switch-like behavior and thus provide a natural computational framework for analyzing transcription regulation.

Linear splines are described by

$$\theta(x,0) = x, \text{ if } x \geq 0 \tag{12}$$
$$= 0, \text{ otherwise}$$

There are two types of splines as shown in **Fig. 2**: $\theta(x - \xi, 0)$ and $\theta(\xi - x, 0)$. The first type is linear in the range $x \geq \xi$, while the second type is linear when $x \leq \xi$. The point ξ where the function changes from being zero to linear is called a knot. Thus, a motif contributes to expression if its count (or, PWM score) is beyond a certain threshold. When only pair-wise interactions are allowed, the spline model for expression looks like:

$$y_g^p = a + \sum_{\mu,i} b_{\mu,i} \theta\left(n_g^\mu - \xi_{\mu,i}, 0\right) + \sum_{\mu,v,i,j} c_{\mu,v,i,j} \theta\left(n_g^\mu - \xi_{\mu,i}, 0\right) \cdot \theta\left(n_g^v - \xi_{v,j}, 0\right) \tag{13}$$

where $\xi_{\mu,i}$ is the *ith* knot for the motif μ. The other type of spline is also considered in the model fitting. The difference between models (11) and (13) is that there are now additional degrees of freedom because of the knots $\xi_{\mu,i}$.

Das et al. *(3)* developed a method called MARSMotif to build the spline model as shown in **Eq. 13** starting from expression data. MARSMotif starts with a large number of motifs and prioritizes them using the Kolmogorov–Smirnov (KS) test, which is a nonparametric test. The MARS *(15,16)* (Multivariate Adaptive Regression Splines) algorithm is then used to build the spline model in **Eq. 13** using the prioritized motifs as input. MARS is a nonparametric and adaptive method. It builds a large number of models using a combination of forward selection and backward elimination. The terms and knots are enumerated by minimizing the residual sum of squares. The final model is selected by minimizing the general cross-validation score (GCV), which controls overfitting:

$$GCV = \frac{\sum_{g=1}^{N} \left[\log(E_g / E_{gC}) - y_g^p \right]^2}{[1 - M/N]^2} \tag{14}$$

where M is the effective number of parameters in the model and N is the total number of genes. M is estimated by cross validation. GCV-based model selection ensures the number of terms in the model is small *(3)*. Interactions involving the same motif are written as a sum of splines in MARS. Thus, $\mu \neq v$ in the third term in **Eq. 13**. MARSMotif works with both motif counts and weight matrix scores. In fact, it can work with a hybrid set of such inputs *(3)*.

4.1. Periodic Regulation of Cell Cycle

We first discuss the differences between a linear model and a MARSMotif model for a single motif and a pair of motifs. When the expression level of a given TF is low, the *cis*-regulatory motif to which it binds is inactive, and the corresponding regression model for this motif must yield $\Delta\chi^2 \approx 0$. On the other hand, when the expression level of the TF is high, its binding *cis*-motif is active (under typical conditions), and its regression model must lead to $\Delta\chi^2 \gg 0$. Because the expression levels of some of the key regulators vary periodically with the cell cycle *(1,2)*, the %RIV for their corresponding binding elements should also vary periodically. This is shown in **Fig. 3A,B** for SCB and MCB elements, respectively, where word counts have been used as inputs. There are actually two cell cycles in these experiments. But, because $\Delta\chi^2 \geq 0$, there are four peaks in these figures instead of two. For a single motif with word count, we notice that the linear and MARSMotif models are almost identical. This is not surprising because a linear model with word counts already has a built-in cutoff as word counts are discrete, and thus in a sense, mimics linear splines. This is not the case for position-weight matrices, as shown in **Fig. 3C**, where we show the time course of the Mcm1 motif. Mcm1 is a very degenerate motif with two conserved dinucleotides, separated by six nucleotides *(2)*. In this case, the periodicity is still retained in the linear model, but peaks are much sharper in the MARSMotif model. We have also shown here the model that uses only a single linear spline. Both in terms of periodicity and sharpness of peaks, this seems to be the optimal choice (*see* **Note 4**). Thus, for a single motif, the analog of a linear model with motif count as input is a linear spline model with PWM score as input. For a pair of motifs, the interactions are important. In this case **(Fig. 3D)**, the periodicity is lost in a linear model, and in the MARSMotif model, it clearly stands out.

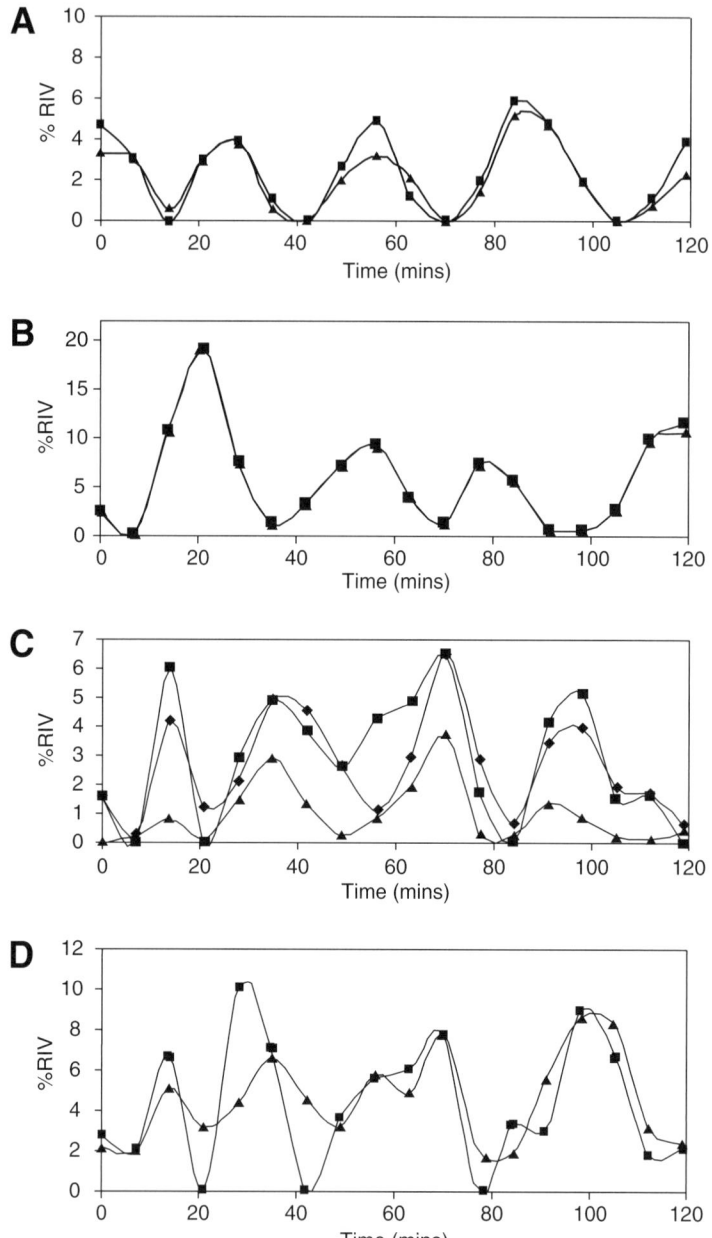

Fig. 3. Time courses of various motif combinations for the alpha-arrest experiments *(1)*: (**A**) SCB, (**B**) MCB, (**C**) Mcm1, and (**D**) Mcm1-SFF pair. Linear models are shown as triangles, MARSMotif models are as squares and the single linear spline model in **C** is shown as diamonds.

4.2. Summary of MARSMotif Results

Das et al. *(3)* applied MARSMotif to the yeast cell cycle data *(1,17)* using six input sets of motifs generated by different *ab-initio* motif-finding algorithms: (1) counts of motifs found by AlignACE *(10)*, a Gibbs-sampling approach, (2) weight matrix scores of motifs from AlignACE *(10)*, (3) counts of motifs discovered by cross-species conservation *(18)*, (4) counts of a curated set of motifs *(3)*, (5) counts of 5–7mer DNA words, which were clustered by their similarity to each other to obtain a nonredundant set, and (6) same as set (5), except that clustering was done using motifs obtained by cross-species conservation *(18)* as templates. MARSMotif yielded a higher %RIV than REDUCE, regardless of which type of motif input was used: 13.9–32.9% on an average, which is about 1.5–3.5 times that of REDUCE. The %RIV is highest for word counts, as in set (5), and worst for set (3). Because REDUCE was done with word counts, true improvement lies toward the upper end of this range. When interactions were included in the model, %RIV increased in 69–88% of the cases, and the fractional increase in %RIV in these cases was 47–96%, depending on which motif set was used. This shows that MARSMotif can suitably model synergistic effects that are widespread in eukaryotic transcription regulation. It is sensitive to which type of motif set is used as input. When both %RIV and modeling of synergistic effects are considered, combination of word counts and cross-species conservation (input set [5] above) is the optimal choice for yeast.

MARSMotif not only led to a higher quantitative accuracy, but also detected several motifs and motif pairs previously known as important regulators of cell cycle. For example, the classical cell cycle-regulatory motifs were found at the correct stages of cell cycle: MCB and SCB in G_1/S phase, Mcm1 and SFF at the G_2/M phase, and Ace2, Swi5, and Ste12 at the M/G_1 phase. Several nonclassical motifs, e.g., Rme1, Adr1, and Rap1, were also identified as significant. Among motif pairs, the well-known Mcm1-SFF pair was identified as functional in the G_2/M phase. Other examples of known cell cycle-regulatory pairs detected by MARSMotif included Mcm1-Ste12 and Ace2-Swi5. The rest of the pairs identified as significant by MARSMotif were either known pairs that participate in processes secondary to the cell cycle (e.g., Alpha2-Mcm1), completely novel (e.g., GCR1-SWI4), or were supported by other computational methods (e.g., Ace2-SFF). An important point is that, in contrast to a method like the EC score approach, MARSMotif can identify the specific phase/time-point where a given motif combination is active. More details are available in **ref. 3**.

5. Summary

In this chapter we have reviewed how regression methods can be used to extract information on transcription regulation from microarray data in eukaryotic systems. Here all genes are fit. So regulatory information of all genes can be obtained, at least in principle. The relative impact of each motif and motif pair on gene expression can be directly quantified as well. Percent reduction in variance of expression log ratios, on the other hand, provides a quantitative estimate of how complete the discovery is. No background sequence sets or any prior system-specific knowledge of transcription are necessary either. In this sense, the methods are quite unbiased. They can work with limited expression data: microarray data from a single time-point and a control set are sufficient to do the analysis. Additionally, regression splines model the underlying bioenergetics and can produce a quantitatively highly accurate model of transcription regulation. Individual motifs and cooperative motif combinations, which are active under a specific condition, can also be very accurately predicted. Apart from modeling energetics, linear splines help to filter noise present in the input motif sets by allowing nonzero contributions only beyond a certain threshold.

Predicting gene expression levels from DNA sequence information and invoking combinatorial logic in this prediction are important topics of current research in modeling gene regulation *(19)*. It is very easy to see from the previous discussions that regression methods allow one to predict expression levels of a gene from sequence data. Combinatorial logic of the type AND, OR, and NOT are also captured in the splines framework. Presence of AND logic is obvious from the product terms in **Eq. 13**. OR logic can be seen from the involvement of terms of type $\theta(S_1 - \xi_1, 0) + \theta(S_2 - \xi_2, 0)$ where S_i is the PWM score of the motif i. There is a finite contribution to expression if $S_1 > \xi_1$ or $S_2 - \xi_2$ or both. An example of NOT logic would be a term like $\theta(S_1 - \xi_1, 0)$, where the knot ξ_1 is very small. That is, this term is finite only when the motif is absent.

Use of cross-species conservation in promoter regions has been shown to improve the performance of regression methods *(14)*. However, conservation is also known to increase the false-negative rate of identifying motifs specific to a given organism *(20)*. Constraints on regulatory elements, e.g., relative orientation, distance from transcription start site, and so on need to be incorporated to obtain a more accurate view of transcription regulation. In this context, application of Bayesian networks is noteworthy *(19)*. Several classification methods have also been applied to the problem of regulatory element identification that we have not reviewed here *(21,22)*.

Regression methods have been applied to expression data from higher eukaryotes as well, e.g., in *Drosophila (23)*, and have now been successfully extended to mammals *(24)*. Additionally, linear splines allow one to predict direct targets of active motif combinations from a small amount of microarray data with high accuracy *(24)*. In conclusion, current developments lead us to believe that regression methods will allow researchers to comprehensively dissect the transcription-regulation process across a wide range of eukaryotic systems even when only a limited amount of microarray data is available.

6. MARSMotif: An Implementation

Here we discuss how to implement the MARSMotif algorithm *(3)*. We first discuss the algorithm for individual motifs, and then for combinations of motifs allowing for interactions.

6.1. MARSMotif for Individual Motifs

Given a set of candidate motifs, we first examined association of each motif with expression using the KS test. It is a nonparametric test that assigns a p-value based on the maximum distance between the two respective cumulative distribution functions. For any given motif, we compared the distribution of expression values for the genes that have the motif with the distribution for genes that do not have that motif. The KS test was implemented using the subroutine given in **ref. 25**. This subroutine works only when $n_e = n_1 n_2/(n_1 + n_2) \geq 4$, where n_1 and n_2 are the number of genes in the two samples. For all other cases, we used the KS test available in S-PLUS.

The top 100 motifs by KS p-value were used in MARS regression. MARS was run iteratively with 40 motifs at a time; at most, top 30 motifs were retained from the previous run where motif ranking is based on the variable importance reported by MARS. This was augmented with additional motifs to make the number up to a maximum of 40. The final run produced the list of significant motifs.

We used the MARS program available from Salford Systems *(26)* (http://www.salford-systems.com/). We ran MARS with basis functions (linear splines and their products) at six times the number of motifs (minimum number of basis functions = 25) and speed=1, allowing for no interactions between distinct motifs (int=1). Speed=1 ensures that the accuracy of the program is highest, although at the expense of speed. We used 10-fold cross validation to obtain the effective number of parameters appearing in the GCV score (**Eq. 14**) (*see* **Note 5**).

6.2. MARSMotif for Pairs of Motifs

For a given set of input motifs, the pairs of motifs were first constructed from the top 100 motifs selected using the KS test for individual motifs (*see* **Subheading 6.1.**). For any given pair of motifs, we compared the expression values of genes that have that pair of the motifs with the expression values of genes that have one or the other motif (but not both) using the KS test. This comparison allowed us to capture the potentially synergistic pairs. KS test was implemented as in **Subheading 6.1.**

The top 200 motif pairs from the KS test were then used in MARS regression. In each MARS iteration, every time a motif was included all of its interacting partners detected via KS test were included as well. We stopped adding motifs to the input set for a given iteration as soon as the number of motifs exceeded 40. MARS was run allowing for pair-wise (int = 2) and third-order (int = 3) interactions separately. Apart from the interactions, the settings for MARS runs were the same as those for the individual motifs (*see* **Subheading 6.1.**).

For each interaction setting, the motifs that were found significant by MARS were then combined with the set of motifs found significant in the MARS run with individual motifs (*see* **Subheading 6.1.**). MARS was then rerun allowing for the same order of interactions (int=2 or 3) in this set. The motifs and motif pairs identified to be important by MARS in this final run were considered as significant.

6.3. Final Model Selection

For each interaction setting, *p*-values of motifs and motif pairs discovered by MARS were computed based on an F-test *(16)* (*see* **Note 6**):

$$F = \frac{(RSS_0 - RSS_1)/(p_1 - p_0)}{RSS_1/(N - p_1 - 1)} \tag{15}$$

where RSS_1 is the residual sum of squares of the final MARS model with $p_1 + 1$ terms, and RSS_0 is the residual sum of squares of the MARS model without a particular motif (or, motif pair) which has $p_0 + 1$ terms in it. N is the number of genes used in the model. The F statistic has an F distribution with $p_1 - p_0$ numerator degrees of freedom and $N - p_1 - 1$ denominator degrees of freedom. The corresponding *p*-value was calculated in S-PLUS. The *p*-values were then corrected for multiple testing *(3)*. Following corrections, if $p > 0.01$ for a motif (or a motif pair), all the basis functions involving that motif (or motif pair) were deleted from the MARS model. This is the final pruned model for a given interaction setting. We then obtain the $\Delta\chi^2$ corresponding to this pruned model. The interaction setting for which the pruned model had $\Delta\chi^2$ as maximum was identified as the optimal model by MARSMotif.

7. Notes

1. The advantage of using ratios of expression levels is that only a few motifs that are different between the test and control samples contribute significantly to the model.
2. Here $n.\varepsilon_1$ represents the total binding free energy owing to the motif under the given condition. Thus, ε_1 is implicitly dependent on the average concentration of the TF binding to this motif.
3. When interactions are included through a more complete modeling via linear splines, this is generally not true. Word counts perform better than the weight matrices in yeast.

4. We think this is because of the noise arising from use of multiple splines in MARS for the case of one motif.
5. Use of a large number of basis functions can unusually slow down the program.
6. Although a third-order combination can be directly inferred from the int=3 model, we decomposed such combinations into pairs because more often experimental evidence for pairs of motifs are reported in the literature.

Acknowledgments

We thank Gengxin Chen for a careful reading of the manuscript. This work was supported by NIH grants HG01696 (M. Q. Z) and GM60513 (M. Q. Z) and CSHL Association Fellowship (D. D.).

References

1. Spellman, P. T., Sherlock, G., Zhang, M. Q., et al. (1998) Comprehensive identification of cell cycle-regulated genes of the yeast Saccharomyces cerevisiae by microarray hybridization. *Mol. Biol. Cell.* **9,** 3273–3297.
2. Bussemaker, H. J., Li, H., and Siggia, E. D. (2001) Regulatory element detection using correlation with expression. *Nat. Genet.* **27,** 167–171.
3. Das, D., Banerjee, N., and Zhang, M. Q. (2004) Interacting models of cooperative gene regulation. *Proc. Natl. Acad. Sci. USA* **101,** 16,234–16,239.
4. Conlon, E. M., Liu, X. S., Lieb, J. D., and Liu, J. S. (2003) Integrating regulatory motif discovery and genome-wide expression analysis. *Proc. Natl. Acad. Sci. USA* **100,** 3339–3344.
5. Djordjevic, M., Sengupta, A. M., and Shraiman, B. I. (2003) A biophysical approach to transcription factor binding site discovery. *Genome Res.* **13,** 2381–2390.
6. Tavazoie, S., Hughes, J. D., Campbell, M. J., Cho, R. J., and Church, G. M. (1999) Systematic determination of genetic network architecture. *Nat. Genet.* **22,** 281–285.
7. Liu, X. S., Brutlag, D. L., and Liu, J. S. (2002) An algorithm for finding protein-DNA binding sites with applications to chromatin-immunoprecipitation microarray experiments. *Nat. Biotechnol.* **20,** 835–839.
8. Carey, M. (1998) The enhanceosome and transcriptional synergy. *Cell* **92,** 5–8.
9. Ptashne, M. and Gann, A. (1997) Transcriptional activation by recruitment. *Nature* **386,** 569–577.
10. Pilpel, Y., Sudarsanam, P., and Church, G. M. (2001) Identifying regulatory networks by combinatorial analysis of promoter elements. *Nat. Genet.* **29,** 153–159.
11. Banerjee, N. and Zhang, M. Q. (2003) Identifying cooperativity among transcription factors controlling the cell cycle in yeast. *Nucleic Acids Res.* **31,** 7024–7031.
12. Kato, M., Hata, N., Banerjee, N., Futcher, B., and Zhang, M. Q. (2004) Identifying combinatorial regulation of transcription factors and binding motifs. *Genome Biol.* **5,** R56.
13. Keles, S., van der Laan, M., and Eisen, M. B. (2002) Identification of regulatory elements using a feature selection method. *Bioinformatics* **18,** 1167–1175.

14. Chiang, D. Y., Moses, A. M., Kellis, M., Lander, E. S., and Eisen, M. B. (2003) Phylogenetically and spatially conserved word pairs associated with gene-expression changes in yeasts. *Genome Biol.* **4**, R43.

15. Friedman, J. H. (1991) Multivariate Adaptive Regression Splines. *Annals of Statistics* **19**, 1–67.

16. Hastie, T., Tibshirani, R., and Friedman, J. H. (2001) *The Elements of Statistical Learning*, Springer Verlag, New York, NY.

17. Cho, R. J., Campbell, M. J., Winzeler, E. A., et al. (1998) A genome-wide transcriptional analysis of the mitotic cell cycle. *Mol. Cell.* **2**, 65–73.

18. Kellis, M., Patterson, N., Endrizzi, M., Birren, B., and Lander, E. S. (2003) Sequencing and comparison of yeast species to identify genes and regulatory elements. *Nature* **423**, 241–254.

19. Beer, M. A. and Tavazoie, S. (2004) Predicting gene expression from sequence. *Cell* **117**, 185–198.

20. Pennacchio, L. A. and Rubin, E. M. (2001) Genomic strategies to identify mammalian regulatory sequences. *Nat. Rev. Genet.* **2**, 100–109.

21. Keles, S., van der Laan, M. J., and Vulpe, C. (2004) Regulatory motif finding by logic regression. Bioinformatics **20**, 2799–2811.

22. Phuong, T. M., Lee, D., and Lee, K. H. (2004) Regression trees for regulatory element identification. *Bioinformatics* **20**, 750–757.

23. Orian, A., van Steensel, B., Delrow, J., et al. (2003) Genomic binding by the Drosophila Myc, Max, Mad/Mnt transcription factor network. *Genes Dev.* **17**, 1101–1114.

24. Das, D., Nahlé, Z., and Zhang, M. Q. (2006) Adaptively inferring human transcriptional subnetworks. *Mol. Syst. Biol.* **2**, 2006. 0029.

25. Press, W. H., Flannery, B. P., Teukolsky, S. A., and Vetterling, W. T. (1992) *Numerical Recipes in C: The Art of Scientific Computing,* Cambridge University Press, Cambridge, UK.

26. Steinberg, D. and Colla, P. (1999) *MARS: An Introduction.* Salford Systems, San Diego, CA.

6

Statistical Framework for Gene Expression Data Analysis

Olga Modlich and Marc Munnes

Summary

DNA (mRNA) microarray, a highly promising technique with a variety of applications, can yield a wealth of data about each sample, well beyond the reach of every individual's comprehension. A need exists for statistical approaches that reliably eliminate insufficient and uninformative genes (probe sets) from further analysis while keeping all essentially important genes. This procedure does call for in-depth knowledge of the biological system to analyze.

We conduct a comparative study of several statistical approaches on our own breast cancer Affymetrix microarray datasets. The strategy is designed primarily as a filter to select subsets of genes relevant for classification. We outline a general framework based on different statistical algorithms for determining a high-performing multigene predictor of response to the preoperative treatment of patients. We hope that our approach will provide straightforward and useful practical guidance for identification of genes, which can discriminate between biologically relevant classes in microarray datasets.

Key Words: Microarray; prognostic classification; algorithm; preoperative chemotherapy; breast cancer.

1. Introduction

The broad application of microarrays during the last years gave an enormous impulse for biomedical research and promoted numerous studies in all fields of the biological and medical disciplines. There are numerous questions being addressed with microarray experiments in this field. One of the most popular of them belongs to diagnostic and prognostic prediction, treatment selection, and individualized medicine. Microarrays have been utilized extensively for the characterization of cancerous tissues in cancer diagnosis *(1,2)*. The underlying assumption is that gene expression profiles might serve as molecular fingerprints allowing a far more accurate classification of the tumor type and fate compared with present day "traditional" marker detection. Although preliminary data published in this area are promising, there is a need for proper validation of the microarray data in the realm of their feasibility. This validation does refer on the

From: *Methods in Molecular Biology, vol. 377, Microarray Data Analysis: Methods and Applications*
Edited by: M. J. Korenberg © Humana Press Inc., Totowa, NJ

one hand to the technology of high multiplex measurement themselves, but even more to the compiled gene lists, which describe certain properties of a training cohort and have to show their power also in independent validation groups.

Because micoarray technology has reached almost industrial standard, today's more problematic aspect of DNA microarray technology is the nonstandardized area of data analysis. This inconsistency does on the one hand reflect the different array platforms used in the scientific community. On the other hand, it reflects the need for an individual adoption of the statistical techniques applied to a certain biological question. Standardization does take place in the generation of raw data values and in the experiment description (e.g., minimum information about a microarray experiment [MIAME] standard). Nevertheless there are many obvious and hidden pitfalls in the microarray data analysis that may lead to erroneous decisions. The success of analysis relies on the right choice of appropriate statistical method and a clear understanding of the subtleties of analysis *(3)*.

The first statistical efforts in the microarray field dealt with such problems as cross-hybridization on the array, normalization between different array experiments and their reproducibility, and automated image analysis for array hybridization experiments. Because the technology became more mature, the preference of problems has changed.

As already mentioned, one of the present problems concerns compatibility of different microarray platforms and data exchange. Microarray technology is evolving rapidly. Laboratories studying global gene expression in samples of similar origin often use different microarray platforms. These platforms differ in deposition technology, design, probe sets, as well as handling protocols. There have been few studies examining the data correlation among different platforms. The results demonstrated both concordance and discordance of different platforms depending on the applied procedures for raw data readout and normalization. Obviously, these technological differences may influence the results of gene expression profiling *(4)*. Nonetheless, the remarkable degree of overlap for results of differential gene expression has been demonstrated in one of the latest studies on "cross platform comparison" for genes commonly represented on Affymetrix, Aglient, and Amersham CodeLink platforms. This study was based on the oligonucleotide reporters used for the different platforms *(5)*.

At the beginning of the last decade, the number of genes whose expression could be examined on the array was limited to several hundreds. Since then, the situation has changed. Although the technology itself allows collection of a huge amount of gene expression data quickly, accurate analysis and the correct interpretation of the data are still a really big problem for many investigators.

The microarray technology relies on mathematical statistics because of the diverse nature of experiments, and the huge number of genes under study *(6)*. Additionally, there are different sorts of questions, which are addressed with microarray experiments. A question of interest requires the appropriate statistical method, which will be applied for analysis. Categories of questions include: (1) search for genes differentially expressed in different classes (time-points, treatment groups, and so on); (2) identification of genes whose expression is correlated with each other; (3) identification of gene sets involved in the same biological processes (pathway or network oriented); and (4) classification of samples based on their gene expression profiles (patients groups, tissues, and so on). Nonparametric methods, such as nonparametric *t*-test, Wilcoxon (or Mann–Whitney) rank sum test, and a heuristic method based on high Pearson correlation are suitable for identification of differentially expressed genes but also for coregulation or coexpression of gene sets *(7)*. Such statistical techniques as regression methods and discriminant analyses have been applied to determine predictive gene sets *(8)*. Nearly all categories of questions can be approached with clustering techniques, which, if they are applied in an unsupervised fashion, can give an overview of the manifold features of a biological system *(9)*. But any of these techniques will lead to a proper result only if the input datasets are carefully chosen to answer that very question, and the overall expression has been "debulked" for genes, which would hinder the identification of a significant classifier. This "debulking" process may not be restricted to genes but can also include samples. It is mandatory to exclude a whole dataset from further analysis if the overall expression or even the signal intensities of certain areas on the microarray surface are affected by artifacts. The impact of such disturbances on the overall data structure may differ between the individual microarray platforms. In order to get the optimum at the end one should raise the bar right from the beginning.

There are also some biological aspects, which make the microarray application to the field of cancer characterization more difficult. Most cancers are heterogeneous diseases. The development of every tumor is a unique event because every gene dysregulation may be highly specific to each individual patient. There can occur DNA amplification and chromosomal rearrangement, loss of whole chromosomes, and aneuploidity. All these factors will have an impact on the overall expression level of a certain tumor sample and on the selection of genes that can be identified as up- or downregulated. Therefore, statistical methods using average gene expression may hide important expression subtypes. Additionally, it is important to remember that tumor samples are typically a mixture of different cell types. Almost in all studies, the tumor sample is treated as homogeneous. However, different compounds of tumor including tumor cells, surrounding stroma, and blood vessels will react in different way when the tumor is under treatment. How important are such interactions within the tumor for the patient's outcome or response to therapy? We believe that it is one of the very important questions to ask. While cell culture systems do offer the chance to monitor drug activity within a certain cell type, it is practically impossible to control and study the different tumor compounds under treatment in vivo. Therefore, almost all research groups working in this area try to use expression levels of genes in pretreatment tumor samples, as individual portraits, which can hide the patient's destiny.

Precise clinicopathological information and an appropriate data analysis are the anchor stones to successfully build up a tumor classification based on transcription profiling. Because the number of tissue samples examined is usually much smaller than the number of genes on a given array, efficient data deconvolution and dimensional reduction is important. Reliable statistical procedures should be able to eliminate most of the unaffected genes from further consideration while keeping essentially all genes whose expressional changes are potentially important for the aim of a study.

The purpose of this report is to describe an analytic statistical framework for a gene expression-based tumor classification scheme that can allow data analysis in a formal and systematic manner. Here we provide a brief outline of a multistep data analysis, which resulted in a predictor set of 59 genes for predicting response to neoadjuvant epirubicin/cyclophosphamide (EC) chemotherapy of breast cancer patients, and a comparison of this predictor with gene sets obtained by appropriate application of other statistical methods.

2. Materials

2.1. Breast Cancer Data

The example database comes from our recent study on prediction of clinical outcome after neoadjuvant chemotherapy in patients with primary breast cancer disease, in which Affymetrix platform (namely GeneChip HG-U133A consisting of 22,283 probe sets) has been used *(10)*. For marker discovery we used a 56 patient training cohort and 5 normal breast tissue samples. An additional 27 samples were used later on as an independent test cohort for validation purposes.

2.2. Software

Expressionist Analyst software (GeneData, Basel, Switzerland) was applied for statistical data analyses. Additionally, partial least squares discriminant analysis (PLS-DA) using SIMCA-P 10.0 software (Umetrics, Umea, Sweden) has been used.

3. Methods

The methods described next outline (1) data filtering; (2) short description of statistical methods applied for the development of predictive gene sets; (3) the discovery and validation of the 59-gene predictor set; (4) the validation of the gene predictor on the independent cohort; (5) partial least squared regression

analysis of expression data from the training cohort and results from the validation on the test cohort; and (6) the description of the alternative statistical analysis
for the development of a multigene predictor gene ranking using ANOVA.

3.1. Data Filtering

The analytical approach used in this study to minimize the gene probe set is
depicted graphically in **Fig. 1**. In brief, raw data from all microarray hybridization experiments were acquired using MicroSuite 5.0 software (Affymetrix)
and normalized to a common arbitrary global expression value (target signal
value [TGT]; TGT=100). All data were imported into GeneData's Expressionist
software package for further detailed statistical analyses.

3.1.1. Selection of Gene Probe Sets Based on Their Signal Quality

In order to get only high-quality signatures we excluded gene probe sets
from the subsequent analysis owing to various reasons.

1. 59 probe sets corresponding to hybridization controls (housekeeping genes, and so
   on) as identified by Affymetrix were removed from the analysis. We kept the information for the 3′ located probe set for the *GAPDH* and β-*actin* genes as indicated
   by the manufacturer.
2. 100 genes, whose expression levels are routinely used in order to normalize
   between HG-U133A and HG-U133B GeneChip versions, were also removed from
   the analysis because their expression levels did not vary over a broad spectrum of
   human tissues.
3. Genes with potentially high levels of noise (81 probe sets), which is frequently
   observed for genes with low absolute expression values (below 30 relative light
   units [RLU] through all experiments), were removed from the dataset.
4. The remaining genes were preprocessed to eliminate those genes (3196), which were
   labeled as "absent" or above a trustful *p*-value of 0.04 by MicroSuite 5.0. To apply a
   higher stringency to the data we eliminated genes whose significance level ($p < 0.04$)
   was only reached in 10% of all breast cancer samples ever analyzed by our institu-
   tions. This further filtering step resulted in the exclusion of 3841 probe sets.

Data for the remaining 15,006 probe sets were used for all subsequent analysis
steps as described in **Subheading 3.1.2.**

3.1.2. Prefiltering of Data Regarding ER Alpha Status and Genes Involved in the Regulation of the Immune System

1. The content of immune cells varies in breast cancer tissue samples to a great
   extent. In addition, it is difficult to clearly decipher the amount and the impact of
   these cells on the overall gene expression. The "immune" genes (1025 probe sets)
   were selected by their biological properties and based on prior published knowledge and excluded from further analysis.

Data pre-processing

1. Analytical Data Set (**22,283** probe sets, TGT 100)
2. Extract: – 59 AFFYX-control probe sets and
           – 100 Normalizing genes
             22,134 probe sets left
3. Extract: – 81 probe sets maximum < 30 RLU
           – 3,196 probe sets <10% present call in > than
             400 breast cancer samples; P of > 0.04 MAS 5.0
             15,006 probe sets left
4. Extract: – 828 ER-dependent genes
           – 1,025 Immune system genes
             13,145 probe sets left

↓

In parallel application of:

1. *t*–Test, Welch, Wilcoxon, Kolmogorov–
Smimov tests to: (I) n=40 PR *vs.* n=8 NC;
                 (II) n=8 pCR *vs.* n=40 PR;   } *P* < 0.05
                 (III) n=8 pCR *vs.* n=8 NC
                 =>2,301 probe sets qualified
2. Additional restrictions: 2-fold change of median expression
and average expression > 30 RLU
                 =>1,512 probe sets qualified
3. Kruscal-Wallis and ANOVA tests to:
n=8 NC *vs.* n=40 PR *vs.* n=8 pCR; *P* < 0.05 for both tests
                 =>414 probe sets qualified

↓

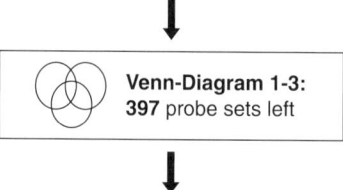 **Venn-Diagram 1-3:**
397 probe sets left

↓

PCA with all pre-defined classes and 397 probe sets:
=> **327** probe sets qualified

↓

Extract genes highly expressed in blood vessels;
adipocytes, muscles, LCM dissected breast cancer tissues
=> **264** probe sets left

Fig.1. Statistical analysis method used in this study. A whole set of probe sets was filtered on signal intensity, regulation fold change, and statistical significance.

2. Genes whose expression is related to ER alpha were also excluded from the final gene lists. It is known that a large number of genes expressed in breast tumors are associated with ER alpha status *(11)*, and the expression signatures of ER-related genes may camouflage additional signatures we desired to identify. Based on our previous analysis on two patient cohorts with positive and negative ER status (100 patients each) we identified 828 Affymetrix probe sets by ANOVA and *t*-test ($p < 0.005$) with a median fold change of 1.2 or more between the two groups. By rejection of the ER alpha-related probe sets, the dataset subsequently used in statistical procedures contained 13,145 probe sets.

3.2. Statistical Analysis

To identify genes differentially expressed in response to chemotherapy we explored several methods including the nonparametric Wilcoxon rank sum test, two-sample independent Student's *t*-test, and two-sample Welch's *t*-statistics *(12)*. A nonparametric Wilcoxon (or Mann–Whitney–U) test is an alternative to the *t*-tests with less power. The Wilcoxon test works better under the assumption that distribution of data under comparison are nonsymmetrical. This test operates on rank-transformed data rather that the raw values *(13)*.

In a next step, the *p*-value for each gene for the null hypothesis that expression values for all experiments are drawn from the same probability distribution and calculated in all tests. For groups with less than 9 samples, the random permutation test has been applied to calculate the *p*-value. Therefore, if the *p*-value is close to zero, than the null hypothesis is probably wrong, and the medians of expression values are significantly different in the two classes. By combining the individual results of these tests with criteria of $p < 0.05$ and median fold change between groups > 2 in a SUM-Rank test we could determine an order of the top performing probe sets in each of the statistical tests applied.

The application of one-way analysis of variance (ANOVA) and Kruskal–Wallis tests appeared to be useful in this study setting because we were dealing with two well-defined sample groups, pCR (complete remission) and NC (no change) as the most extreme response patterns to chemotherapy, and with a third group of partial responders (PR), which was expected to show features of the other two. The Kruskal–Wallis test is a nonparametrical version of the ANOVA *(14)*. It uses the ranks of the data, and is an extension of the Wilcoxon test to more than two groups. If all classes under comparison have at least five samples, the distribution of discriminatory weights can be approximated by a χ^2 distribution. Then, if the *p*-value is close to zero it suggests that the null hypothesis is wrong, and the median of expression levels for at least one group of samples is significantly different from the others.

Principal components analysis (PCA) was most prominently used for data display and structural analysis but in certain steps of the identification process also for dimensional (probe set) reduction *(15)*. Principal components are the orthogonal linear combinations of the genes showing the greatest variability among the cases. Using principal components as predictive features provides a reduction in the dimension of the expression data. However, the PCA has two limitations. First of all, the principal components are not necessarily good predictors. Second, utilization of such principal components as a predictor requires measuring expression of all genes in the particular dataset to classify. This makes the PCA unsuitable for routine clinical applications. For the subsequent classification process and the mandatory cross-validation procedures we selected the rather robust k-nearest neighbors (k-NN) algorithm *(16)*. All these different tools were used as implemented in the Gene Data Expressionist Analyst software package and were only modified by selection of starting parameters and appropriate distance weight matrices.

PLS-DA is a partial least squares regression of one set of binary variables on the other set of predictor variables. This technique is specially suited to deal with a much larger number of predictors than observations and with the multicollineality, which are two of the main problems encountered when analyzing microarray data. PLS is known as a "supervised" method because it uses the independent (expression levels) as well as the dependent variables (classes). The multivariate statistical methods, soft independent modeling of class analogy, and partial least squares modeling with latent variables (PLS) allow all variables to be analyzed simultaneously.

When PLS is applied to microarray data, it is a better method than PCA *(17)*. PCA finds the directions in multivariate space and is capable of identifying common variability rather than distinguishing "among-classes" variability. PLS-DA finds a model that discriminates among classes of objects on the basis of their N variables *(18)*. Additionally, PLS-DA provides a quantitative estimation of the discriminatory power of each descriptor by means of VIP (variable importance for the projection) parameters. VIP values represent an appropriate quantitative statistical parameter ranking descriptors (gene expression values) according to their ability to discriminate different sample classes (tumor types).

The ability to successfully distinguish between tumor classes using gene expression data is an important aspect of cancer classification. Feature selection, as an important step in the process of PLS-DA, is used to identify genes that are differentially expressed among the classes. So far several variations in the algorithms based on linear discriminant analysis (LDA) have been published and used on data from microarray studies for class prediction. One of those is the LDA, which is a classical statistical approach for classifying samples of unknown classes, based on training samples with known classes *(19)*. Fisher's LDA is an oldest form of linear discriminant, but it performs well only if the number of selected genes is small compared with the number of samples. Sparse discriminant analysis is a special case of Fisher's discriminant analysis, which makes it possible to analyze many genes when the number of samples is small *(20)*.

Support vector machines (SVMs) are well suited for two-class or multiclass pattern recognition *(21)*. A SVMs algorithm implements the following idea: it maps the input vectors, i.e., samples into a high-dimensional feature space (variables or genes) and constructs an optimal separating hyperplane, which maximizes the distance (margin) between the hyperplane and nearest data-points of each class in the space. It is important to mention that SVMs can handle large feature spaces while effectively avoiding overfitting and can automatically identify a small subset of informative data-points. The classification of biological samples and thereby the identification of a neoplastic lesion as well as the response of such lesion to therapeutic agents based on gene expression data is often a multiclass classification task.

k-NN as a nonparametric pattern recognition approach is one of the suitable algorithms to opt for when predicting class membership. The method of *k*-NN proposed by T. M. Cover and P. E. Hart *(22)* is quite easy and efficient. Partly because of its perfect mathematical theory, the NN method has developed into several variations. As we know, if we have infinitely many sample points then the density estimates converge to the actual density function. The classifier becomes the Bayesian classifier if samples on a large scale are provided. But in practice, given a small number of samples, the Bayesian classifier usually fails in the estimation of the Bayes error especially in a high-dimensional space, which is called the disaster of dimension. Therefore, the method of *k*-NN has a great disadvantage that the number of training samples must be large enough.

In *k*-NN classification, the training data set is used to classify each member of a "target" dataset. The structure of the data is that there is a classification (categorical) variable of interest (e.g., "responder" (CR) or NC), and a number of additional predictor variables (gene expression values). Generally speaking, the algorithm works as follows:

1. For each sample in the dataset to be classified, locate the *k*-NN of the training data set. A Euclidean distance measure can be used to calculate how close each member of the training set is to the target sample being examined.
2. Examine the *k*–NN; which classification do most of them belong to? Assign this category to the sample being examined.
3. Repeat this procedure for the remaining samples in the target set.

Of course the computing time goes up as *k* goes up, but the advantage is that higher values of *k* provide smoothing that reduces vulnerability to noise in the training data. In practical applications, typically, *k* is in units or tens rather than in hundreds or thousands. The distance to the "NN" in higher dimensional space may also be determined. The *k*-NN method gathers the nearest *k* neighbors and lets them vote; the class with highest number of neighbors wins. Theoretically, the more neighbors we consider, the smaller the error rate. Ben-Dor et al. *(23)* and Dudoit et al. *(24)* compared several simple and complex methods on several public datasets, both have found that *k*-NN classification generally performed as well as or better than other methods *(21,22)*.

3.3. Discovery and Validation of 59 Genes Predictor Set

3.3.1. Discovery of Multigene Predictor Set

1. The training cohort of 56 cases with known response was used to develop and train our predictors (**Fig. 1**). 8 of the training cases experienced a pathologically confirmed pCR, 40 cases experienced PR, and 8 experienced stable or progressive disease (NC). In order to identify the most significant genes determining each group's properties we considered the following comparisons for the training set: (I) n=40 PR vs n=8 NC; (II) n=8 pCR vs n=40 PR, and (III) n=8 pCR vs n=8 NC. These comparisons were made by nonparametric *t*-test, Welch, Wilcoxon, and Kolmogorov–Smirnov tests. We reported as significant only those genes that reached significance at the level $p < 0.05$ in all tests. Altogether, 2301 probe sets were qualified.

2. Because such statistical filtering does not take signal strength or factor of gene regulation in the individual groups into account, we applied the following restrictions: at least twofold change of median expression level and average expression more than 30 RLU for all three groups were under comparison. Only 1512 probe sets were qualified for further analyses following this independent filtering step.

3. In parallel, statistical significance in the comparison of all three response classes (n=8 pCR vs n=40 PR vs n=8 NC) was measured with the Kruskal–Wallis and one-way ANOVA tests. For this study we assumed that those tumors with a mediocre response to chemotherapy but at least a reduction of the tumor mass of 25% (PR) may represent an individual gene signature. For the three-group tests we applied a cutoff of $p < 0.05$. Only 414 probe sets passing this filter were identified. Based on Venn diagram analysis of the three gene sets derived from previous individual analyses we qualified 397 probe sets to go on with. These genes do combine the requested features of appropriate signal intensity, regulation fold change, and statistical significance.

4. PCA using all predefined tissue classes, normal tissue (collection of > 100 different tissue/cell types; NT), normal breast tissue (NB), pCR, cCR (good clinical response), PR, and NC, was applied to the 397 probe sets, to filter based on the major components (eigengenes). In our particular case the separation of pCR and cCR tumors on the one hand and of NC samples on the other was defined by only two most distinguishing components. We applied a cutoff on the correlation matrix of the PCA and filtered for genes at < -0.4 and > 0.4. This removed 72 and left 325 probe sets.

5. Because a further gene reduction of the predictor set was mandatory for ease of usability later on, we performed filtering for genes based on biological knowledge.

We filtered out probe sets highly expressed in blood vessels, adipocytes, and muscle tissue vs expression profiles obtained from individual tumor cells dissected by laser capture microdissection from breast cancer tissue samples. Besides this attempt to filter out nontumor-specific gene expression, we identified two genes (*FHL1* and *CLDN5*) as highly discriminative between most "normal" tissue samples and all breast cancer samples analyzed. We combined the two genes with the 57 genes identified before as top ranked in a SUM-Rank test for all samples and with respect to the 13,145 genes.

3.3.2. Cross Validation

The model discovery process is depicted graphically in **Fig. 2**. Cross-validation was performed for the training set and for classes NB, pCR, PR, and NC using the *k*- NN with *k*=3 and 59 probe sets (57 filtered probe sets and 2 genes, which can distinguish between normal and cancerous breast tissue). Thus, each sample was represented by a pattern of expression that consisted of 59 genes. Each sample was then classified according to the class memberships of its *k*-NN, as determined by the Euclidean distance in higher dimensional space. Training error was determined using "leave-25%-out" cross-validation method. Cross-validation removes randomly each time 25% of observations in turn, constructs the classifier, and then computes whether this classifier correctly classifies the removed test fraction. Finally, a *k*-NN model was built using all 56 training cases (with no samples left out), which was then used to predict classification of the test cases. The specificity of the best performing classifier on the training set was 99% for normal breast tissue, approx 90% for pCR, 80% for PR, and 25% for NC.

3.3.3. Optimization of the Gene Classifier Using Decision Tree

This classifier could be subdivided into three groups of genes. These contain genes/probe sets, which are able to distinguish:

1. Normal breast vs breast cancer tissues (two genes).
2. pCR or cCR (collectively, CR) cases vs the nonfavorable outcomes PR or NC (31 probe sets or "good response signature").
3. NC vs PR (26 probe sets or "poor response signature").

We expected that both signatures, good and poor, would effectively recognize expression patterns corresponding to those that it was trained on. It is necessary to admit that the fuzziness of the ultrasound imaging applied for tumor-size determination prior to chemotherapy, compared with the rather accurate measurement by a pathologist on the resected tumor margins, has introduced an undesirable error in true response status and, subsequently, in the further statistical analysis. Therefore, the developed model may have lower sensitivity (i.e., predict many NC cases as PR and vice versa), which is reflected in low prediction accuracy for NC cases (*see* above).

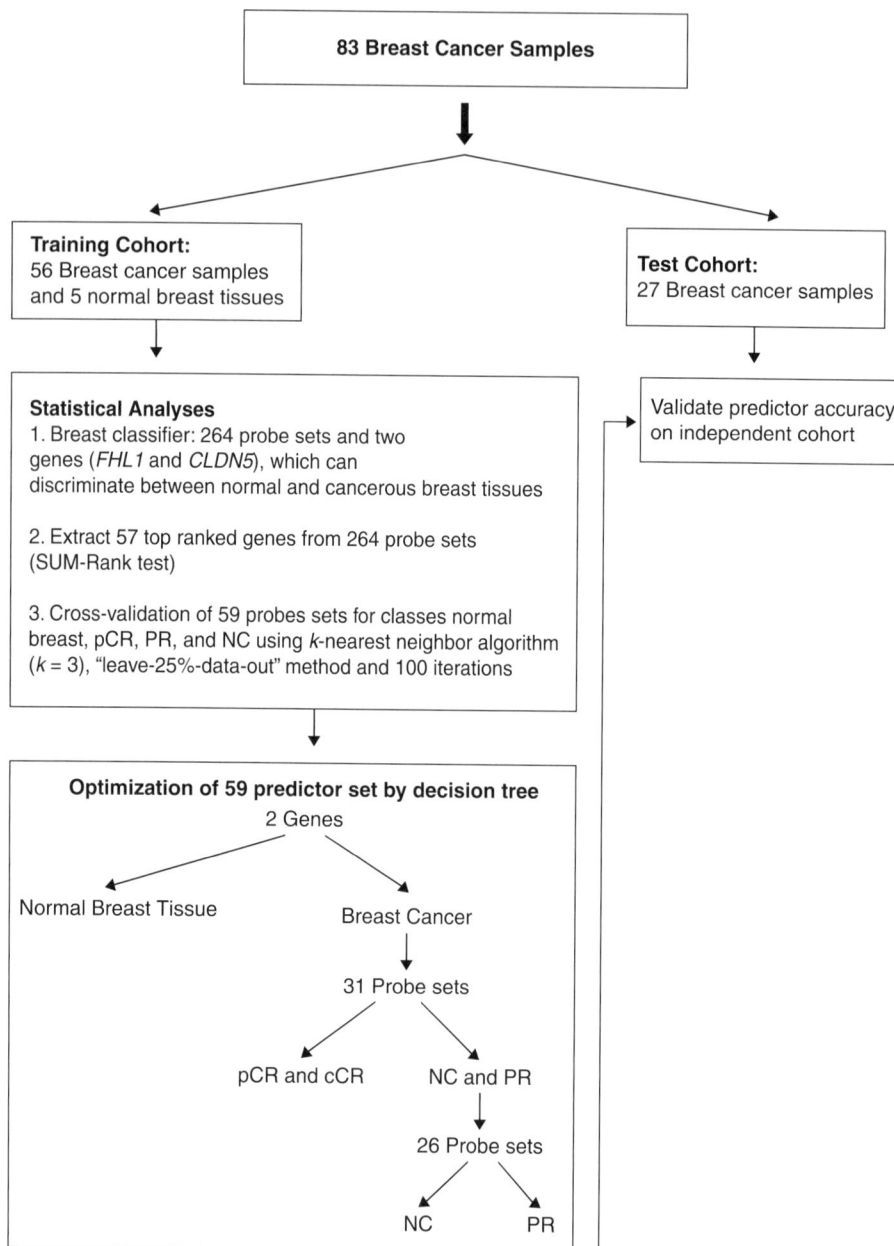

Fig. 2. A supervised learning approach to develop multigene predictors of clinical outcome. pCR, complete tumor remission; cCR, good clinical response; PR, partial response; NC, no change or progressive disease.

3.4. Validation on Independent Cohort

The classifier was tested on an independent test cohort (n = 27; 3 pCR and 1 ccR, 4 NC, 19 PR) as follows (**Fig. 2**). Classification was performed by *k*-NN algorithm (*k*=3) in three steps of a decision tree algorithm using the 59 genes mentioned in **Subheading 3.3.1.**

1. All 27 tumor samples were correctly qualified as cancerous tissues using the two-gene signature (*FHL1* and *CLDN5*).
2. Using the genes from the "good response signature" a group of 7 tumor samples was classified as CR, and the remaining 20 tumors as other (i.e., NC and PR together).
3. The latter 20 tumors were classified as either NC or PR by use of the "poor response signature."

All CR and NC cases were correctly predicted. Results of classification for the test cohort are shown in **Table 1**.

3.5. PLS-DA Using Training Set and Results Validation on the Test Set

1. Following the experimental setup as described herein, the training and test cohorts consisting of 56 and 27 samples, respectively, have been used. PLS-DA was carried out first with those 13,145 probe sets that passed the quality control filtering process (*see* **Subheading 3.1.**). Although this leads to an over-parameterized model with poor prediction properties, it provides a first assessment about the most important discriminant variables. We let the algorithm work on two independent starting models each consisting of two classes: model 1: class 1 – pCR, class 2 – NC (PR cases were excluded); and model 2: class 1 – pCR, class 2 – NC and PR together. Another model, with three classes (pCR, NC, and PR), demonstrated rather poor prediction power because it strongly depends on the definition of PR, which may often be rather controversial.
2. Three and four components were defined by PLS-DA in models 1 and 2, respectively. Then those variables satisfying criteria of having expression levels more than 60 RLU (as a mean value in at least one of each sample group, pCR and NC), ratio (pCR/NC) > 1.9 or < 0.55, and VIP of > 1.9 were retained. We performed a second iteration of the PLS-DA of model 1 with the selected 96 probe sets and model 2 with 90 probe sets. **Figure 3A,B** shows a scatter plot of samples in the training set grouped according to the two components for either PLS in the model 1 (96 probe sets; **Fig. 3A**) or in the model 2 (90 probe sets; **Fig. 3B**). The pCR and NC samples are clearly discriminated, although results of permutation tests for both models (data not shown) demonstrated that both reduced models were still over-parameterized.
3. Thus, we retained the 20 probe sets deduced from model 1 (pCR vs NC) and 20 probe sets from model 2 (pCR vs NC and PR) with highest VIP values. A reassessment of the performance of both second iteration models is shown in **Fig. 4A,B**.

Table 1
Comparison of Predicted and Pathological Response in Test Cohort

Case	Tumor reduction (%)	Response, pathologic	Predicted response k-NN. Decision treeAlgorithm (59 genes) (**Subheading 3.4.**)	Predicted response PLS-DA model 1: pCR vs NC; without PR (**Subheading 3.5.**)	Predicted response PLS-DA model 2:pCR vs NC&PR (**Subheading 3.5.**)	Predicted response k-NN (63 probe sets) (**Subheading 3.6.**)
N1	0	NC	NC	NC	PR	PR
N2	0	NC	NC	NC	NC	PR
N3	0	NC	NC	NC	PR	NC
N4	10	NC	NC	NC	PR	PR
N5	100	pCR	CR	CR	CR	PR
N6	100	pCR	CR	CR	CR	PR
N7	100	pCR	CR	CR	CR	PR
N8	100	cCR	CR	CR	CR	CR
N9	40	PR	PR	CR	CR	PR
N10	47	PR	PR	NC	NC	NC

N11	40	PR	PR	CR	CR	PR
N12	90	PR	CR	CR	CR	PR
N13	80	PR	PR	NC	NC	NC
N14	92	PR	PR	PR	PR	PR
N15	0	PR	PR	PR	CR	PR
N16	0	PR	NC	NC	NC	NC
N17	40	PR	PR	PR	PR	PR
N18	62	PR	NC	PR	PR	PR
N19	22	PR	NC	NC	PR	PR
N20	10	PR	NC	NC	NC	NC
N21	33	PR	PR	NC	NC	PR
N22	50	PR/NC	PR	NC	NC	PR
N23	0	PR/NC	NC	CR	CR	NC
N24	68	PR	CR	NC	NC	PR
N25	5	NC	NC	NC	NC	NC
N26	25	PR	NC	NC	NC	NC
N27	85	PR	CR	CR	CR	PR

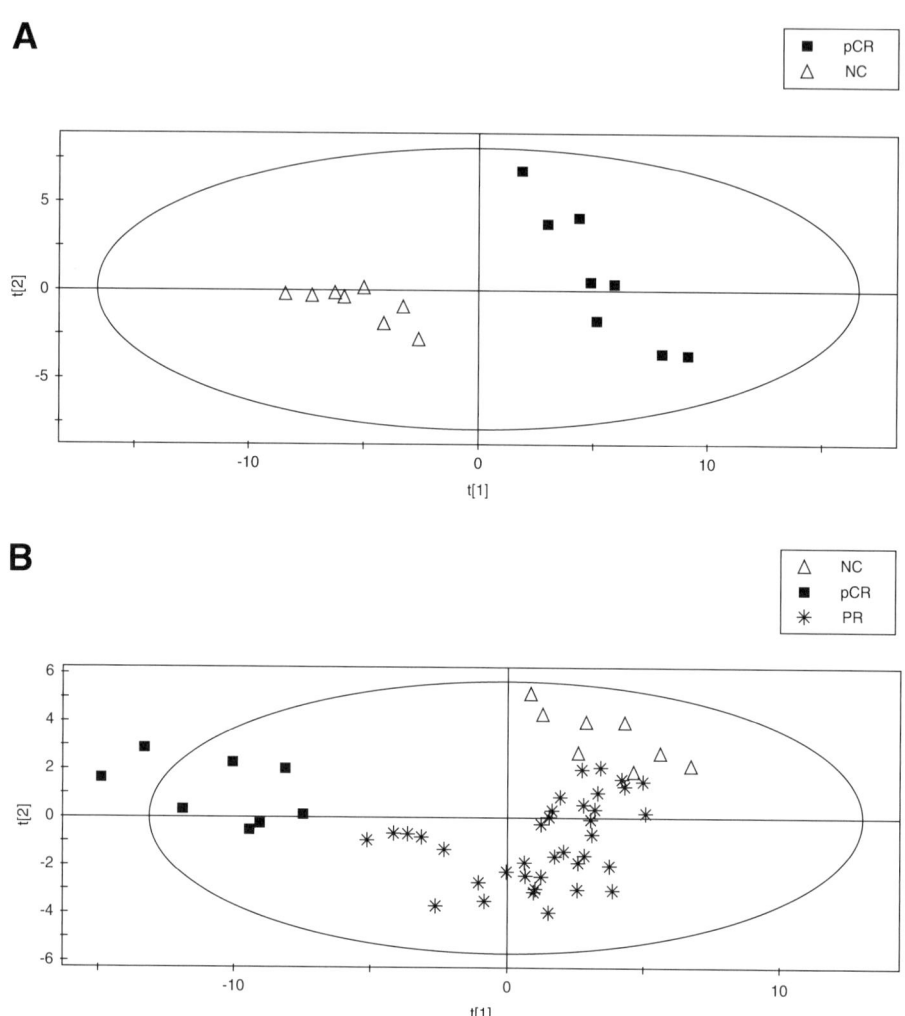

Fig. 3. PLS discrimination according to tumor response class using the variables selected by PLS (*VIP* > 1.9) and ratio (pCR/NC) > 1.9 or < 0.55. (**A**) Model 1 (PR cases were deleted; class 1 – pCR, *black boxes*; class 2 – NC, *open triangles*); 96 probe sets (cDNAs) retained. (**B**) Model 2(class 1 – pCR, *black boxes*; class 2 – NC, *open triangles* and PR, *stars*); 90 probe sets retained.

In both cases, models performed much better than expected by chance. Both groups of selected probe sets were compared and nine probe sets were found to be represented in both lists. The combined list of unique probe sets we used for model validation contained 31 probe sets.

4. For an independent validation, a group of 27 tumor samples was used in order to test the discriminative power of the final gene list. The results are presented in

A

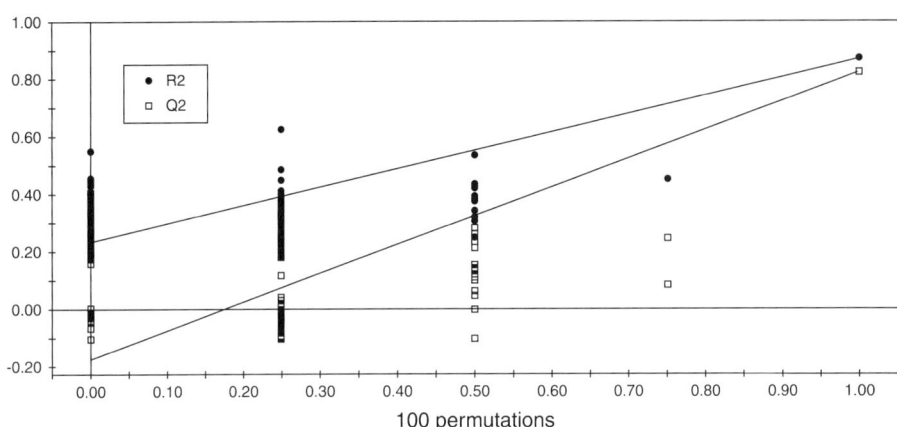

B

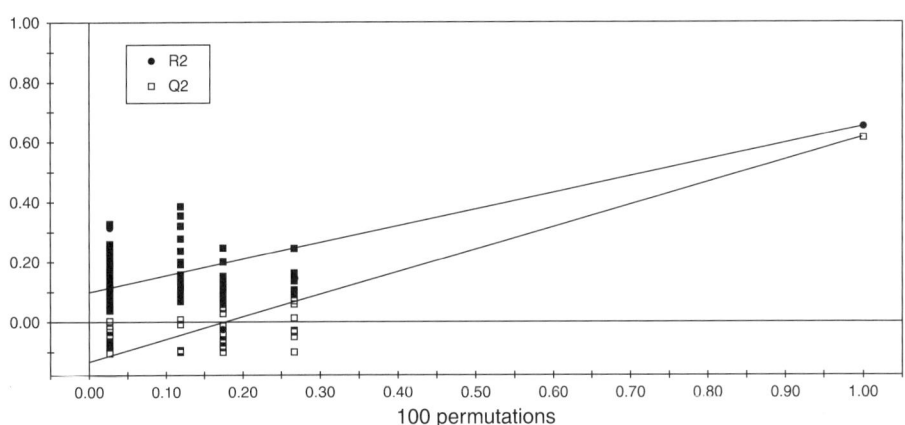

Fig. 4. Validation of PLS discrimination analysis by permutation. **(A)** Model 1 (class 1 – pCR; class 2 – NC; PR cases were deleted) using 20 selected from 96 probe sets. **(B)** Model 2 (class 1 – pCR; class2 – NC and PR together) using 20 selected from 90 probe sets. The *horizontal axis* shows the correlation between the permuted and actual data, the *vertical axis* is the value of R^2 (the variance explained in fitting) and Q^2 (the variance explained in predicting). The two values on the right hand corner $r=1$ correspond to the values of R^2 and Q^2 for the actual data. Each *symbol* represents a permutation result, R^2 is shown by *black dots* and Q^2, by *squares*.

Table 1. It is obvious that true pCR cases are correctly predicted in both models, while NC cases are better predicted in model 1. Nevertheless, it was interesting to see that also by PLS-DA as before by statistical tests partial responding tumors with either good (>60% tumor shrinkage) or very poor response to therapy were predicted to show potentially complete response (e.g., N12, N24, N27) or no change of tumor (e.g., N22, N25, N26), respectively. This observation indicates that for further studies the monitoring of tumor shrinkage during preoperative systemic chemotherapy is pivotal to correctly judge the final pathological response classification.

A comparison of results obtained by applying the two statistical approaches in microarray data analysis, one that resulted in a 59-gene EC predictor set, and the other resulting in 96- and 90-gene sets identified by PLS-DA, showed that 19 genes were identified by both statistical approaches. However, the PLS-DA itself had overall worse predictive ability in comparison to the first multistep analysis combined with a k-NN classification at the end.

3.6. Gene Ranking—ANOVA

Additionally, for the purpose of comparison, we applied Fisher's and Sparse LDA, SVMs, and k-NN classification in a gene ranking procedure to find genes that are significant for response to EC chemotherapy in our training set.

Unfiltered data consisting of all 22,283 probe sets were used. The minimal misclassification rate for each of three algorithms, Sparse LDA, SVMs, and k-NN together corresponded to the gene set containing 63 probe sets. The predictive accuracy demonstrated by such a gene set in a cross validation (k-NN) was high: 100% for CR, 61% for PR, and 88% for NC. Unfortunately, an independent validation on the test cohort was less successful than by use of the 59-gene classifier. Validation results are shown in **Table 1**. Only one case in each group, CR and NC, was correctly predicted.

This problem is also known as "overfitting" the training set. We had so many parameters that they could fit all of the random variations well. Therefore, all tests have found predictors, which fit the model in the training set very well, but provided inaccurate predictions for the independent test cohort.

4. Conclusions

This statistical approach offers a possibility for successful expression data filtering and analyses concerning the development of a multigene predictor set. We have applied several simple but very effective steps for the data reduction and prefiltering. Statistical methods described here provide improved approaches to microarray data analysis. After applying a proposed model, a predictive probe set was selected, which could be successfully validated on the independent cohort of samples. The data reduction and an appropriate statistical analysis algorithm are crucially important for the identification of new molecular markers for response prediction.

References

1. Olson, J. A., Jr. (2004) Application of microarray profiling to clinical trials in cancer. *Surgery* **136**, 519–523.
2. Jain, K. K. (2004) Applications of biochips: from diagnostics to personalized medicine. *Curr. Opin. Drug Discov. Devel.* **7**, 285–289.
3. Simon, R. (2003) Diagnostic and prognostic prediction using gene expression profiles in high-dimensional microarray data. *Br. J. Cancer* **89**, 1599–1604.
4. Hardiman, G. (2004) Microarray platforms—comparisons and contrasts. *Pharmacogenomics* **5**, 487–502.
5. Shippy, R., Sendera, T. J., Lockner, R., et al. (2004) Performance evaluation of commercial short-oligonucleotide microarrays and the impact of noise in making cross-platform correlations. *BMC Genomics* **5**, 61.
6. Quackenbush, J. (2001) Computational analysis of microarray data. *Nat. Rev. Genet.* **2**, 418–427.
7. Troyanskaya, O. G., Garber, M. E., Brown, P. O., Botstein, D., and Altman, R. B. (2002) Nonparametric methods for identifying differentially expressed genes in microarray data. *Bioinformatics* **18**, 1454–1461.
8. Liu, Y. and Ringner, M. (2003) Multiclass discovery in array data. *BMC Bioinformatics* **5**, 70.
9. Shannon, W., Culverhouse, R., and Duncan, J. (2003) Analyzing microarray data using cluster analysis. *Pharmacogenomics* **4**, 41–52.
10. Modlich, O., Prisack, H-B., Munnes, M., Audretsch, W., and Bojar, H. (2005) Predictors of primary breast cancers responsiveness to preoperative Epirubicin/Cyclophosphamide chemotherapy: translation of microarray data into clinically useful predictive signatures. *J. Transl. Med.* **3**, 32.
11. Gruvberger-Saal, S. K., Eden, P., Ringner, M., et al. (2004) Predicting continuous values of prognostic markers in breast cancer from microarray gene expression profiles. *Mol. Cancer Ther.* **3**, 161–168.
12. Walpole, R. E. and Myers, R. H. (eds.) (1985) *Probability and Statistics for Engineers and Scientists, 3rd ed.,* Macmillan, New York, NY.
13. Wilcoxon, F. (1945) Individual comparisons by ranking methods. *Biometrics* **1**, 80–83.
14. Kruskal, W. H. and Wallis, W. A. (1952) Use of ranks in one-criterion variance analysis. *J. Amer. Statist. Assoc.* **47**, 583–621.
15. Manly, B. F. J. (ed.) (2004) *Multivariate Statistical Methods: A Primer, 3rd ed.,* Chapman Hall, London, UK.
16. Duda, R. O. and Hart, P. E. (1973) *Pattern Classification and Scene Analysis,* Wiley, New York, NY.
17. Datta, S. (2001) Exploring relationships in gene expressions: a partial least squares approach. *Gene Expr* **9**, 249–255.
18. Wold, S., Sjöström, M., and Erikson, L. (1998) PLS in chemistry. In: *The Encyclopedia of Computational Chemistry* (Schleyer, P.v.R., Schreiner, P. R., Allinger, N. L., et al, eds.), John Wiley & Sons, Chichester, UK, pp. 2006–2020.

19. Johnson, R. A. and Wichern, D. W. (eds.) (1982) *Applied Multivariate Statistical Analysis*. Prentice-Hall, Englewood Cliffs, NJ.

20. Cox, J. (2002) Comparative study of classification algorithms and gene selection methods for the discrimination of cancer tissue samples using microarray expression data. 3rd Intl. Conf. on Systems Biology, Stockholm, Sweden.

21. Vapnik, V. (ed.) (1998) *Statistical Learning Theory*. Wiley, New York, NY.

22. Cover, T. M. and Hart, P. E. (1967) Nearest neighbour pattern classification. *IEEE Transactions on Information Theory* **13**, 21–27.

23. Ben-Dor, A., Bruhn, L., Friedman, N., Nachman, I., Schummer, M., and Yakhini, Z. (2000) Tissue classification with gene expression profiles. *J. Comput. Biol.* **7**, 559–583.

24. Dudoit, S., Fridlyand, J., and Speed, T. P. (2002) Comparison of discrimination methods for the classification of tumors using gene expression data. *J. Am. Stat. Assoc.* **97**, 77–87.

7

Gene Expression Profiles and Prognostic Markers for Primary Breast Cancer

Yixin Wang, Jan Klijn, Yi Zhang, David Atkins, and John Foekens

Summary

Genome-wide measures of gene expression have been used to classify breast tumors into clinically relevant subtypes, as well as provide a better means of risk assessment on an individual basis for lymph node-negative (LNN) breast cancer patients. We have applied Affymetrix GeneChips of 22,000 transcripts to analyze total RNA of frozen tumor samples from 286 LNN breast cancer patients in order to identify a gene signature for identification of patients at high risk for distant recurrence.

Key Words: Microarray; gene expression; node-negative breast cancer; prognosis.

1. Introduction

Genome-wide measures of gene expression provide a powerful approach to identify gene expression patterns that are correlated to tumor behaviors. Several reports in colon, breast, lung, and lymphoma cancers suggest that the new approach could be complementary to clinical or pathological examination (*1,2*). Based on gene expression patterns, breast tumors could be classified as those with different clinically relevant subtypes (*3–6*), different prognosis, and response to chemotherapy (*7–17*). Determining prognosis of the breast tumor for an individual patient requires careful assessment of multiple clinical and pathological parameters; however, traditional prognostic factors are not always sufficient to predict patient outcomes accurately. It is important to identify those patients at high risk for relapse and who definitely need adjuvant systemic therapy after primary surgery, instead of giving adjuvant therapy to all lymph node-negative (LNN) patients, resulting in over-treatment. There have been many attempts to find novel gene or protein markers for breast cancer progression. Few have been implemented in routine practice. A possible reason for these individual marker candidates is that breast cancer progression is a complex

From: *Methods in Molecular Biology, vol. 377, Microarray Data Analysis: Methods and Applications*
Edited by: M. J. Korenberg © Humana Press Inc., Totowa, NJ

function of multiple molecular events that may arise within the malignant tumor cells or may be induced by stromal events. Genome-wide measurements allow us to perform a more comprehensive assessment of these molecular events in LNN primary breast cancer *(18)*.

2. Materials
2.1. Patient Samples

1. Frozen tumor specimens from LNN patients treated during 1980–1995, but untreated with systemic adjuvant therapy, were selected from our tumor bank at the Erasmus Medical Center (Rotterdam, Netherlands).
2. The tumor samples were originally collected from 25 regional hospitals. The guidelines for primary treatment were the same for all hospitals.
3. 436 invasive tumor samples were screened for inclusion to the study. Patients with a poor, intermediate, and good clinical outcome were included. Samples were rejected based on insufficient tumor content (53), poor RNA quality (77), and poor chip quality (20) leaving 286 samples eligible for further analysis.
4. The study was conducted according to the approved protocol by the institution's Medical Ethical Committee (MEC no. 02·953).
5. Median age of the patients at the time of surgery was 52 yr (range, 26–83 years).
6. Prior to inclusion, all the 286 tumor samples were confirmed to have sufficient (>70%) and uniform involvement of tumor in H&E-stained, 5-μm sections cut from the frozen tumors.
7. Estrogen receptor (ER) (and progesterone receptor [PgR]) levels were measured by ligand-binding assay or enzyme immunoassay *(19)* or by immunohistochemistry (in nine tumors). The cutoffs used to classify tumors as ER or PgR positive were >10 fmol/mg protein or >10% positive tumor, respectively.
8. Patient followup involved examination every 3 mo during the first 2 yr, every 6 mo for year 2 to 5, and annually from year 5 of the followup period.
9. Date of distant metastasis was defined as the date of confirmation of metastasis after complaints and/or clinical symptoms, or at regular followup.
10. Of the 286 patients included, 93 (33%) showed evidence of distant metastasis within 5 yr and were counted as failures in the analysis of distant metastasis-free survival.

2.2. Reagents

1. Total RNA Isolation. Life Technologies Trizol Reagent Total RNA Isolating (Invitrogen).
2. cDNA synthesis.
   a. 50 μ*M* GeneChip T7-Oligo(dT) Promoter Primer kit, 5′–GGCCAGT GAATTGTAATACGACTCACTATAGGGAGGCGG-(dT)24–3′, HPLC-purified (Affymetrix, P/N 900375).
   b. SuperScript™ II (Invitrogen Life Technologies, P/N 18064-014) or SuperScript Choice System for cDNA synthesis (Invitrogen Life Technologies).
   c. *Escherichia coli* DNA ligase (Invitrogen Life Technologies, P/N 18052-019).

    d. *E. coli* DNA polymerase I (Invitrogen Life Technologies).

    e. *E. coli* RNaseH (Invitrogen Life Technologies).

    f. T4 DNA polymerase (Invitrogen Life Technologies, P/N 18005-025).

    g. 5X Second-strand buffer (Invitrogen Life Technologies, P/N 10812-014).

    h. 10 m*M* dNTP (Invitrogen Life Technologies).

    i. 0.5 *M* EDTA.

3. Sample Clean-Up (Sample Cleanup Module; Affymetrix).
4. Synthesis of biotin-labeled cRNA (Enzo BioArray HighYield RNA Transcript Labeling kit *(10)*; Enzo Life Sciences).
5. cRNA fragmentation (all from Sigma-Aldrich): Trizma base, magnesium acetate (P/N M2545), potassium acetate, and glacial acetic acid.
6. Target hybridization.

    a. Water (molecular biology grade, BioWhittaker Molecular Applications/ Cambrex).

    b. 50 mg/mL Bovine serum albumin solution (Invitrogen Life Technologies).

    c. Herring sperm DNA (Promega Corporation).

    d. GeneChip eukaryotic hybridization control kit (Affymetrix) contains control cRNA and control Oligo B2.

    e. 3 nM Control Oligo B2 (Affymetrix).

    f. 5 *M* NaCl (RNase-free, DNase-free) (Ambion).

    g. MES hydrate (SigmaUltra; Sigma-Aldrich).

    h. MES sodium salt (Sigma-Aldrich).

    i. 0.5 *M* Solution EDTA disodium salt (100 mL) (Sigma-Aldrich).

    j. DMSO (Sigma-Aldrich).

    k. 10%Surfact-Amps 20 (Tween-20) (Pierce Chemical).

7. Washing, staining, and scanning.

    a. Water (molecular biology grade; BioWhittaker Molecular Applications/Cambrex).

    b. Distilled water (Invitrogen Life Technologies).

    c. 50 mg/mL Bovine serum albumin solution (Invitrogen Life Technologies).

    d. R-phycoerythrin streptavidin (Molecular Probes).

    e. 5 *M* NaCl (RNase-free, DNase-free; Ambion).

    f. PBS, pH 7.2 (Invitrogen Life Technologies).

    g. 20X SSPE: 3 *M* NaCl, 0.2 *M* NaH$_2$PO$_4$, 0.02 *M* EDTA (BioWhittaker Molecular Applications/Cambrex).

    h. Goat IgG (reagent grade; Sigma-Aldrich).

    i. Anti-streptavidin antibody (goat) (biotinylated; Vector Laboratories).

    j. 10%Surfact-Amps 20 (Tween-20) (Pierce Chemical).

3. Methods

3.1. RNA Extraction

1. Homogenize tissue samples in 1 mL of Trizol reagent per 50–100 mg of tissue using the disperser/homogenizer (Ultra-turrax T8 dispersers/homogenizers; IKA Works). Wash the stainless-steel probe with the following solutions in sequence: absolute

ethanol, RNase-free water, RNase away, and RNase-free water twice. Then dry the probe with Kim wipes. Repeat this between processing two samples.

2. Incubate the homogenized samples for 5 min at room temp to permit the complete dissociation of nucleoprotein complexes. Add 200 µL of chloroform per 1 mL of TRIzol reagent. Cap sample tubes securely and shake vigorously by hand for 15 s. Incubate them at room temp for 2–3 min. Centrifuge the samples at no more than 12,000g for 15 min at 4°C. Following centrifugation, the mixture separates into a lower red, phenol–chloroform phase, an interphase, and a colorless upper aqueous phase. RNA remains exclusively in the aqueous phase. The volume of the aqueous phase is about 60% of the volume of TRIzol reagent used for homogenization.

3. Transfer the aqueous upper phase to a fresh tube. Precipitate the RNA from the aqueous phase by mixing with isopropyl alcohol. Use 500 µL of isopropanol per 1 mL of TRIzol reagent used for the initial homogenization. Incubate samples at –20°C for 30 min and centrifuge at no more than 12,000g for 10 min at 4°C. The RNA precipitate, often invisible before centrifugation, forms a gel-like pellet on the side and bottom of the tube.

4. Remove the supernatant from **step 3**. Wash the RNA pellet once with 75% ethanol (in DEPC water), adding at least 1 mL of 75% ethanol per 1 mL of TRIzol reagent used for the initial homogenization. Mix the sample by vortexing and centrifuge at no more than 7500g for 5 min at 4°C. The RNA precipitate can be stored in 75% ethanol at 4°C for at least 1 wk, or at least 1 yr at –20°C.

5. Briefly dry the RNA pellet (air-dry for 5–10 min). Be careful not to let the pellet dry completely as this will decrease solubility. Add Rnase-free water (how much depends on the size of the pellet and how concentrated or dilute you want your sample), vortex, and heat the sample at 55–60°C for 10 min.

6. If using microcuvet (pathlength of 0.5 cm), make a 1:5 dilution in a volume of 10 µL (8 µL of water + 2 µL RNA) in a fresh tube. Take absorbance readings using the Hewlett Packard spectrophotometer at 260 and 280 wavelengths. Calculate the 260/280 ratio. A ratio of <1.6 indicates the sample is only partially dissolved. A260 × 40 × 2 (cuvet pathlength is adjusted to 1 cm) × 5 (dilution factor)/ 1000 = µg/µL.

7. Assess the integrity of total RNA samples on an Agilent 2100 Bioanalyzer. For a high-quality total RNA sample, two well-defined peaks corresponding to the 18S and 28S ribosomal RNAs should be observed, similar to a denaturing agarose gel, with ratios approaching 2:1 for the 28S to 28S bands. The sum of percent areas under the 18S and 28S ribosomal RNAs should be more than 15.

3.2. Gene Expression Analysis

1. Biotinylated targets were prepared using published methods (Affymetrix) *(20)* and hybridized to Affymetrix oligonucleotide microarray U133a GeneChip. Arrays were scanned using the standard Affymetrix protocol.

2. Expression values for each gene were calculated using Affymetrix GeneChip analysis software MAS 5·0.

3. In order to normalize the chip signals, all probe sets were scaled to a target intensity of 600 and scale mask files were not selected. Chips were rejected if average intensity was less than 40 or if the background signal exceeded 100.

3.3. Statistical Analysis

1. Gene expression data was filtered to include genes called "present" in two or more samples. 17,819 genes passed this filter and were used for hierarchical clustering.
2. Each gene was divided by its median expression level in the patients. This standardization step minimized the effect of the magnitude of expression of genes, and grouped together genes with similar patterns of expression in the clustering analysis. Average linkage hierarchical clustering was performed on both the genes and the samples using GeneSpring 6·0.
3. To identify gene markers that best discriminate between patients who developed a distant metastasis and those who remained metastasis free within 5 yr, we used supervised class prediction approaches.
4. The patients were first placed into one of the two subgroups stratified by ER status. Each patient subgroup was then analyzed separately in order to select markers. The patients in the ER-positive subgroup were divided into a training set of 80 patients and a testing set of 129 patients. The patients in the ER-negative subgroup were divided into a training set of 35 patients and a testing set of 42 patients. The selection of the patients into the training and the testing set was entirely random.
5. As a quality control step, Kaplan–Meier survival curves *(21)* of the training and the testing set were evaluated to ensure that there was no significant difference and no bias was introduced by the random selection of the training and the testing set. The training set was used for gene selection and the testing set was used for independent validation.
6. The sample size of the training set was determined by a resampling method to ensure its statistical confidence level. Briefly, the number of patients in the training set started at 15 patients and was increased by 5 at a time. For a given sample size, 10 training sets with randomly selected patients were made. A gene signature was constructed from each of training sets and then tested in a designated testing set of patients using receiver operating characteristic (ROC) curve analysis using distant metastasis within 5 yr as the defining point. The mean and the coefficient of variation of the area under the curve (AUC) for a given sample size were determined. A minimum number of patients required for the training set were chosen at the point that the average AUC plateaued and the coefficient of variation of the 10 AUCs was below 5%.
7. Univariate Cox proportional hazards regression was used to identify genes whose expression levels (on $\log_2$ scale) were correlated with the length of distant metastasis-free survival.
8. To reduce the effect of multiple testing and to test the robustness of the selected genes, the Cox model was performed with bootstrapping of the patients in the training set *(22)*. Briefly, 400 bootstrap samples of the training set were constructed, each containing 80 patients randomly chosen with replacement. The Cox model

was run on each of the bootstrap samples. A bootstrap score was created for each gene by removing the top and bottom 5% *p*-values and then averaging the inverses of the remaining bootstrap *p*-values. This score was used to rank the genes.

9. To construct a multiple gene signature, combinations of gene markers were tested by adding one gene at a time according to the rank order. ROC analysis using distant metastasis within 5 yr as the defining point was performed to calculate AUC for each signature with increasing number of genes until a maximum AUC value was reached.

10. The Relapse Score (RS) was used to determine each patient's risk of distant metastasis. The score was defined as the linear combination of weighted expression signals with the standardized Cox regression coefficient as the weight:

$$RS = A \cdot I + \sum_{i=1}^{60} I \cdot w_i x_i + B \cdot (1 - I) + \sum_{j=1}^{16} (1 - I) \cdot w_j x_j$$

Here A and B are constants, and $I = 1$ if ER level > 10 and otherwise $I = 0$. The w_i and w_j are the standardized Cox regression coefficients for ER+ and ER− markers respectively, and x_i and x_j are the expression values of ER+ and ER− markers, respectively, in $\log_2$ scale.

11. The threshold was determined from the ROC curve of the training set to ensure 100% sensitivity and the highest specificity. The values of the constants A of 313.5 and B of 280 were chosen to center the threshold of RS to zero for both ER positive and ER negative patients.

12. Patients with positive RS scores were classified into the poor prognosis group and patients with negative RS scores were classified into the good prognosis group. The gene signature and the cutoff were validated in the testing set.

13. Kaplan–Meier survival plots and log-rank tests were used to assess the differences in time-to-distant metastasis of the predicted high- and low-risk groups.

14. Sensitivity was defined as the percent of the distant metastasis patients within 5 yr that were predicted correctly by the gene signature, and specificity was defined as the percent of the patients free of distant recurrence for at least 5 yr that were predicted as being free of recurrence by the gene signature.

15. Odds ratio was calculated as the ratio of the odds of distant metastasis between the predicted relapse patients and relapse-free patients.

16. All statistical analyses were performed using S-Plus 6·1 software (Insightful, Seattle, WA).

3.4. Pathway Analysis

1. The list of Affymetrix probe set IDs was used as the input to search for the biological networks built by the software. A functional class was assigned to each of the genes in the prognostic signature. Pathways analysis was performed using the Ingenuity 1.0 software (Ingenuity Systems, Redwood City, CA).

2. Biological networks identified by the program were then confirmed by using general functional classes in gene ontology classification. Pathways that have two or more genes in the prognostic signature were selected and evaluated.

References

1. Ntzani, E. and Ionnidis, J. P. A. (2003) Predictive ability of DNA microarrays for cancer outcomes and correlates: an empirical assessment. *Lancet* **362,** 1439–1444.

2. Wang, Y., Jatkoe, T., Zhang, Y., et al. (2004) Gene expression profiles and molecular markers to predict recurrence of Dukes' B colon cancer. *J. Clin. Oncol.* **22,** 1564–1571.

3. Perou, C. M., Sørlie, T., Eisen, M. B., et al. (2000) Molecular portraits of human breast tumors. *Nature* **406,** 747–752.

4. Sørlie, T., Perou, C. M., Tibshirani, R., et al. (2001) Gene expression patterns of breast carcinomas distinguish tumor subclasses with clinical implications. *Proc. Natl. Acad. Sci. USA* **98,** 10,869–10,874.

5. Sørlie, T., Tibshirani, R., Parker, J., et al. (2003) Repeated observation of breast tumor subtypes in independent gene expression data sets. *Proc. Natl. Acad. Sci. USA* **100,** 8418–8423.

6. Korkola, J. E., DeVries, S., Fridlyand, J., et al.: Differentiation of lobular versus ductal breast carcinomas by expression microarray analysis. *Cancer Res.* **63,** 7167–7175.

7. Van't Veer, L., Dai, H., Van de Vijver, M. J., et al. (2002) Gene expression profiling predicts clinical outcome of breast cancer. *Nature* **415,** 530–536.

8. Van de Vijver, M. J., Yudong, H. E., Van't Veer, L., et al. (2002) A gene expression signature as a predictor of survival in breast cancer. *N. Engl. J. Med.* **347,** 1999–2009.

9. Ahr, A., Kam, T., Solbach, C., et al. (2002) Identification of high-risk breast-cancer patients by gene-expression profiling. *Lancet* **359,** 131–132.

10. Huang, E., Cheng, S. H., Dressman, H., et al. (2003) Gene expression predictors of breast cancer outcomes. *Lancet* **361,** 1590–1596.

11. Sotiriou, C., Neo, S. -Y., McShane, L. M., et al. (2003) Breast cancer classification and prognosis based on gene expression profiles from a population-based study. *Proc. Natl. Acad. Sci. USA* **100,** 10,393–10,398.

12. Woelfle, U., Cloos, J., Sauter, G., et al. (2003) Molecular signature associated with bone marrow micrometastasis in human breast cancer. *Cancer Res.* **63,** 5679–5684.

13. Ma, X. -J., Salunga, R., Tuggle, J. T., et al. (2003) Gene expression profiles of human breast cancer progression. *Proc. Natl. Acad. Sci. USA* **100,** 5974–5979.

14. Ramaswamy, S., Ross, K. N., Lander, E. S., et al. (2003) A molecular signature of metastasis in primary solid tumors. *Nat. Genet.* **33,** 1–6.

15. Chang, J. C., Wooten, E. C., Tsimelzon, A., et al. (2003) Gene expression profiling for the prediction of therapeutic response to docetaxel in patients with breast cancer. *Lancet* **362,** 362–369.

16. Sotiriou, C., Powles, T. J., Dowsett, M., et al. (2003) Gene expression profiles derived from fine needle aspiration correlate with response to systemic chemotherapy in breast cancer. *Breast Cancer Res.* **4,** R3.

17. Hedenfalk, I., Duggan, D., Chen, Y., et al. (2001) Gene-expression profiles in hereditary breast cancer. *N. Engl. J. Med.* **344,** 539–548.
18. Wang, Y., Klijn, J., Zhang, Y., et al. (2005) Gene-expression profiles to predict distant metastasis of lymph-node-negative primary breast cancer. *Lancet* **365,** 671–679.
19. Foekens, J. A., Portengen, H., van Putten, W. L. J., et al. (1989) Prognostic value of estrogen and progesterone receptors measured by enzyme immunoassays in human breast tumor cytosols. *Cancer Res.* **49,** 5823–5828.
20. Lipshutz, R. J., Fodor, S. P., Gingeras, T. R., et al. (1999) High density synthetic oligonucleotide arrays. *Nat. Genet.* **21,** 20–24.
21. Kaplan, E. L. and Meier, P. (1958) Non-parametric estimation of incomplete observations. *J. Am. Stat. Assoc.* **53,** 457–481.
22. Efron, B. (1981) Censored data and the bootstrap. *J. Am. Stat. Assoc.* **76,** 312–319.

8

Comparing Microarray Studies

Mayte Suárez-Fariñas and Marcelo O. Magnasco

Summary

We present a practical guide to some of the issues involved in comparing or integrating different microarray studies. We discuss the influence that various factors have on the agreement between studies, such as different technologies and platforms, statistical analysis criteria, protocols, and lab variability. We discuss methods to carry out or refine such comparisons, and detail several common pitfalls to avoid. Finally, we illustrate these ideas with an example case.

Key Words: Microarray; meta-analysis; crossplatform comparisons.

1. Introduction

In the past few years a profusion of research has dealt with comparisons of different microarray studies, both to cross-validate or integrate different studies as well as to assess the differences between platforms *(1–9)*. The latter regale newcomers to the field with a bewildering array of orthogonal conclusions— some conclude that different platforms generate largely incompatible data, whereas others conclude that laboratory variability is in general greater than that from the platform, and so on. Upon closer inspection it is seen that these studies assess equality or difference in dramatically different ways, that some comparisons are dramatically less fairer than others, and some may be downright incorrect. In light of the profusion of different "comparison technologies," the aim of this chapter is to introduce the reader to the problems and issues that arise in comparing high-throughput experiments in general and microarray studies in particular.

The traditional notion of equality or equivalence (of a measurement) is predicated on overlap: the measurements of two objects occupy the same footprint. This is so because all measurements are inaccurate, so we must measure several times to estimate a probability distribution, and then compare the distributions. We say that two things weigh the same if, when we weigh each many times,

From: *Methods in Molecular Biology, vol. 377, Microarray Data Analysis: Methods and Applications*
Edited by: M. J. Korenberg © Humana Press Inc., Totowa, NJ

there is substantial overlap between the measurements of both, as assessed, for example, by checking that the difference between the means is smaller than the standard errors (a *t*-test).

But in high-dimensional spaces (when we are measuring many things at the same time) overlap vanishes exponentially fast. Imagine we repeat 101 times an experiment with a microarray probing 10,000 transcripts in a given tissue. Just like 2 points determine a line and 3 points determine a plane, these 101 microarray measurements determine a 100-dimensional hyperplane in the 10000-dimensional space of gene expression values. The overlap between geometrical objects is dictated by their dimensionalities; if the sum of the dimensions of the objects is smaller than the dimension of the ambient space, the objects are unlikely to intersect at all. If we now repeat the experiment, treating the tissue with some factor, the 100-dimensional planes corresponding to our two experiments have a zero chance of intersecting—even if the treatment did not do anything. Thus, comparing microarray experiments (or any other high-dimensional measurement) cannot be done as with low-dimensional experiments by comparing the measurements *as probability distributions in the space of the measurement.*

Some further modeling is required to compare experiments, and may be implicit in the form the comparison takes. For instance, assuming that individual gene expression values are statistically independent the probability distribution factors into a product; 10,000 independent tests can be applied to assess differences. This comparison (and the *p*-values obtained thereof) is only valid under the (likely incorrect) assumption of statistical independence. In a radically different vein one may coarse grain over sets of genes, e.g., participating in given pathways or having common gene ontology classifications. *All such comparisons have an implicit model and may give incorrect results if the model is too far off the mark in the particular case at hand.*

To keep the discussion as practical and how-to as possible, we shall explain, not only how to do something, but a number of things to avoid, which we collect in **Subheading 2.** (an "anti-methods" if you will), and also notes are provided. We present as example one case study.

2. Pitfalls

2.1. Correlation of Absolute Expression Values Against Relative Values

Not infrequently cross-platform comparisons are carried out through the correlation of signal intensities for individual transcripts (*4,6,7*). We follow here an argument given in **ref.** *10* demonstrating that such comparisons are misleading because they are adversely affected by "probe effects:" probe-specific and platform-specific multiplicative factors that have a large variability (*11–13*).

The advantage of relative expression over absolute expression can be easily understood if the following model is considered.

$$Y_{ijk} = \theta_i + \phi_{ij} + \varepsilon_{ijk}$$

where Y_{ijk} is the k-th measurement of expression (in log scale) on a gene i by platform j, θ_i is the real expression of a gene, ϕ_{ij} is the platform-specific probe/spot effect and ε is a random error in the measurement, and all effects in the model are independent random variables with variances σ_θ^2, σ_ϕ^2 and σ_ε^2, respectively. In Affymetrix arrays the probe-effect variability σ_ϕ^2 is larger than the variance of the expression level, σ_θ^2 *(14)*. The within-platform correlation is given by

$$corr(Y_{ij1}, Y_{ij2}) = \frac{\sigma_\theta^2 + \sigma_\phi^2}{\sigma_\theta^2 + \sigma_\phi^2 + \sigma_\varepsilon^2} \qquad (1)$$

and is usually near one because σ_ε^2 is much smaller than $\sigma_\theta^2 + \sigma_\phi^2$. The across-platform correlation can be written as:

$$corr(Y_{i1k}, Y_{i2k}) = \frac{\sigma_\theta^2}{\sigma_\theta^2 + \sigma_\phi^2 + \sigma_\varepsilon^2} \qquad (2)$$

and it is smaller than the within-platform correlation because the probe effect is not common to both platforms, so the term σ_ϕ^2 does not appear in the numerator. The probe effect can be calibrated *(11)* so absolute mRNA concentrations can be estimated, but to do so nominal concentrations of spiked-in mRNAs must be provided. A simpler solution to avoid the probe-effect problem is to consider only relative expression values. Usually in microarrays we are comparing control vs condition. If we consider Y_{ijk}^A, Y_{ijk}^B, the absolute expression values for samples A and B, the relative expression value can be modeled as:

$$M_{ijk} = Y_{ijk}^A - Y_{ijk}^B = d_i + \varphi_{ij} + \eta_{ijk} \qquad (3)$$

where d_i is the true amount of differential expression (in log-fold change). The terms ϕ_{ij} should be the same for sample A and B so the probe effect cancels out. As in practice this is not removed completely, the term φ_{ij} in **Eq. 3** is included to represent a platform-dependent bias. The within- and across-platform correlation of the M-values are respectively:

$$corr(M_{ij1}, M_{ij2}) = \frac{\sigma_d^2 + \sigma_\varphi^2}{\sigma_\theta^2 + \sigma_\varphi^2 + \sigma_\eta^2} \text{ and } corr(M_{i1k}, M_{i2k}) = \frac{\sigma_d^2}{\sigma_\theta^2 + \sigma_\varphi^2 + \sigma_\eta^2}$$

but now the term σ_φ^2 is much smaller than σ_ϕ^2. The confirmation of this theoretical effect can be checked in the results of **ref.** *15* and our examples.

2.2. Preprocessing Steps

In all microarray technologies, a good amount of preprocessing follows image analysis. Various groups have shown the impact of normalization and background correction procedures on downstream analysis in Affymetrix *(16)* and cDNA *(17)*. As an example, we compute expression values for Affymetrix' Spike-in experiments (HGU133a chips) using four of the most popular algorithms. **Figure 1** shows the magnitude of the differences between each pair of algorithms' outcomes. Note that for a substantial number of genes the difference can be bigger than twofold changes. It is then not surprising that those discrepancies can affect agreement across platforms, as shown in **ref.** *5*, where correlation of M-values between Affymetrix and cDNA Agilent platform varies from 0.6 to 0.7 when RMA *(16)*, MAS5, and dChip *(14)* algorithms are used to compute expression values. Most of the authors *(2,3,7–9,18)* use the default algorithm provided by the array manufacturer's software to preprocess the data. Although analytical software provided by manufacturers require very little input from the user, there are alternatives developed by the academic community shown to have better performance. **Reference** *15* clearly shows how the agreement within and between platforms can be increased by proper use of available alternative algorithms.

2.3. Annotations

Agreement between platforms can be affected by the identification of common genes as **refs.** *3* and *9* suggested in their studies. The selection of the identifiers is a difficult issue because none of them maps genes one-to-one. For example, the number of common genes for the three experiments in our case study is almost 8000 using Unigene identifiers, but around 15,000 using Locus Link identifiers. Sequence-matched probes can increase cross-platform correlation between M-values as reported in **ref.** *5* (*see* **Note 2**). However, matching the sequences could be a hard procedure especially if more than two studies and a large number of genes are involved. One solution is to take the intersection of various identifiers, i.e., genes matched by two or more identifiers, which can improv the cross-platform agreement *(15)*.

2.4. Statistical Protocol

Although researchers are quite aware that experimental results are sensitive to the protocol used, it is not unusual that studies using different statistical approaches are compared on the same basis. This is particularly delicate if we are trying to assess platform reproducibility. For example, **refs.** *2*, *4*, and *18* based their comparisons on the agreement of lists generated from different statistical criteria. There is only one solution to this problem: the data must be reanalyzed using the same statistical approach.

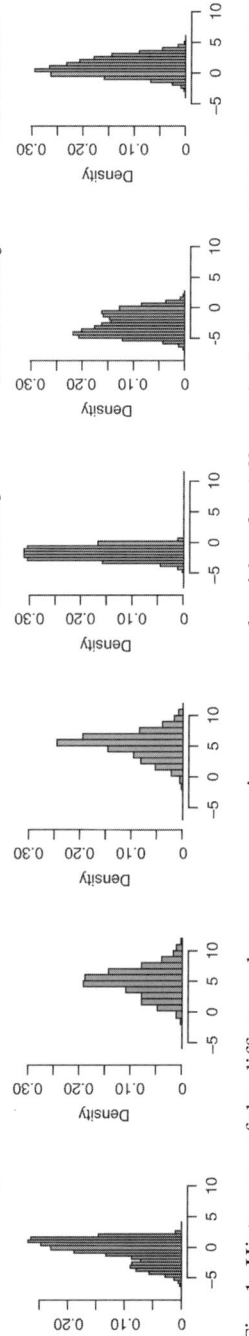

Fig. 1. Histogram of the difference between expression measures algorithm for Affymetrix Spike In data (HGU133a chips).

Yet, is the intersection of individual lists a good strategy to measure concordance? Even with the same statistical protocol some studies concluded that a small amount of genes lay in the intersection *(7,8)*. The caveat here is that the intersection between lists should be considered as a compound statistical test, whereby the null hypothesis (for the intersection) is rejected only when all three null hypothesis (for the individual lists) are rejected. The false-positive rate of the intersection is thus the product of the individual false-positive rates; however the true positive (TP) rate is also the product of the three individual TP rates, and as these are also smaller than one, their product could be quite small. As a result, the *p*-values which generate adequate lists with good false-positive control will be inadequate for intersecting (too few TPs at the intersection). *In order to use list intersection as a criterion for comparing studies, care should be taken that the* p-*values for the individual lists should be chosen so as to give good numbers at the intersection, not on individual lists.*

2.5. Lab Effect

Gene expression is the nervous system of cells, easily imprinted by anything in the environment; gene expression is affected by the way the laboratory sets up the experiment—sample collection methods in the case of tumors, culture system variables for cultured tissues, and animal feeding and maintenance protocols. For example, a study *(3)* found poor correlations between M-values using samples of cancer cell lines, but it was carried out independently in two different laboratories, and variations that may have arisen from independent cell culturing, RNA isolation, and purification were not controlled. The influence of the lab effect on cross-platform agreement was pointed out in **ref. 8**, where the sample variability was revealed to be the main source of data variation, and was confirmed in **ref. 15**.

3. Methods

We now present some ideas as to how comparisons can be carried out soundly; we anticipate that better methods will be devised in the future as our understanding evolves, and the practitioner in the field should try to keep aware of the latest techniques. We shall first outline the basic overall flow and then the pieces.

3.1. Overall Flow

1. Get the raw data for all studies.
2. Use uniform data preprocessing steps.
3. Identification of the common genes for all studies.
   a. Obtain up-to-date annotations for all the studies.
   b. Try to match the sequence or use more than one identifier to match them.
4. Further reduce the scope of the comparisons by eliminating genes that might have been erratically affected; e.g., by the integrated correlations approach (*see* **Subheading 3.3.**).

5. Use the same statistical methodology to define differential expression for the individual studies and the most powerful available tests (*see* **Note 4**).
6. Create a list of common differentially expressed genes; by list intersection or by summary statistics *(see* **Subheading 2.5.**). The *p*-value of the *intersection* has to be set, which is the product of individual *p*-values, if intersection is considered (*see* **Note 5**).

3.2. Raw Data

The researcher attempting comparative studies should procure the rawest level of data possible from all sources. There is no current standard as to how to analyze data in the field, and studies are published with vastly different analysis methods. It is thus *imperative* to redo the analysis from scratch. For Affymetrix GeneChip data, the raw ".CEL" files should be procured; these contain data from individual probes and permit execution of quality control algorithms diagnosing, e.g., the quality of the hybridization or presence of blemishes *(19)*, as well as usage of other summarization methods besides the "closed-box" Affymetrix MAS5 algorithms. In the case of cDNA-like techniques, the database containing foreground and background intensities can be used, assessing the quality trough, e.g., arrayMagic *(20)*. Nevertheless, for a better standardization in the case of studies where the image analysis software used different feature extraction criteria, processing the original image will be a plus.

3.3. Integrated Correlations Approach

To reduce the "lab effect," it would be advisable to identify and eliminate from the analysis genes that which appear to be affected erratically across labs; for example, many in vivo studies are afflicted by immunity genes flaring up because of some flu or other condition affecting a litter of animals. Integrative correlation analysis was introduced to validate agreement across studies and to select genes that exhibit a consistent behavior across them *(21)* by examining all pair-wise correlations of gene expression.

Define x_g to be the expression profile for a gene g, and $\rho_p^s = corr(x_{g_1}, x_{g_2})$, the correlation for the pair of genes p=(g_1,g_2) in the study *s*. Based on ρ_p^s we can assess both overall reproducibility between studies and gene-specific reproducibility. The integrated correlation $I(s,s') = corr(\rho_p^s, \rho_p^{s'})$, quantifies the reproducibility between studies. If this expression is calculated considering only the pairs containing a gene g, then we have a measure of the gene-specific reproducibility between two studies, that is $R^{s,s'}(g) = corr(\rho_p^s, \rho_p^{s'})$, where p=(g,j). When more than two studies are involved, the average over all s and s' is used as a reproducibility score for a gene g,

$$R_g = \sum_{s=1}^{n} \sum_{s'>s} \frac{R^{s,s'}(g)}{\binom{n}{2}}.$$

3.4. Coinertia

Another technique to measure agreement between studies is the coinertia analysis (COIA). Initially developed in the ecological area, it was applied to perform cross-platform comparisons in **ref. 22**. It does not require cross referencing the annotation of the transcript or statistically based filtering of data prior to cross-platform analysis, but it is only possible if both experiments have exactly the same amount of arrays with the same sample. Furthermore, it does not offer a way to identify common differentially expressed genes.

The idea of COIA is to produce for each study a new representation of the arrays in a gene hyperspace where the two new representations maximize the square covariance (of the arrays) between the two studies. This produces a set of axes, one from each dataset, where the first pair of axes is chosen so as to be maximally covariant and represent the most important joint trend in the two datasets. The second pair of axes is chosen so as to be maximally covariant but orthogonal to the first pair, and so on for the rest of the axes. Once the new representations are obtained, the similarity is measured either as the correlation between the data-points projected on the first corresponding axes for each study or by the RV-coefficient, a multivariate extension of the Pearson correlation.

3.5. List Intersection or Summary Statistics

We discussed briefly the problems with list intersection previously mentioned (*see* **Subheading 1.4.**). Here we reiterate that intersecting lists is an extremely valuable technique that can potentially refine results by considering what has been seen to happen repeatably in many experiments; the fundamental trick is to recreate the lists from scratch because the cutoff criteria that give good control over the false-positive and -negative rates in an intersection are far from those that give good control to each individual list *(23)*.

Another approach *(24)* is the use of summary statistics based on individual p-values. The p-values for each gene in each study is obtained (normal single study). The meta-analysis consists in defining the summary statistic $S = -2\sum \log(p_i)$. The distribution of S is obtained by simulation and a "summary P" for gene g is defined as P(S>Sg). This technique is potentially sensitive to outliers because a single large value in one of the studies can place a gene on the list, and would be best to use with a robust approach.

3.6. Discussion: A Case Study

We illustrate the practical use of these procedures through a study *(23)* carried out to compare three different studies of human embryonic stem cells (HESC). Each *(25–27)* study concluded with a list of genes that are upregulated in stem cells, but the three lists of significantly upregulated genes, as published, are quite

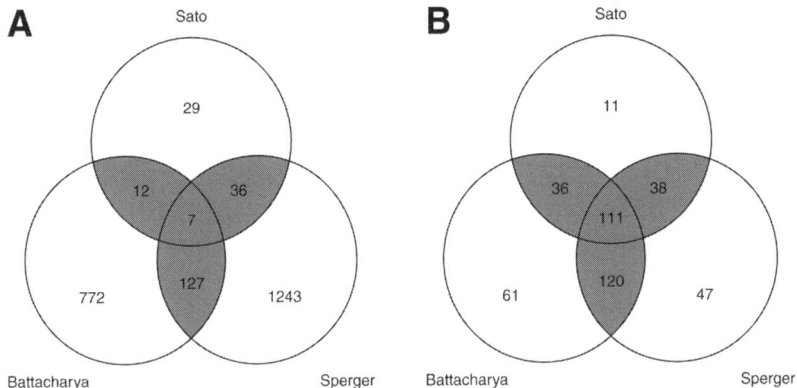

Fig. 2. Intersection between the published lists of upregulated genes for each study. (**A**) As published. (**B**) After reanalysis.

different. Their intersection is shown in **Fig. 2A**: seven genes appear in all three studies out of the 2226 total genes in the union. This is particularly troublesome because all three studies appear to be technically reliable and each study has good reproducibility between replicates. After we carried out the procedure described next we obtained a much more significant level of agreement illustrated in **Fig. 2B**.

The Bhattacharya study *(25)* has 6 cDNA chips (8 × 4,23 × 23 design) where different HESC lineages were hybridized to the red channel (Cy5), and control samples hybridized to the green channel (Cy3) were isolated from a collection of adult human tissues. The Sperger study *(26)* used a 12 × 4, 30 × 30 design of cDNA chips, also hybridizing individual lineages; the control samples were also a common reference pool of mRNA. The Sato study *(27)* had six Affymetrix HGU133A chips, three replicates of H1 cells, and three replicates of "nonlineage-directed differentiation."

We then proceeded as follow:

1. We carried out the analysis using the open-source R language version 2.0 *(28)* and packages provided in *Bioconductor* project *(29)*.
2. For **cDNA arrays**: we used the same image analysis criteria to exclude low-quality spots for cDNA arrays. Transcripts with excessive numbers of low-quality spots across the set of arrays were excluded from the analysis. The *marray* package from the Bioconductor suite was used for preprocessing. Normalization was executed in two steps, first within-print-tip-group location-dependent intensity normalization followed by a within-print-tip group scale normalization using median absolute deviation. Single-channel normalization of two-color cDNA was done as proposed by ref. *30*, using quantile normalization.
3. For **Affymetrix chips**, the GCRMA algorithm was used to summarize data as proposed in ref. *31*. This algorithm improves the widely used RMA *(16)* by including

an extra step to adjust for nonspecific binding, and computing the sequence-specific affinities between probes as described *(13)*.

4. We verified that within-platform reproducibility is fairly good in all the studies, even noting that Battacharya's and Sperger's designs contain different lineages of HESC rather than true replicates of a single lineage.

5. Annotations were obtained with the raw data from each study. For both Bhattacharya and Sperger studies, annotations were obtained from SOURCE from the Stanford microarray data homepage (http://source.standford.edu). For Affymetrix data, annotations packages from Bioconductor were used. The IMAGE clone IDs and the Affymetrix probes were matched using Unigene Cluster Annotation. Genes with no Unigene identifier were eliminated and duplicated probes/spots were averaged together.

6. After this process there are 7373 genes common to all three studies. We filtered for evidence of variation across samples, reducing our set of interesting genes to 2463.

7. Within this universe of 2463 genes we executed an integrated correlation approach. The integrated correlation coefficients between studies were extremely small (0.13 in the best case) and inspection of correlation between M-values indicates poor general agreement between studies. For each pair of studies, the two-dimensional density of the pair-wise correlations (data not shown) suggests that we can find many "negatively coherent" pairs of genes, positive correlated in one study and negatively correlated in the other, and in any such pair, one must be inconsistent.

   Figure 3 paints a much more hopeful picture. The histograms of the coherence scores between study pairs (shown as marginal distributions around the two-dimensional densities in **Fig. 3**) reveal the existence of a group of genes with high coherence scores in all study pairs. The bivariate density of the coherence score between pairs of studies shows that despite variations, there is a group of genes where scores between Bhattacharya–Sperger are similar to the score of Bhattacharya–Sato, those that have higher values in both are part of the coherent set. The histogram of the average pair-wise reproducibility **(Fig. 4)** shows a bimodal distribution, with an apparently clear-cut distinction between two groups of genes, one of them having positive reproducibility scores ("coherent") and the other one close to zero ("erratics") or negative ("incoherents"). So the general poor agreement observed between the studies is a result of averaging over a set of genes with both positive and negative coherences. We decided to keep for further analysis the 739 genes in the top 30% of the gene-coherence distribution. Eliminating erratic genes enormously improves the general agreement between the studies, with integrated correlation value of 0.78 in the worst case.

8. Within the set of coherent genes, we study those that are up- or downregulated in stem cells vs their differentiated controls in *each one* of the studies. *Exactly* the same statistical tests and criteria were applied to all three studies, with a strict cutoff value selection based *both* on a *p*-value and a positive log of the odds (that a gene is differentially expressed) *(32)*. We used the moderated *t*-statistics as proposed in **ref. 33** and the false discovery rate procedure was used to adjust the *p*-values for multiple hypothesis *(34)*. *P*-value cutoff was set at 0.01, which implies than the probability of error is 10^{-4} in the pair-wise comparison and 10^{-6} when the three studies are considered.

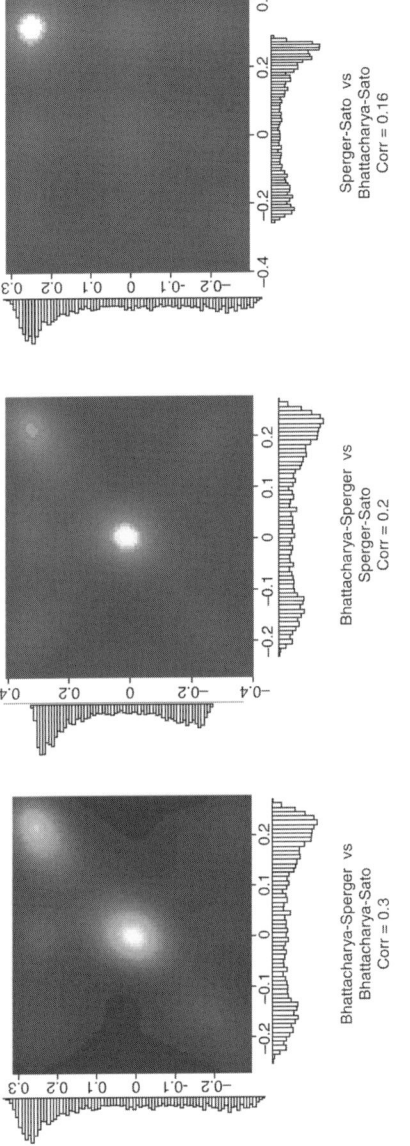

Fig. 3. Bivariate densities of the coherence score.

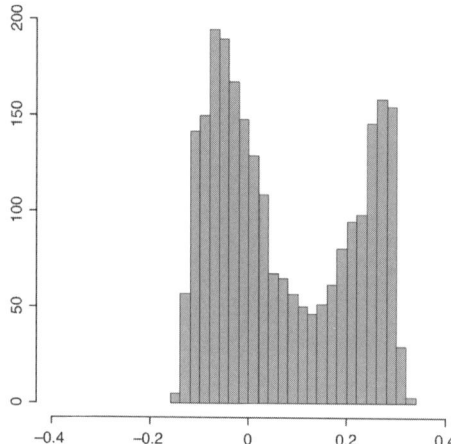

Fig. 4. Gene-coherence score (average over all three comparisons).

The intersection between the lists is now quite a bit larger and statistically significant (*see* **Fig. 2B**), 111 genes were found to be upregulated genes common to all three studies (95 downregulated) against 3 expected by chance. Notice that the 111 upregulated genes in this list are *not* necessarily the "most" upregulated for any individual study; yet they are significantly upregulated for *each* study.

4. Notes

1. Different ways to measure the agreement. Some comparisons are solely based on correlation of signal or correlation of M-values, others are based on intersection of the list or alternatives analysis such as COIA. As an example, **refs.** *3* and *22* used the same panel of 60 cell lines from the National Cancer Institute to compare Affymetrix and cDNA arrays, arriving at different conclusions.

2. Although in a recent study *(4)* it was concluded that verification of sequence identity appears to play only a small role in the improvement of the result, the study was limited to the analysis of baseline quantitation of biological replicates and does not compare the arrays ability to detect changes.

3. The agreement between platform can be affected by slow signal *(1,3)*, cross-hybridization *(18)*, and GC-content *(3)*.

4. *t*-test: with small number of replicates, the variance is easily underestimated and hence significance can be overestimated. Recently proposed solutions to this problem include the moderated *t*-test *(33)*.

5. *p*-values: do not forget that expression of genes is a coordinated business, and hence adjustments for multiple hypothesis should be made, e.g., as described in **ref.** *34*.

References

1. Barczak, A., Rodriguez, M. W., Hanspers, K., et al. (2003) Spotted long oligonucleotide arrays for human gene expression analysis. *Genome Res.* **13,**1775–1785.

2. Kothapalli, R., Yoder, S. J., Mane, S., and Loughran, T. P., Jr. (2002) Microarray results: how accurate are they? *BMC Bioinformatics* **3**, 22.

3. Kuo, W. P., Jenssen T. K., Butte A. J., et al. (2002) Analysis of matched mRNA measurements from two different microarray technologies. *Bioinformatics* **18**, 405–412.

4. Mah, N., Thelin, A., Lu, T., et al. (2004) A comparison of oligonucleotide and cDNA-based microarray systems. *Phys. Genom.* **16**, 361–370.

5. Mecham, B. H., Klus, G. T., Strovel, J., et al. (2004) Sequence-matched probes produce increased cross-platform consistency and more reproducible biological results in microarray-based gene expression measurements. *Nucl. Acids Res.* **32**, e74.

6. Rogojina, A., Orr, W. E., Song, B. K., and Geisert, E. E., Jr. (2003) Comparing the use of Affymetrix to spotted oligonucleotide microarrays using two retinal pigment epithelium cell lines. *Mol. Vision* **9**, 482–496.

7. Tan, P., Downey, T. J., Spitznagel, E. L., Jr., et al. (2003) Evaluation of gene expression measurements from commercial microarray platforms. *Nucl. Acids Res.* **31**, 5676–5684.

8. Yauk, C., Beendt, M. L, Williams, A., Douglas, G. R., et al. (2004) Comprehensive comparison of six microarray technologies. *Nucl. Acids Res.* **32**, e124.

9. Yuen, T., Wurmbach, E., Pfeffer, R. L., Ebersole, B. J., and Sealfon, S. C. (2002) Accuracy and calibration of commercial oligonucleotide and custom cDNA microarrays. *Nucl. Acids Res.* **30**, e48.

10. Irizarry, R. A., Warren, D., Spencer, F., et al. (2005) Multiple-laboratory comparison of microarray platforms (vol 2, pg 345, 2005). *Nat. Methods* **2**, 477–477.

11. Hekstra, D., Taussig, A. R., Magnasco, M., and Naef, F. (2003) Absolute mRNA concentrations from sequence-specific calibration of oligonucleotide arrays. *Nucl. Acids Res.* **31**, 1962–1968.

12. Naef, F., Lim, D. A., Patil, N., and Magnasco, M. (2002) DNA hybridization to mismatched templates: A chip study. *Phys. Rev. E.* **65** (040902).

13. Naef, F. and Magnasco, M. (2003) Solving the riddle of the bright mismatches: Labeling and effective binding in oligonucleotide arrays. *Phys. Rev. E.* **68** (011906).

14. Li, C. and Wong, W. (2001) Model-based analysis of oligonucleotide arrays: Expression index computation and outlier detection. *Proc. Natl. Acad. Sci. USA* **98**, 31–36.

15. Irizarry, R. A., et al. (2004) *Multiple Lab Comparisons of Microarray Platforms*, in *Dept. of Biostatistics Working Papers, Johns Hopkins University.*

16. Irizarry, R. A., Bolstad, B. M., Collin, F., Cope, L. M., Hobbs, B., and Speed, T. P. (2003) Summaries of Affymetrix GeneChip probe level data. *Nucl. Acids Res.* **31**, e15.

17. Yang, Y. H., Dudoit, S., Luu, P., et al. (2002) Normalization for cDNA microarray data: a robust composite method addressing single and multiple slide systematic variation. *Nucl. Acids Res.* **30**, e15.

18. Li, J., Pankratz, M., and Johnson, J. (2002) Differential gene expression patterns revealed by oligonucleotide versus long cDNA arrays. *Toxicol. Sci.* **69**, 383–390.

19. Suárez-Fariñas, M., Haider, A., and Wittkowski, K. M. (2005) "Harshlighting" small blemishes on microarrays. *BMC Bioinformatics* **6**, 65.

20. Buness, A., Huber, W., Steiner, K., Sultmann, H., and Poustka, A. (2005) arrayMagic: two-colour cDNA microarray quality control and preprocessing. *Bioinformatics* **21**, 554–556.

21. Parmigiani, G., Garrett-Mayer, E. S., Anbazhagan, R., and Gabrielson, E. (2004) A cross-study comparison of gene expression studies for the molecular classification of lung cancer. *Clin. Cancer Res.* **10**, 2922–2927.

22. Culhane, A., Perriere, G., and Higgins, D. (2003) Cross-platform comparison and visualisation of gene expression data using co-inertia analysis. *BMC Bioinformatics* **4**, 1600–1608.

23. Suárez-Fariñas, M., Noggle, S., Heke, M., Hemmati-Brivanlou. A, and Magnasco, M. O., et al. (2005) How to compare microarray studies: The case of human embryonic stem cells. *BMC Genomics* **4**.

24. Rhodes, D., Barrette, T. R., Rubin, M. A., Ghosh, D., and Chinnaiyan, A. M. (2002) Meta-analysis of microarrays: Interstudy validation of gene expression profiles reveals pathway dysregulation in prostate cancer. *Cancer Res.* **62**, 4427–4433.

25. Bhattacharya, B., Miura, T., Brandenberger, R., et al. (2004) Gene expression in human embryonic stem cell lines: unique molecular signature. *Blood* **103**, 2956–2964.

26. Sperger, J. M., Chen, X., Draper, J. S., et al. (2003) Gene expression patterns in human embryonic stem cells and human pluripotent germ cell tumors. *Proc. Natl. Acad. Sci. USA* **100**, 13,350–13,355.

27. Sato, N., Sanjuan, I. M., Heke, M., et al. (2003) Molecular signature of human embryonic stem cells and its comparison with the mouse. *Dev. Biol.* **260**, 404–413.

28. Available from: http://www.r-project.org. Last accessed: 10/19/2006.

29. Available from: http://www.bioconductor.org. Last accessed: 10/19/2006.

30. Yang, Y. H. and Thorne, N. (2003) Normalization for two-color cDNA microarray data. In: *Science and Statistics: A Festschrift for Terry Speed*, (Goldstein, D. R., ed.) IMS Lecture Notes Monograph Series, vol 40, pp. 403–418.

31. Wu, Z., Irizarry, R. A., Gentleman, R., Martinez-Murillo, F., and Spencer, F. (2004) A model based background adjustement for oligonucleotide expression arrays. *J. Amer. Stat. Assoc.* **99**, 909–917.

32. Lonnstedt, I. and Speed, T. (2002) Replicated microarray data. *Statistica Sinica* **12**, 31–46.

33. Smyth, G. K. (2004) Linear models and empirical Bayes methods for assessing differential expression in microarray experiments. *Stat. Appl. Genet. Mol. Biol.* **3**, Article 3.

34. Dudoit, S., Shaffer, J., and Boldrick, J. (2003) Multiple hypothesis testing in microarray experiments. *Stat. Sci.* **18**, 71–103.

9

A Pitfall in Series of Microarrays
The Position of Probes Affects the Cross-Correlation of Gene Expression Profiles

Gábor Balázsi and Zoltán N. Oltvai

Summary

Using *Escherichia coli* cDNA microarray slides and Affymetrix GeneChips, we study how the relative position of probes on microarrays affects the cross-correlation of gene expression profiles. We find that in cDNA arrays, every spot located within the same block is affected by a similar, experiment-specific bias. As a result, the cross-correlation between some gene expression profiles is significantly altered, depending on the similarity between these "block-dependent" biases through the series of cDNA microarray experiments. In addition, the position of probes within the blocks can also contribute to the measured gene expression. We outline the necessary steps to computationally identify and correct these biases.

Key Words: Microarray; bias; gene expression; cross-correlation; position; block; probe; spot; correction.

1. Introduction

Microarray technology is used to simultaneously monitor the mRNA expression levels of all genes within a given organism *(1,2)* and has become an indispensable tool in cell biology *(3)*. Following the experiments and data collection, possible avenues of microarray data analysis range in complexity from the identification of significantly affected genes *(4)* to the application of sophisticated computational methods to cluster, classify, and interpret the observed gene expression patterns *(5–7)*. Unfortunately, in addition to biological variations in gene expression *(8)*, microarray data are also affected by a large number of technological factors *(9–13)*. To gain a better understanding of the assayed biological phenomena, it is crucial to identify the source of technological biases

From: *Methods in Molecular Biology, vol. 377, Microarray Data Analysis: Methods and Applications*
Edited by: M. J. Korenberg © Humana Press Inc., Totowa, NJ

and to develop computational or experimental methods for their reduction or elimination *(10–15)*.

By using *Escherichia coli* gene expression data collected from in-house-printed cDNA arrays *(16)*, we show that the relative position of probes on microarrays affects their coexpression, and can have important consequences on gene coexpression measurement and clustering of expression profiles. We outline the steps necessary to identify and reduce such errors in existing microarray data. No significant bias was found using Affymetrix GeneChip *(17)* data.

2. Software

For data processing, identification, and correction of position-dependent bias, we used Microsoft Excel and Matlab® by The Mathworks, Inc.

3. Methods

3.1. Microarray Platforms

We used two types of microarray platforms to identify the effect of relative position on gene coexpression *(see* **Note 1**): in-house-printed cDNA arrays *(16)* and commercial Affymetrix GeneChips *(17)*.

All the steps needed to construct the in-house-printed cDNA array have been described in detail before *(16)*. Therefore, we omit discussing the construction procedure, and focus on the identification and correction of biases instead.

The custom-built cDNA array slides contained three copies of the *E. coli* genome in a total of 24 blocks of spots (8 blocks per genome, *see* **Fig. 1**). Each block contained 26 columns and 23 rows, or a total of $N_B = 576$ spots in a rectangular array (the last row of spots was incomplete, containing only four spots). The total number of spots per genome was 4608 and there were 13,824 spots per slide. The spreadsheets for data analysis were generated with the GenePix Pro 4.0 software (Axon Instruments), and contained 14,352 entries (including the 22 empty spots from the last row of each block).

The second type of microarray was the commercially available Affymetrix *E. coli* GeneChip *(17)*. The chips contained duplicate probe sets ("perfect match" and "mismatch") for 7312 locations on the *E. coli* chromosome (including intergenic regions).

3.2. Data Processing–cDNA Microarrays

We generated expression data tables in Microsoft Excel, containing the following information for each of the 14,352 entries: block *(B)*, column *(X)*, and row *(Y)* number, red foreground (R_F) and background (R_B), green foreground (G_F) and background (G_B) intensity. The position of each probe within a block, P is defined by the pair of integers *(X,Y)*.

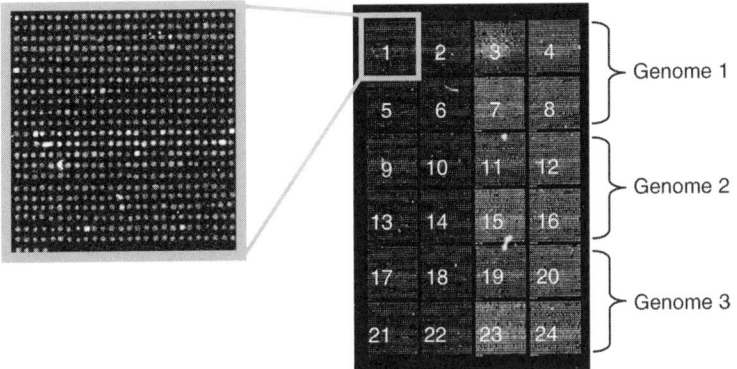

Fig. 1. Geometry of the custom-built cDNA microarray.

We subtracted the background intensities from the foreground intensities, and then used the corrected red (R_C) and green (G_C) intensities to calculate the $\log_{10}$ ratios, or lg ratios (E) of gene expression:

$$E = \lg\left(\frac{R_C}{G_C}\right) = \lg\left(\frac{R_F - R_B}{G_F - G_B}\right) \tag{1}$$

In some cases, when the intensity of the background was higher than or equal to the intensity of the foreground, the resulting lg ratios became complex or infinity. These values were eliminated using the find, imag, and isfinite functions in Matlab® (*see* **Note 2**).

3.3. Block-Dependent Biases–cDNA Microarrays

Using the find, nanmean, and nanstd functions in Matlab, we calculated averages and standard deviations within each block for each of the foreground and background intensities, as well as the corrected values and the lg ratios. In the absence of block-dependent biases, one would expect the average corrected log ratios

$$\langle E \rangle = \frac{1}{N_B} \sum_{i=1}^{N_B} E_i \tag{2}$$

to be around 0 and show no systematic differences. However, as **Fig. 2** indicates, systematic differences between blocks are present even after the background subtraction and calculation of lg ratios. These systematic differences (biases) originate in the biases of the original red and green foreground and background intensities, and background subtraction or other global normalization methods are not sufficient to eliminate them (*see* **Note 1**).

Balázsi and Oltvai

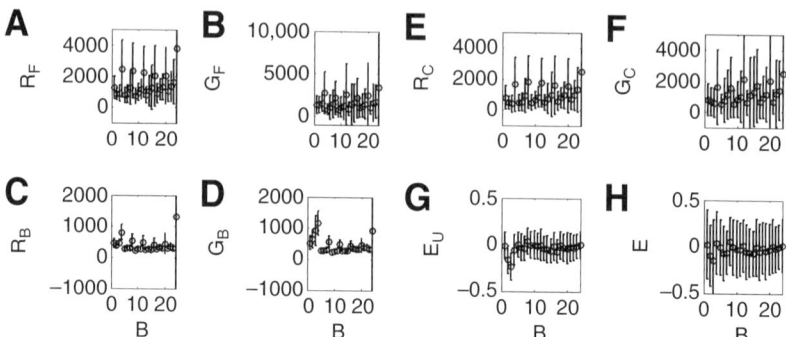

Fig. 2. The average red foreground (R_F, plot A), green foreground (G_F, plot B), red background (R_B, plot C), green background (G_B, plot D), red corrected (R_C, plot E), green corrected (G_C, plot F), uncorrected lg ratio (E_U, plot G) and corrected lg ratio (E, plot H) within each of the 24 blocks on the cDNA microarray slide.

3.4. Position-Dependent Biases of Higher Order

To identify position-dependent biases of higher order, biases of order 0 have to be corrected (*see* **Subheading 3.6.** and **Note 3**). If the 0th order-corrected lg ratio values E_0, plotted in the sequential order of rows and columns contain systematic trends, the microarray data are affected by higher order bias.

Experimental noise can in general be reduced by increasing the number of experiments and averaging the results of repeated experiments. Contrary to expectation, the effect of position-dependent biases on the cross-correlation between gene expression profiles *increases* instead of decreasing with the length of the experiment series *(12)*. This is an important problem because cross-correlation is the most frequently used distance metric in hierarchical clustering *(5)*.

To illustrate the adverse effect of multiple experiments, we plot in **Fig. 3** the average cross-correlation <ρ(P_1,P_2)> between probes as a function of their relative distance within blocks $d(P_1,P_2)$, defined as

$$d(P_1,P_2) = \sqrt{\left(X_1 - X_2\right)^2 + \left(Y_1 - Y_2\right)^2} \qquad (3)$$

The cross-correlation between probes, ρ(P_1,P_2) is defined as

$$\rho(P_1,P_2) = \frac{\left\langle E_1 E_2 \right\rangle - \left\langle E_1 \right\rangle \left\langle E_2 \right\rangle}{\sigma_1 \sigma_2}, \qquad (4)$$

where the averages are taken over the experiment series and the standard deviation over N experiments is

$$\sigma = \sqrt{\frac{1}{N-1} \sum \left(E - \left\langle E \right\rangle\right)^2} \qquad (5)$$

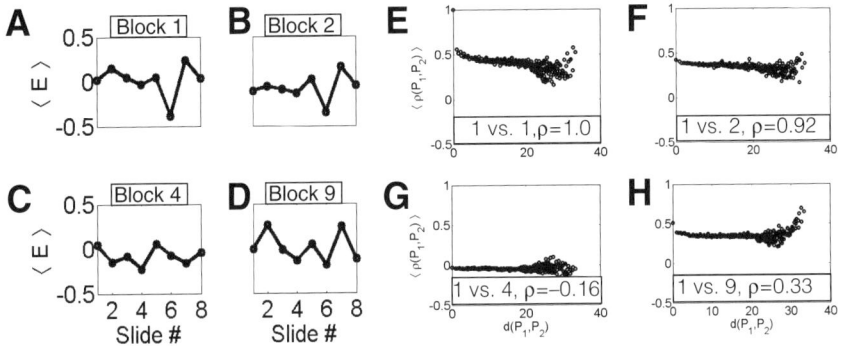

Fig. 3. Average expression values (biases) within blocks 1, 2, 4, and 9 in the series of 8 microarray experiments (**A, B, C, D**), and the average cross-correlation $\rho(P_1,P_2)$ between various pairs of blocks as a function of the relative distance $d(P_1,P_2)$ of spots within the blocks (**E, F, G, H**). The cross-correlation values ρ shown on the bottom of graphs **E, F, G, H** were calculated between the bias on graph **A**, and graphs **A, B, C, D**, respectively.

Selecting all probes located at the same relative distance $d(P_1,P_2)$ (within blocks), their average cross-correlations as a function of distance indicate the strong contribution of biases to the cross-correlation (**Fig. 3**).

Ideally, all correlations should be around 0 (as in **Fig. 3G**), except when the genes probed by the spots are identical, when $\rho(P,P)$ should be 1. This would happen, for example, within block 1, when $d = 0$ (*see* the first data-point in **Fig. 3E**), or within two blocks, B_1 and B_2 for which $B_2 = B_1 + 8k$ (such as blocks 1 and 9, 2 and 18, etc.), while cross-correlation for any pair of spots for which $d \neq 0$ or for which $B_2 \neq B_1 + 8k$ should be around 0. For example, cross-correlations between blocks 1 and 2 should be nonsignificant because they probe different genes printed by different tips. However, as **Fig. 3** indicates, spots in blocks 1 and 2 are affected by very similar biases (**Fig. 3A,B**, $\rho = 0.92$), and the result is a strong cross-correlation (**Fig. 3F**). On the other hand, one would expect a strong correlation between spots in blocks 1 and 9, especially for $d = 0$, as they probe the same set of genes and have been printed by the same tip (**Fig. 1**). Nevertheless, the biases affecting blocks 1 and 9 (**Fig. 3A,D**) are more different ($\rho = 0.33$) than the ones affecting blocks 1 and 2 (**Fig. 3A,B**, $\rho = 0.92$), and the result is a reduced cross-correlation (**Fig. 3H**). Notice that the cross-correlation between spots probing identical genes (located at $d = 0$ within blocks 1 and 9—first data-point on **Fig. 3H**) is slightly higher than for the rest of the spot pairs, but is far less than 1, the value expected in the absence of bias and noise.

3.5. Position-Dependent Biases (Affymetrix GeneChips)

The position (X,Y) of probe sets containing the perfect match and mismatch sequences was defined as the pair of averages $(<X>,<Y>)$ over all probes within

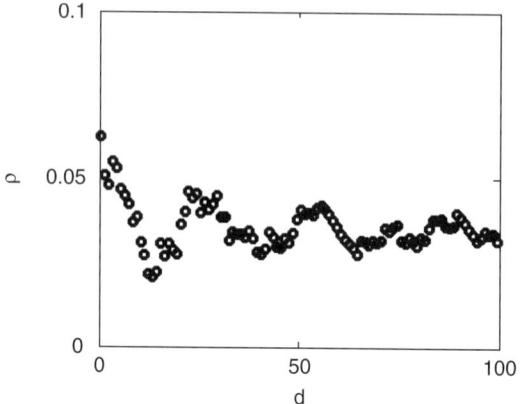

Fig. 4. Position-dependent bias in Affymetrix GeneChips.

the probe set. We used MAS5 (Microarray Suite Software) normalized data from 24 Affymetrix GeneChip experiments to study position-dependent bias.

A method similar to the one described in **Subheading 3.4.** was applied to determine the effect of relative probe set position on the corresponding cross-correlation. The formulas used to calculate d and ρ were identical to **Eqs. 3**, **4**, and **5**, except X was replaced by $<X>$ and Y was replaced by $<Y>$. Cross-correlations $\rho(P_1, P_2)$ of probe set pairs (P_1, P_2) were averaged over increasing distances d ranging from $k < d < k + 10$, $k = 0,1,2,3,\ldots$.

As **Fig. 4** indicates, probe sets located at certain distances tend to have a slight (nonsignificant) increase in cross-correlation. As a result, the cross-correlation seems to fluctuate as a function of the distance d between probe set pairs.

3.6. Correction of the Biases

Ideally, the value of the block-dependent biases should be 0. It is straight forward to achieve this by calculating the bias $<E>$ for each block and subtracting it from all the individual expression values within the block:

$$E_{i,0} = E_i - \langle E \rangle = E_i - \frac{1}{N_B} \sum_{j=1}^{N_B} E_j \qquad (6)$$

Even after the correction of position-dependent biases of order 0, biases of higher order often persist within blocks, visible as column- or row-dependent trends (*see* **Fig. 5**). To better visualize them, lg ratios can be averaged over columns and/or rows and plotted as a function of row number Y and column number X, respectively. Higher order biases can be eliminated by linear interpolation within individual columns and/or rows (using the fit function in Matlab), and subtraction of the linear trend (*see* **Note 3**).

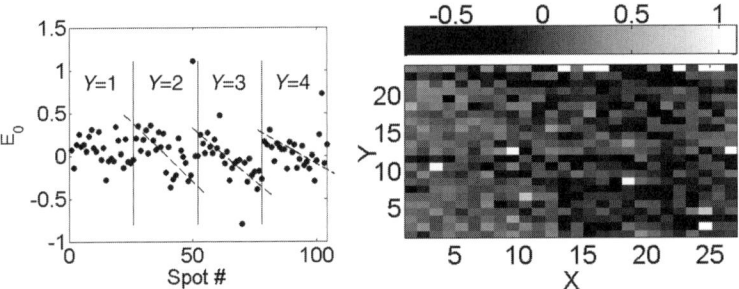

Fig. 5. Column-dependent bias within block 5 of the first cDNA microarray remaining after first-degree correction. The decreasing trends within rows (left) correspond to a "darkening" tendency of the lg ratios toward the right-hand side of block 5 (right).

Although position-dependent biases can be reduced computationally, it might be safer to prevent them by appropriate experimentation (*see* **Notes 4** and **5**).

4. Notes

1. Microarray data are frequently preprocessed and subject to other normalization techniques *(14,15)*. Nevertheless, it is important to check for the presence of position-dependent biases, which often remain present after global normalization methods affecting all the spots.
2. Before identifying position-dependent biases, data from all "flagged" spots should be replaced with "NaN" (not a number). This assures that appropriately measured intensities are used to identify and correct biases.
3. Position-dependent biases of higher order are column- or row-dependent trends that can occasionally be nonlinear. In this case, linear interpolation is not appropriate to remove them. A quadratic or higher order polynomial or other functions can then be used instead of the linear fit to remove the remaining biases.
4. The origin of position-dependent biases in cDNA arrays is unclear. Incorrect estimation of background intensities within spots might play an important role *(10)*. Because different blocks are affected differently, it is likely that the volume and the concentration of the material deposited are print-tip and print-time dependent. One possible method to diminish position-dependent biases experimentally is to randomize the position of probes on the microarrays, so that nearby spots become distant from slide to slide.
5. The intensity of the green (Cy3) channel is affected by a spot-localized contaminating fluorescence, which can be reduced by allowing the slides to dry before printing *(10)*. Also, using a hyperspectral scanner to obtain the Cy3 and Cy5 intensities *(13)* will likely improve data quality and might result in the complete elimination of position-dependent biases. However, at the present it is not known if spot-localized contaminating fluorescence is the main source of position-dependent bias.

Acknowledgments

The authors thank John W. Campbell and Krin A. Kay for designing the cDNA microarrays, performing the experiments, and providing the data, as well as the Applied BioDynamics Laboratory (Boston University) for providing the Affymetrix data. This work was supported by the U.S. Department of Energy, the National Institutes of Health and the National Science Foundation.

References

1. Schena, M., Shalon, D., Davis, R. W., and Brown, P. O. (1995) Quantitative monitoring of gene expression patterns with a complementary DNA microarray. *Science* **270**, 467–470.
2. Wodicka, L., Dong, H., Mittmann, M., Ho, M. H., and Lockhart, D. J. (1997) Genome-wide expression monitoring in *Saccharomyces cerevisiae*. *Nat. Biotechnol.* **15**, 1359–1367.
3. Brown, P. O. and Botstein, D. (1999) Exploring the new world of the genome with DNA microarrays. *Nat. Genet.* **21**, 33–37.
4. Wei, Y., Lee, J. M., Richmond, C., Blattner, F. R., Rafalski, J. A., and LaRossa, R. A. (2001) High-density microarray-mediated gene expression profiling of Escherichia coli. *J. Bacteriol.* **183**, 545–556.
5. Eisen, M. B., Spellman, P. T., Brown, P. O., and Botstein, D. (1998) Cluster analysis and display of genome-wide expression patterns. *Proc. Natl. Acad. Sci. USA* **95**, 14,863–14,868.
6. Salmon, K., Hung, S. P., Mekjian, K., Baldi, P., Hatfield, G. W., and Gunsalus, R. P. (2003) Global gene expression profiling in Escherichia coli K12. The effects of oxygen availability and FNR. *J. Biol. Chem.* **278**, 29,837–29,855.
7. Alter, O., Brown, P. O., and Botstein, D. (2003) Generalized singular value decomposition for comparative analysis of genome-scale expression data sets of two different organisms. *Proc. Natl. Acad. Sci. USA* **100**, 3351–3356.
8. Kaern, M., Elston, T. C., Blake, W. J., and Collins, J. J. (2005) Stochasticity in gene expression: from theories to phenotypes. *Nat. Rev. Genet.* **6**, 451–464.
9. Kerr, M. K. and Churchill, G. A. (2002) Experimental design for gene expression microarrays. *Biostatistics* **2**, 183–201.
10. Martinez, M. J., Aragon, A. D., Rodriguez, A. L., et al. (2003) Identification and removal of contaminating fluorescence from commercial and in-house printed DNA microarrays. *Nucleic Acids Res.* **31**, e18.
11. Yang, Y. H., Dudoit, S., Luu, P., Lin, D. M., Peng, V., Ngai, J., and Speed, T. P. (2002) Normalization for cDNA microarray data: a robust composite method addressing single and multiple slide systematic variation. *Nucleic Acids Res.* **30**, e15.
12. Balázsi, G., Kay, K. A., Barabási, A. L., and Oltvai, Z. N. (2003) Spurious spatial periodicity of co-expression in microarray data due to printing design. *Nucleic Acids Res.* **31**, 4425–4433.
13 Timlin, J. A., Haaland, D. M., Sinclair, M. B., Aragon, A. D., Martinez, M. J., and Werner-Washburne, M. (2005) Hyperspectral microarray scanning: impact on the accuracy and reliability of gene expression data. *BMC Genomics* **6**, 72.

14. Cui, X., Kerr, M. K., and Churchill, G. A., Transformations for cDNA microarray data. (2003) *Stat. Appl. Gen. Mol. Biol.* **2,** Article 4.
15. Quackenbush J. (2002) Microarray data normalization and transformation. *Nat. Genet.* **32,** 496–501.
16. Tong, X., Campbell. J. W., Balázsi, G., et al. (2004) Genome-scale identification of conditionally essential genes in E. coli by DNA microarrays. *Biochem. Biophys. Res. Commun.* **322,** 347–354.
17. Selinger, D. W., Cheung, K. J., Mei, R., et al. (2000) RNA expression analysis using a 30 base pair resolution Escherichia coli genome array. *Nat. Biotechnol.* **18,** 1262–1268.

10

In-Depth Query of Large Genomes Using Tiling Arrays

Manoj Pratim Samanta, Waraporn Tongprasit, and Viktor Stolc

Summary

Identification of the transcribed regions in the newly sequenced genomes is one of the major challenges of postgenomic biology. Among different alternatives for empirical transcriptome mapping, whole-genome tiling array experiment emerged as the most comprehensive and unbiased approach. This relatively new method uses high-density oligonucleotide arrays with probes chosen uniformly from both strands of the entire genomes including all genic and intergenic regions. By hybridizing the arrays with tissue specific or pooled RNA samples, a genome-wide picture of transcription can be derived. This chapter discusses computational tools and techniques necessary to successfully conduct genome tiling array experiments.

Key Words: Tiling array; oligonucleotide array; maskless array synthesizer; transcriptome; human genome; mammalian genome.

1. Introduction

Microarray is a powerful technology that combines the complementary base pairing properties of the DNA molecules with microfabrication techniques of the electronics industry to quantitatively measure cellular transcript levels at a global scale *(1)*. Innovative applications of this experimental tool have revolutionized biology over the last decade. Such novel uses include monitoring of differential gene expressions under multiple conditions *(2–4)*, locating transcription factor binding sites on the chromosomes *(5)*, identification of chromosomal replication sites *(6)*, and monitoring of small RNA and miRNA expressions *(7)*. In all of the examples, the real power of the array technology is in its ability to conduct high-throughput measurements, and in providing quantitative rather than qualitative estimates of the transcript levels. The success of the DNA arrays has inspired development of other array techniques for measuring protein and glucose levels that respectively utilize antigen–antibody- and glucose–lectin-binding properties *(8,9)*.

From: *Methods in Molecular Biology, vol. 377, Microarray Data Analysis: Methods and Applications*
Edited by: M. J. Korenberg © Humana Press Inc., Totowa, NJ

Innovative applications of the array technologies mentioned in the previous paragraph were mostly on model organisms, such as yeast *S. cerevisiae*, worm *C. elegans*, and fruit fly *D. melanogaster*, with well-annotated genome structures. In recent years, chromosomes of several other eukaryotic organisms have been partially or fully sequenced. With rapidly rising sequencing capabilities around the globe, it is anticipated that many more genomes will be decoded in the near future. A key challenge therefore is to quickly develop comprehensive annotations for the new genomes, so that further downstream experiments can be conducted *(10)*. Unfortunately, this process has become a large bottleneck in the postgenomic era. Conventional approaches for annotating new genomes have several shortcomings. Computational gene prediction algorithms *(11)* often produce many erroneous genes, and the results must be empirically verified before further use. On the other hand, traditional experimental approaches for genome annotation, such as sequencing of expressed sequence tags and full-length cDNAs, are not comprehensive enough, but biased toward detecting the highly expressed genes.

Genome tiling array is a relatively new application of the microarray technology to comprehensively identify the transcribed regions of large complex genomes *(12–21)*. This powerful, but relatively inexpensive, technology can substantially reduce the gap between sequencing and annotation. In this method, millions of oligonucleotide probes are chosen uniformly from the entire genome, synthesized on proper substrates, and hybridized with biotinylated RNA samples extracted from the tissues under study. Strong signals on consecutive probes matching a segment of the genome suggest transcription of the corresponding genomic region. Therefore, by properly mapping all observed probe signals back to the genome, genome-wide transcriptional activities can be identified in a comprehensive manner.

Recent emergence of genome tiling arrays is not accidental, but closely linked to the continual improvements of semiconductor fabrication technologies guided by Moore's law. Several alternatives currently exist for developing high-density arrays necessary for genome tiling studies. In one approach, commercialized by Affymetrix™, chromium-based masks are used to pattern chemicals on the arrays *(12,13)*. The second approach, marketed by Nimblegen™, utilizes a maskless array synthesizer, where an optical virtual mask is used to guide patterning of the nucleotides on the arrays *(14–17,22,23)*. In a third alternative, Agilent™ builds their arrays applying a modified ink-jet technology. We should note here that for initial annotations of the large genomes, maskless technologies offer more flexibilities than the mask-based technologies. Huge fixed costs of designing the masks in the mask-based technologies prohibit reselecting locations or lengths of the probes for further refinements of the annotations.

Irrespective of the underlying array technology, the computational challenges faced by a researcher in designing of whole-genome tiling array experiments and data analysis remain similar *(21)*. This chapter describes the general methods for successfully completing the task. In this context, one should keep in mind that significant differences exist between the goals of traditional array-based projects for differential gene expression monitoring, and the tiling arrays applied for genome-wide identification of the transcripts. Therefore, the design and the data analysis techniques differ substantially between the two projects. Also, tiling array technology is an emerging and active area of research, and some of the theoretical issues for analysis are not fully settled yet.

2. Materials

1. Latest release of the genome sequence.
2. Latest annotation of the genome, if any. Annotations for all protein-coding genes, small RNAs, and other genomic features should be considered.

3. Methods

A typical situation often faced by the scientists leading the genome sequencing projects is as follows: (1) draft genome sequence of the organism of interest is assembled, (2) a number of genes are computationally predicted from the draft sequence, but many of those genes do not have homologs in any other organism, (3) several large segments of the genome do not contain any predicted gene. The key questions that need to be answered at this point are: how do we know for sure, whether the predicted genes are real? Is the list of genes exhaustive or are there other transcribed regions on the genome?

Genome tiling array experiments can answer the previous questions in a comprehensive but cost-effective manner. In this approach, oligonucleotide probes are chosen uniformly from the entire genome (**Fig. 1**) and hybridized with either pooled RNA from many tissues or RNA extracted from selected individual tissues. **Figure 2** shows the steps necessary to successfully conduct tiling array experiment on a large genome. Computational skills and resources are necessary at two stages: (1) in design of the experiment and more specifically in selecting the probes that are being placed on the arrays, (2) in analysis of the array data after completion of the hybridization. The following three subsections discuss the steps in further details.

3.1. Design of Experiment

1. Conducting tiling array experiment on a large genome is expensive (although relatively cheaper than other alternatives to achieve the same goal of transcription mapping), and many of the important decisions need to be made at an early stage. Once the probes are selected and hybridized on the arrays, correction of any earlier error

Fig. 1. Probe selection in a tiling array experiment. Oligonucleotide probes chosen uniformly from the entire genome are synthesized on a substrate and hybridized with mRNA extracted from any cell line. This provides a comprehensive and unbiased map of the entire transcriptome.

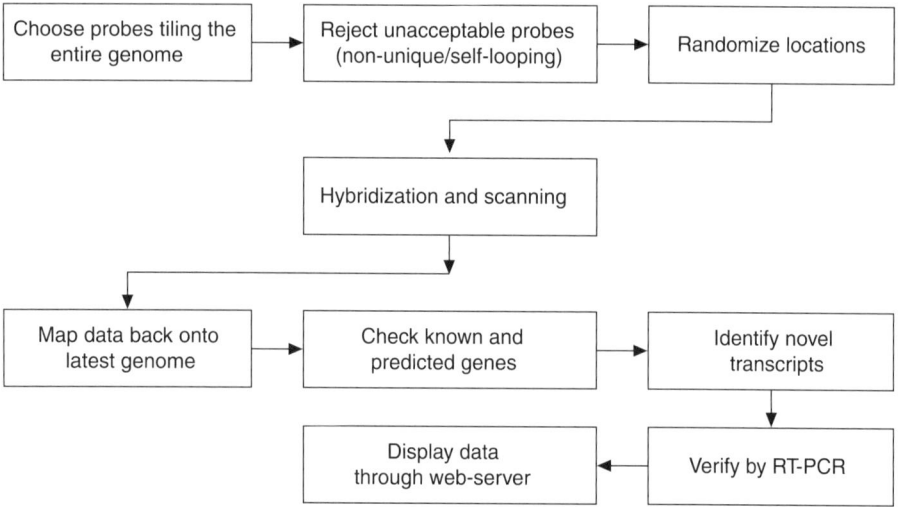

Fig. 2. Flowchart showing all components of a tiling array experiment. Steps necessary in successful completion of a genome-wide tiling array experiment.

becomes very costly. Therefore, several parameters need to be thoughtfully chosen at the design phase to obtain most information about the genome.

2. The first parameter in designing an array experiment is the choice of the optimal probe size. Very short probes (<20 nucleotides) are not specific enough and are likely to cross-hybridize with different RNAs in the sample. Very long probes, on the other hand, may miss smaller genes and exons. Also, the cost of synthesizing longer probes can be considerably higher, especially if mask-based technologies are used. Currently published studies in the literature have used probe sizes of 25 or 36 nucleotides. 36-mer probes are favored by the maskless designs because they provide additional sensitivities than the 25-mer probes without significant cost increase *(24)*.

3. Density of the selected probes on the given genome is the next important design parameter. Probes can be chosen overlapping each other, end-to-end, or with gaps of few bases between the consecutive ones. The previous choice depends on the available resources, genome size, and the expected lengths of the exons and introns. On a tighter budget, the best option is to choose probes tiling only the predicted gene regions. For a complete transcriptional profiling of an entire genome, choosing probes end-to-end or with small interprobe gap (~1/3 of the probe size) has been sufficient. A pilot study on *Arabidopsis thaliana* found that choosing overlapping probes did not add many additional values *(13)*.

4. The third design criterion is the type of RNA sample to be used and the number of replicates necessary to successfully conduct the array experiment. The goal of a genome tiling project is to identify all transcribed regions within the entire genome. To complete this task in a comprehensive manner for an unexplored genome, the optimal strategy is to first use pooled RNA samples from several tissues and measure only one replicate. A followup experiment to probe tissue-specific activities can be conducted by synthesizing the detected expressed segments within one array, and hybridizing multiple replicates of the array with RNA from different tissues. Single replicate for the genome tiling study (first experiment above) has proved to be sufficient *(16,24)*.

3.2. Probe Selection and Placement

1. The latest assembled version of the genome is obtained and the repeat regions are masked using repeat-masker software. This step is very important for large mammalian genomes, where up to 40% of the sequence may contain repeat regions.

2. Assuming that the probe size (N) and the density has been decided, the simplest way to select the probes would be to start from one end of the genome and continue choosing N-mers uniformly from both strands until the other end is reached. This approach is generally followed, although with certain modifications to ensure that the potential cross-hybridizing probes are excluded.

3. To properly account for the cross-hybridization effect, the following method is used. The entire genome is split into all overlapping 17-mers, and the genome-wide frequencies of the 17-mers are counted. Subsequently, for each 36-mer within the genome, an "average frequency parameter" is computed by averaging the frequencies of all 17-mers within it. 36-mers with large "average frequency parameters" (>5) are more likely to hybridize with multiple regions of the genome and therefore they are excluded. This description assumes a 36-mer probe size, but the same approach can easily be extended to other sizes.

4. Additional filtering criteria include (1) discarding self-looping probes, (2) removing probes with unusually large AT or GC content, (3) filtering out low complexity sequences, and (4) removing probes that require too many synthesis cycles.

5. Probes are chosen uniformly from the rest of the 36-mers. However, instead of selecting probes with uniform spacing, a possible improvement would be to slightly vary the distances between them and ensure that the melting temperatures of all probes in an array lie within a range.

6. In addition to the previous oligonucleotides chosen from the entire genome, each array contains two sets of probes to facilitate the data analysis. The first category consists of probes that do not match any other region of the genome. Those probes are used as negative controls in the analysis. The second category contains a set of randomly chosen genomic probes, and they are placed in each array. They are used to ensure proper normalization of data between the arrays.

7. Locations of all probes are randomized before being placed on the arrays, so that the probes from neighboring genomic locations do not lie next to each other in the arrays. This helps one avoid any possible spatial bias during the hybridization or the scanning stages.

8. As a ballpark estimate, a large genome of size approx ~120 Mb *(Arabidopsis thaliana)* requires 13 arrays, each with approx 400,000 features. This estimate is based on the choices of a probe length of 36 bases and 10 base interprobe distances.

3.3. Analysis of Data

1. All probes are mapped onto the latest version of the genome sequence. This step is often necessary for large genomes because between the time when the probes are selected and when the array hybridization is completed, an additional draft of the assembled genome may be released.

2. Data from different arrays are normalized to reduce any array-to-array experimental variation. The simplest way to normalize is to divide the intensity of each probe from an array by the median signal of the entire array.

3. The normalization scheme only equates the medians (location parameters) of distributions from all arrays, but does not match their standard deviations (scale parameters). A more sophisticated approach to fully match the distributions of all arrays is to convert the array data to percentile or quantile scores (*see* **Note 1**). The quantiles can further be mapped to the average of all array distributions *(25)*. Normalization of data is verified by comparing the signals on common probes that were placed in all arrays for that purpose.

4. At this point, transcription of any gene of interest can be detected by visual inspection of the normalized data mapped back on to the genome and presented in a graphical format (**Fig. 3**). From **Fig. 3**, one can clearly identify the intron and exon boundaries of the gene. Similar plots for other genomic regions of *Arabidopsis thaliana* and several other organisms are available from http://www.systemix.org/At/.

5. It is not possible to visually inspect every gene and find whether it is expressed. Therefore, computer algorithms are developed to perform this task. Most algorithms first derive the threshold level between the expressed and the nonexpressed probes. The threshold is determined based on the specially placed probes not matching any other genomic region, or in their absence, based on the probes from the promoter regions of the previously verified genes *(13,16,20)*. Choice of the threshold ensures that 95% of either the nonmatching or the promoter probes have signals below it, implying that the expected false-positive rate is only 5%.

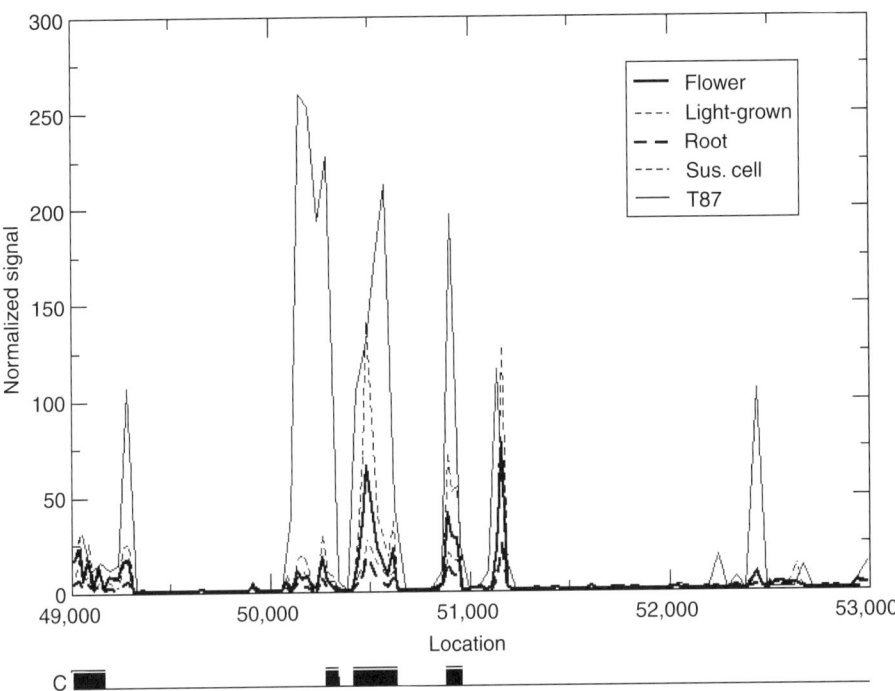

Fig. 3. Tiling array data for a known gene. The activities near the gene *At1g01100* of *Arabidopsis thaliana* are shown. Introns and exons of the gene on the Crick strand are displayed below the figure. A colored version of the same plot and similar plots for other genomic regions is available from http://www.systemix.org/At.

6. Whether an annotated gene is expressed or not is determined from the signals on all probes located fully within its exon regions. If a statistically significant number of the exons have signals above the threshold level, the gene is considered to be expressed (*see* **Notes 2** and **3**). The simplest approach is to determine the median signal of all exonic probes and check whether it is above the predetermined cutoff *(13,16,20)*.

7. A modified method was developed by Bertone et al. that did not rely on either the promoter region or the normalization between the arrays. They checked whether a statistically significant number of probes located within a gene had signals above the median of the array (also *see* **Notes 4** and **5**). By definition, only half in a set of randomly chosen probes were expected to have activities above the median. Therefore, nonrandom activities of probes within a gene could be determined based on a binomial distribution *(15)*, because the probability of k or more probes out of N to have above-median signals by chance is:

$$p = (0.5)^N \sum_{i=k}^{N} \binom{N}{i}$$ (1)

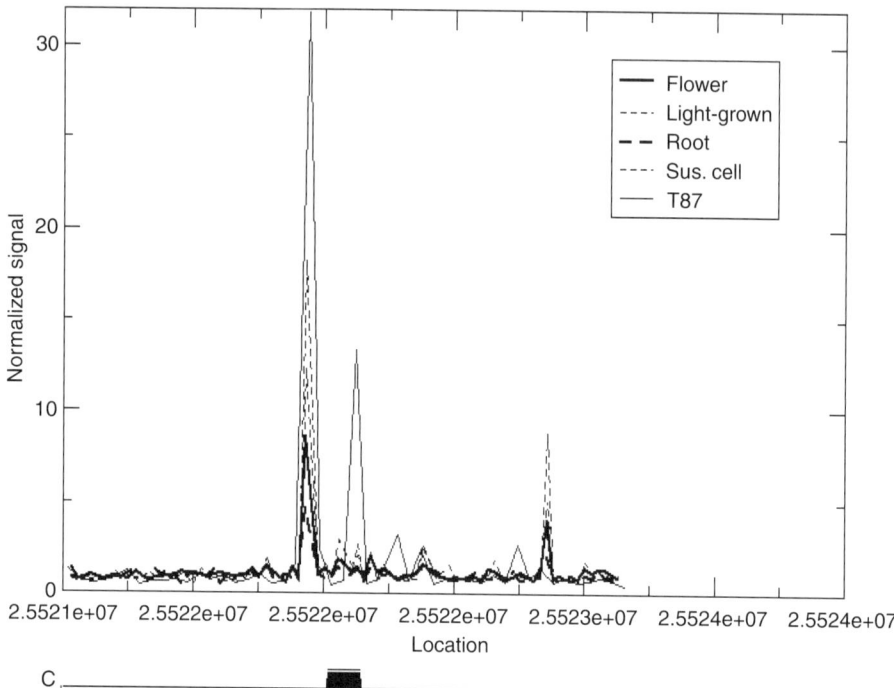

Fig. 4. Probe signals for a known miRNA. The activities around ath-MIR166g are shown. Signal near the peak is observed by both 25-mer and 36-mer based studies, and is therefore likely to be real transcript. The smaller peak observed by 36-mer based study is unconfirmed by the 25-mer probes measuring four different cell-lines. However, this difference may not be noise, and can also be explained by higher sensitivity of the 36-mer probes.

8. A second component of the analysis is to identify novel transcribed regions within the genome (also *see* **Notes 6** and **7**). Both methods previously discussed for verifying known genes can be extended to perform this task. In the simpler approach, all possible open-reading frames on the genome with lengths >50 amino acids are determined and the median signals for probes within them are computed. If the median is higher than the threshold, the open-reading frame is likely to represent potential exon. The method of Bertone et al. is extended by taking signals on 10 consecutive probes (or any appropriate number depending on the average exon size of the organism), which are tested using **Eq. 1** for their transcription by chance. For a large genome, Bonferroni correction to **Eq. 1** needs to be made to avoid too many false positives *(13)*.

9. Further confirmation of the identified novel transcribed regions can be made using homologies with other organisms, identification of a polyadenylation signal, matching with the expressed sequence tag databases, and so on. Owing to the

incompleteness of above alternatives, most studies experimentally confirm the new transcripts using RT-PCR technique.

4. Notes

1. Normalization procedures in **Subheading 3.3.** assume that the data from all arrays have similar distributions. This may not be true for large genomes. In typical tiling array designs, probes for different chromosomal segments are placed on different arrays. If one chromosome is gene rich and another one is gene poor, then it may not be correct to assume the arrays to have identical distributions. The simplest solution of randomizing all probes from the entire genome among all arrays poses some practical difficulties. Typical mammalian genomes require measurements on a hundred or more arrays. If probes are randomized over all arrays, it is not possible to monitor the hybridization quality until experiments on all arrays are completed. A good workable compromise between the two ends is to mix a gene-rich chromosome with a gene-poor chromosome, and then randomize the probes for the combined set among a group of arrays.

2. Tiling arrays can also be used to monitor alternate splicing of genes under different conditions. In this case, additional probes bridging the splice junctions need to be chosen. **Reference *14*** demonstrated one example in *D. melanogaster.*

3. Owing to preferential priming of the mRNAs from their 3′ ends in some experimental methods, signals for the probes near the 3′ end of a gene could be stronger than the 5′ end. Therefore, the algorithm to decide whether an annotated gene is expressed needs to be modified accordingly.

4. Probe-to-probe variation within a gene is a matter of great concern in analyzing tiling array data *(15)*. Mismatch probes are used in some designs *(26,27)* to account for this effect. Such mismatch probes function only partially, and do not eliminate the noise *(15)*. It is important to keep in mind that including a mismatch probe for every probe on the array would reduce the extent of the genome covered by half. The cost vs benefit tradeoff in including the mismatch probes also depends on the probe size and other technological factors.

5. Probe-to-probe variations can also be reduced by applying appropriate smoothing techniques. An example is provided in **ref. *26***, where probe signals within each 100 nucleotide sliding window are replaced by their Hodges–Lehman estimators.

6. One of the surprising observations of tiling array-based studies is the presence of antisense activities for many known genes *(13,16)*. The biological reason of such effect is not clear.

7. In addition to protein-coding genes, tiling arrays also show activities for other noncoding RNAs *(16)*. In **Fig. 4**, signals around a known miRNA of *Arabidopsis* is shown.

Acknowledgments

This work was partly supported by grants to V. Stolc from the NASA Center for Nanotechnology, the NASA Fundamental Biology Program, and the CICT programs (contract NAS2-99092).

References

1. Schena, M., Shalon, D., Davis, R. W., and Brown, P. O. (1995) Quantitative monitoring of gene expression patterns with a complementary DNA microarray. *Science* **270**, 467–470.
2. Chu, S., DeRisi, J., Eisen, M., et al. (1998) The transcriptional program of sporulation in budding yeast. *Science* **282**, 1421.
3. Spellman, P. T., Sherlock, G., Zhang, M. Q., et al. (1998) Comprehensive identification of cell-cyle regulated genes of the yeast Saccharomyces cerevisiae by microarray hybridization. *Mol. Biol. Cell* **9**, 3273–3297.
4. White, K. P., Rifkin, S. A., Hurban, P., and Hogness, D. S. (1999) Microarray analysis of Drosophila development during metamorphosis. *Science* **286**, 2179–2184.
5. Lee, T. I., Rinaldi, N. J., Robert, F., et al. (2002) Transcriptional regulatory networks in *Saccharomyces cerevisiae. Science* **298**, 799–804.
6. Raghuraman, M. K., Winzeler, E. A., Collingwood, D., et al. (2001) Replication dynamics of the yeast genome. *Science* **294**, 115–121.
7. Lu, J., Getz, G., Miska, E. A., et al. (2005) MicroRNA expression profiles classify human cancers. *Nature* **435**, 834–838.
8. Zhu, H., Bilgin, M., Bangham, R., et al. (2001) Global analysis of protein activities using proteome chips. *Science* **293**, 2101–2105.
9. Pilobello, K. T., Krishnamoorthy, L., Slawek, D., and Mahal, L. K. (2005) Development of a lectin microarray for the rapid analysis of protein glycopatterns. *Chembiochem* **6**, 985–989.
10. Roberts, R. J. (2004) Identifying protein function—a call for community action. *PLoS Biol* **2**, E42.
11. Zhang, M. Q. (2002) Computational prediction of eukaryotic protein-coding genes. *Nat. Rev. Genet.* **3**, 698–709.
12. Shoemaker, D. D., Schadt, E. E., Armour C. D., et al. (2001) Experimental annotation of the human genome using microarray technology. *Nature* **409**, 922–927.
13. Yamada, K., Lim, J., Dale, J. M., et al. (2003) Empirical analysis of transcriptional activity in the Arabidopsis genome. *Science* **302**, 842–846.
14. Stolc, V., Gauhar, Z., Mason, C., et al. (2004) A gene expression map for the euchromatic genome of *Drosophila melanogaster. Science* **306**, 655–660.
15. Bertone, P., Stolc, V., Royce, T. E., et al. (2004) Global Identification of Human Transcribed Sequences with Genome Tiling Arrays. *Science* **306**, 2242–2246.
16. Stolc, V., Samanta, M. P., Tongprasit, W. et al. (2005) Identification of novel transcribed Sequences in *Arabidopsis thaliana* using high-resolution genome tiling arrays. *Proc. Nat. Acad. Sci. USA* **102**, 4453–4458.
17. Stolc, V., Samanta, M. P., Tongprasit, W., and Marshall, W. (2005) Genome-wide transcriptional analysis of flagellar regeneration in *Chlamydomonas reinhardtii* identifies orthologs of ciliary disease genes. *Proc. Natl. Acad. Sci. USA* **102**, 3703–3707.
18. The ENCODE Project Consortium (2004) The ENCODE (ENCyclopedia of DNA Elements) Project. *Science* **306**, 636–640.

19. Johnson, J. M., Edwards, S., Shoemaker, D., and Schadt, E. E. (2005) Dark matter in the genome: evidence of widespread transcription detected by microarray tiling experiments. *Trends Genet.* **21,** 93–102.

20. Mockler, T. C., Chan, S., Sundaresan, A., Chen, H., Jacobsen, S. E., and Ecker, J. R. (2005) Applications of DNA tiling arrays for whole-genome analysis. *Genomics* **85,** 655.

21. Royce, T. E., Rozowsky, J. S., Bertone, P., et al. (2005) Issues in the analysis of oligonucleotide tiling microarrays for transcript mapping. *Trends Genet.* **21,** 466–475.

22. Singh-Gasson, S., Green, R. D., Yue, Y., et al. (1999) Maskless fabrication of light-directed oligonucleotide microarrays using a digital micromirror array. *Nat. Biotechnol.* **17,** 974.

23. Nuwaysir, E. F., Huang, W., Albert, T. J., et al. (2002) Gene expression analysis using oligonucleotide arrays produced by maskless photolithography. *Genome Res.* **12,** 1749.

24. Samanta, M. P., Tongprasit, W., and Stolc V. (unpublished).

25. Bolstad, B. M., Irizarry R. A., Astrand, M., and Speed, T. P. (2003) A comparison of normalization methods for high density oligonucleotide array data based on bias and variance. *Bioinformatics* **19,** 185–193.

26. Kampa, D., Cheng, J., Kapranov, P., et al. (2004) Novel RNAs identified from an in-depth analysis of the transcriptome of human chromosomes 21 and 22. *Genome Res.* **13,** 331–342.

27. Cheng, J., Kapranov, P., Drenkow, J., et al. (2005) Transcriptional maps of 10 human chromosomes at 5-nucleotide resolution. *Science* **308,** 1149–1154.

11

Analysis of Comparative Genomic Hybridization Data on cDNA Microarrays

Sven Bilke and Javed Khan

Summary

We present a detailed method to analyze DNA copy number data generated on cDNA microarrays. A web interface is made available for those steps in the workflow that are not typically used in gene expression analysis so that these steps can be carried out online. The end result of the analysis is a list of p-values for the presence of genomic gains or losses for each sample individually or an average p-value, which we show is useful to identify recurrent genomic imbalances.

Key Words: Microarray; comparative genomic hybridization; cancer; disease diagnosis; disease prognosis.

1. Introduction

cDNA microarrays are becoming increasingly popular for applications in comparative genomic hybridization aiming to detect genomic imbalances. Gains or losses of specific DNA regions are frequently observed in tumors *(1)*. Cancers of different diagnostic types often have characteristic genomic alteration profiles, and some profiles are predictive of aggressive behavior *(2)*. Therefore, considerable efforts have been taken to map these genomic alterations for specific cancers in order to identify the genes responsible for the aggressive phenotype. "Traditional" methods to observe DNA copy number changes include metaphase comparative genomic hybridization and fluorescent *in situ* hybridization. Although these are very powerful tools, both have intrinsic limitations. For example, metaphase comparative genomic hybridization has a relatively low spatial resolution (on the order of 10–20 Mbp), as well as a low sensitivity. Fluorescent *in situ* hybridization, on the other hand, provides a good spatial resolution, however, the coverage of the genome is limited to a small number of locations.

From: *Methods in Molecular Biology, vol. 377, Microarray Data Analysis: Methods and Applications*
Edited by: M. J. Korenberg © Humana Press Inc., Totowa, NJ

Array-based comparative genomic hybridization (acGH) combines both a high spatial resolution as well as broad coverage of the genome *(3–5)*. Within a wide range, these parameters are limited only by the number of spots on the array. In this way, acGH partially overcomes the limitations of older methods *(3,4,6–8)*. Different sources of DNA are currently being used for immobilization on the glass carrier, each of which has its own strength. Bacterial artificial chromosomes (BACs), genomic DNA amplified in bacteria, provides probably the highest level of sensitivity as a result of the fact that BAC–DNA sequences are much longer than the sequences used in competing approaches *(3)*. Unfortunately, it is still very laborious to generate BACs and BAC libraries are not widely available. A major advantage of oligonucleotide arrays is the almost complete control over the spotted sequences and the availability from several commercial sources. The popularity of cDNA arrays for gene expression analysis makes these arrays available in many laboratories. The probably biggest advantage for cDNA acGH is the fact that the very same type of chips can not only be used to analyze DNA copy number changes but also for expression analysis. This makes cDNA arrays a superior tool for the investigation of causal links between DNA copy number changes and changes of transcript levels via gene dosage effects *(6,7)*. However, the signals observable with cDNA arrays tend to be relatively weak for small DNA copy number changes *(7,8)* because cDNA sequences with typically around 1000 bp are shorter than the average BAC clone (but longer than oligosequences). Furthermore, cDNA sequences were cloned from mature mRNA and may therefore often differ from the corresponding genomic DNA because of removal of introns and as a result of splicing.

One important problem in the analysis of acGH data generated on cDNA arrays, and to a decreasing extend also for oligonucleotide and BAC arrays, is a reduction of noise in order to be able to detect the lowest levels of gains or losses. This chapter deals mostly with aspects of noise reduction. In the next section we briefly describe the theoretical background of the material covered in this chapter. For the practitioner it is not absolutely necessary to go through this in every detail, the "hands on" description is sufficient to execute the described analysis. Nevertheless it is helpful to understand some of the basics in order to be able to make educated decisions about parameters.

1.1. Theoretical Background

The principle of detecting genomic alterations in CGH data is simple: if the fluorescent ratio of a DNA probe exceeds a specific threshold Θ in comparison to normal DNA, the probe is said to be *gained* (or *lost*, if the ratio falls below a threshold). To do so with statistical confidence it is necessary that the signal induced by the change of the DNA copy number is sufficiently stronger than the inherent noise. With cDNA and oligoarrays (and to an extent also for BAC arrays), the signal-to-noise ratio for single probes is in most cases not large enough to detect the lowest level copy number changes with sufficient statistical confidence. It is not uncommon to find that the noise level (root mean square amplitude of the noise) is of the same magnitude or even larger than the signal intensity. Noise reduction, therefore, is a crucial step for a detection of low level DNA copy number changes.

Sources of measurement uncertainties can broadly be categorized into *systematic* and *stochastic* errors. Stochastic noise is a purely random fluctuation of observed values around the true value. Because this type of noise is undirected (the average signal contributed in many repeated experiments is zero) the noise level can be reduced arbitrarily by repeating experiments. Systematic errors differ from that in that they induce a bias, a constant difference between the observed signal compared with the true value. Consequently, this type of error cannot be reduced by a repetition of experiments. It turns out that both error sources significantly reduce the sensitivity of a CGH chips.

1.1.1. Stochastic Noise

Repeating experiments sufficiently often will eventually reduce stochastic noise below the level that is required to detect genomic alterations with sufficient confidence. In practice, a repetition of hybridizations is rarely used for this purpose because the number of necessary repeats makes this approach too expensive; for example, with the cDNA arrays used by the authors one would need up to 30 repeats to detect single copy losses. Instead "in-slide" noise reduction strategies are commonly used, often combined with a breakpoint detection *(9,10)*. In the biology-related literature, variants of the "running average" smoothing filter are the most frequently used filters to reduce stochastic noise in aCGH data. This algorithm calculates the average observation for a certain number W of consecutive probes in genomic order. The idea is that adjacent probes, within a region of a constant DNA copy number, provide repeated estimates of the same DNA copy number. The result of the averaging is assigned to the respective center position for a "sliding window" moving across the entire genome. In this way each location (with the exception of a few boundary locations) gets assigned a noise-reduced estimate of the local DNA copy number ratio. A factor f parameterizes the level of noise reduction, e.g., $f = 1/2$ indicates a reduction of noise by 50%. When using a *running average* smoothing kernel this factor f is determined by the window's size. Under reasonable assumptions *(11)* for "well-behaved" noise (that is, following an approximately normal distribution) one finds that

$$f(W) = \frac{1}{\sqrt{W}} \quad \text{or} \quad W(f) = f^{-2} \tag{1}$$

the noise factor f shrinks with the inverse square-root of the window size W. Inversely, the required window size grows quadratically with f. For example, to reduce the root mean square stochastic noise level by a factor $f = 1/2$, the window size needed is $W \geq 4$. To reduce the noise level further by another factor 2, that is $f = 1/4$, the window size is already $W \geq 16$. This quadratic growth rapidly reduces the effective spatial resolution; after noise reduction the individual probes are no longer independent from neighboring probes. This does no harm as long as the entire window of probes is within a region of a constant DNA copy number. However, imbalance regions considerably smaller than W cannot be detected because the signal gets dampened by the probes outside of the imbalance region, eventually making signals undetectable. Also, the exact location of a genomic breakpoint, the boundary of the imbalance region, is blurred. Theoretically, one finds that after applying a running average of size W, the distance to the next effectively uncorrelated (as defined by the integrated autocorrelation time *[12]*) probe is $W/2$. This implies that from the N probes on an array one has

$$N_{eff} = \frac{2N}{W},\qquad(2)$$

effectively uncorrelated measurements for $W > 2$. The relevance of this number is that it allows estimating the effective spatial resolution of a aCGH measurement after reduction of stochastic noise. For example, if one assumes that the probes are homogeneously distributed over the genome, one finds in the human genome with approx 3×10^9 nucleotides a resolution

$$Res = 3 * 10^9 \frac{W}{2N}\qquad(3)$$

nucleotides per effectively independent probe. The observation that the resolution decreases with the window size may lead one to choose a small W. In fact, this *resolution-driven* choice is frequently used in the literature, typically with $W = 5...10$. Although this strategy is perfectly valid, it is important to keep in mind that choosing W sets a limit on the sensitivity via the right expression in **Eq. 1**. Consequently low level genomic alterations may not be detectable when the resulting noise reduction factor f is not small enough. Therefore, if the primary concern is the ability to detect low level alterations, a *sensitivity-driven* choice for W is advisable; first estimate the necessary level of noise reduction f and only then choose the window size $W(f)$ using **Eq. 1**.

But, what is an appropriate level of noise reduction? This depends on how statistically certain one wants to be about the results. Two parameters are necessary to discuss this: the false-positive rate α (*see* **Note 1**) and the false-negative rate β (*see* **Note 2**). In essence, the reduced noise level $\sigma' = f\sigma$ must be small enough such that threshold Θ used to define gain or loss is several standard variations away from both the null level (to avoid false positives) and the average signal level (to avoid false negatives). In mathematical terms, this can be expressed as

$$W = \left[\frac{(a+b)\sigma}{\Gamma S}\right]^2 \quad \text{with} \quad \begin{aligned} a &= \sqrt{2}\,\mathrm{erfc}^{-1}(\alpha) \\ b &= \sqrt{2}\,\mathrm{erfc}^{-1}(\beta). \end{aligned} \tag{4}$$

as a function of the raw noise level σ, the signal Γ for a one copy change, the lowest level *S* of DNA copy number change to be observed. Variable *a* measures the threshold Θ in units of the reduced level of noise σ′, while *b* uses the same units to parameterize the distance between the expected signal for the fluorescent ratio <Γ*S*> and the threshold Θ. This leads to an expression

$$\Theta = \Gamma S \frac{a}{a+b} \tag{5}$$

for this threshold.

1.1.2. Systematic Noise

Systematic errors cannot be reduced by a repetition of experiments. Different from the case for stochastic noise, repetitions may even rather increase the relative importance of a bias, as their level remains constant while the amplitude of the random noise gets smaller. This is particularly problematic when one is dealing with the weak signals typical for cDNA-aCGH data analysis. It is not uncommon that the amplitude of the bias reaches the level of the true biological signal. In **ref. *11*** it was shown that a part of the bias varies slowly across the genome leading to a significantly increased false-discovery rate, i.e., regions were labeled as "genomic imbalance" even though they are truly unaltered.

In principle, it would be quite inexpensive to remove a bias by a simple subtraction if one knew the exact magnitude of a bias. However, the systematic error is hardly ever known *a priori*. Algorithmic approaches to bias reduction typically make quite strong assumptions about the nature of the bias, estimate relevant parameters from the data, and subsequently subtract the estimate from the signal. One example is the LOWESS *(13,14)* algorithm that reduces intensity-dependent effects in log-ratio data. Background signals are typically estimated and removed by the image analysis software packages. Print-tip normalization *(15)* reduces a bias introduced in the printing process of the microarrays. Although these and other methods were originally developed in the context of expression analysis they are in most cases also beneficial for aCGH data analysis. However, despite their effectiveness in removing those false signals that follow the basic assumptions of the algorithms it is not uncommon that a significant residue remains reducing the sensitivity of the system.

Our approach to bias removal, which we found to be very effective in context with CGH analysis *(11)*, uses data from so-called "self–self" hybridizations to estimate a bias. In this type of additional experiments the same DNA is split into two groups, labeled with Cy3 or Cy5, respectively, merged again and hybridized on an array. The interesting point of such experiments is that, in principle, the result of measurements is known and trivial, namely a constant fluorescent ratio of one for the entire chip. Reproducible patterns resulting from systematic errors are thus easy to identify and remove.

2. Methods

2.1. Plan Ahead

Understand what the biological question under investigation requires in terms of resolution and sensitivity. This step should always be done *before* any hybridization experiments take place. As discussed in **Subheading 1.1.1.**, it is not possible, for a given technological platform, to choose sensitivity *and* resolution independently. Noise reduction increases the sensitivity for the detection of lower level genomic alterations, however, at the expense of spatial resolution.

It should be decided if the biological problem requires a detection of low level genomic alterations. If the primary focus is on the localization of breakpoints, no or only weak filtering (with $W \leq 5$) should be used. The use of the full resolution provided by the array using all probes is typically only possible for amplicons, more than 10 extra copies of DNA in amplified regions. As a rule of thumb, amplifications can be detected with an *n-fold* analysis without extra noise reduction if the signal exceeds an adjustable threshold Θ the corresponding genomic locus is said to be amplified. For this type of analysis it is sufficient to execute the steps described in **Subheading 2.2. and 2.5.).** To reduce problems related to outliers it is common practice to require that at least two adjacent probes indicate an amplification.

If the biological question requires the detection of low level genomic alterations more planning is necessary. In **Subheading 1.1.1.** it was discussed that an increase of sensitivity via noise reduction typically leads to a loss of spatial resolution; low level genomic alterations can only be detected if the affected region is covered by several probes. At the same time, the precise location of the breakpoint is blurred by this process. Consequently, it is generally not possible for a given dataset to choose both sensitivity *and* resolution. We found it very helpful to acquire a few extra hybridization experiments for an optimization of noise reduction. Self–self hybridizations (*see* **Note 3**) allow us to estimate the level σ of stochastic noise by calculating the variance of the ratio data. These experiments (at least two) can furthermore be used to reduce systematic errors (*see* **Note 4**).

2.2. Image Analysis, Quality Control

Image analysis, the translation of the scanned fluorescent images to the set of numbers used in the subsequent numerical analysis, is the first step in every analysis pipeline. The procedure for CGH arrays does not differ significantly from the steps familiar from gene expression analysis. If your scanning application makes it possible to label bad spots based on the fluorescent intensity, keep in mind that a loss of both copies of DNA may (and should) reduce the intensity in the *signal* channel to numbers close to zero. The option to flag low-intensity spots as low quality may be counterproductive because this could remove regions with DNA loss from the subsequent analysis. It is safe, though, to remove spots with too low intensity in the reference channel, which typically reflects normal DNA copy numbers.

Most image analysis software will label bad spots automatically based on image pathologies. Nevertheless, it is a good practice to eyeball the scanner images individually for obvious pathologies. Although it is practically impossible to identify every, however small, pathology, this step assures that major artifacts do not negatively impact the statistical power of the entire dataset. The overall number of spots marked as "bad" on the different slides may be a good indicator to identify problematic arrays that should either be repeated or removed.

In order to use the software on our website, the result of the image analysis needs to be stored as a flat text file in tab-delimited format. It is expected that the first row contains alpha-numerical column descriptors, typically experiment identifiers. The subsequent rows represent the data for one clone each and the expected format is

id <tab> R/G1 <tab> [R1 <tab> G1 <tab> Q1 <tab>] R/G2....

The row's first column contains the clone (or UniGene) identifier, the second column the ratio of fluorescent intensities for microarray one and (when needed) in columns three to five the red, green intensities and quality, where zero indicates a bad spot and one perfect quality. This data is optional and can be used for an intensity-dependent normalization.

2.3. Normalization

Several physical constants, such as the fluorescent efficiency of the dyes, affect how the numerical scanner value corresponds to the quantity of interest, the (relative) concentration of DNA molecules. Many of the relevant constants are unknown and may vary from experiment to experiment. Consequently, one cannot expect that the ratio of the raw fluorescent signals resulting from *red* and *green* DNA molecules in the same concentration is equal to the expected value one. Adjustment for these unknown parameters, commonly called normalization, is an essential step for CGH array analysis. In general it is safe to use the normalization schemes the reader is accustomed to (*see* **Note 5**). The algo-rithms in later step, the probabilistic detection of gains and losses (**Subheading 2.7.**) partially deals with the potential problems described in **Note 6**.

Our website currently supports LOWESS as well as global normalization. A tab-delimited file in the format described in **Subheading 2.2.** can be uploaded to that website. The user can choose the normalization method and (when available) whether to use the simplified format with only ratio data or the format containing intensity as well as quality information. The normalized file can be downloaded and is formatted in the correct format for the subsequent step described in **Subheading 2.5.**

2.4. Check for Systematic Errors

If at least two self–self hybridizations are available it is now easy to test the importance of systematic errors on the specific array platform used by calculating Pearson correlation coefficients between these experiments. Remember that the expected correlation coefficient for *perfect* self–self hybridizations is $r = 0$ because the experiments should lack any correlated patterns. In our experience it is not uncommon to observe $r \geq 0.5$ indicating that more than 50% of the data variability are of systematic origin.

2.5. Gene Sorting

To facilitate the interpretation of the data it is helpful (and in fact necessary for the noise reduction) to sort the ratio data in genomic order. Our website provides sorting service where tab-delimited files can be uploaded, the expected format is

id <tab> D1 <tab> ... Dn [<tab> Q1 <tab> ....Qn]

with an id-column, *n* data columns, and (optional: mirroring the order of the data columns) quality indicators ranging from zero (unusable) to one (perfect) for each data column.

Our website currently supports *unigene* and *image clone* identifiers for the *id*-field, which need to be selected accordingly. If the data contains quality information (*see* **Note 7**) the checkbox *contains quality data* needs to be selected. If the option *use quality information* is checked, poorly measured clones are either removed or substituted with reasonable values; if excessively many hybridizations (defined by the option *maximal number of bad spots*) are flagged as *bad* the entire clone is removed, whereas with fewer bad spots the values for bad spots are substituted with the average of the remaining spots not marked as bad. If the option *merge clones* is checked, the program averages the values for replicates of the same clone on the microarray or values for distinct clones mapping to the same genomic location into one measurement. This option must be checked if the user wishes to use our algorithm for the detection.

In our studies we frequently remove the data for the X and Y chromosomes because these gender-dependent chromosomes tend to confuse subsequent analysis steps by an apparent change of the DNA copy number. After removing these chromosomes it is generally a good idea to repeat the normalization step (**Subheading 2.3.**).

2.6. Parameter Estimation

The noise-reduction algorithm featured on our website requires the user to adjust how aggressively noise should be removed. As this step works at the expense of spatial resolution one wants to have an estimate about what level is absolutely necessary. First one has to decide what levels of statistical significance are required, namely the false-discovery rate α and the false-negative rate β. In our analysis we typically set $\alpha = \beta = 0.05$. Note that these numbers are *not* adjusted for multiple comparison, nevertheless these values turned out to be sufficient for the detection of the biologically relevant *recurrent* regions (**Subheading 2.8.**).

Besides these user-adjustable thresholds it is necessary to estimate the level of stochastic noise. If you did perform self–self hybridizations, use your favorite statistics program to estimate the variance σ for all samples and calculate the average overall self–self hybridizations. Without self–self hybridizations one can instead use hybridizations where by visual inspection one does not observe a strong signal for a (pessimistic) estimate of σ.

Another important factor is the sensitivity of the microarray platform: how much does the *observed* fluorescent ratio change with a change of the DNA copy number? One way to estimate this parameter is to analyze well-characterized cell lines: extract the data for all probes within regions of known DNA copy number, calculate the median fluorescent ratio within these regions, and then estimate by linear regression the response coefficient Γ. Alternatively, one can use normal, diploid DNA with distinct number of copies of the X chromosome (the details of the biochemistry is outside the scope of this chapter and we refer to **refs. 4** and **16** for details) for the estimation of Γ (*see* **Note 8**). One advantage of this choice is that the autosome data of these hybridizations can substitute the self–self hybridizations discussed previously because in the autosome both signal and background represent the same constant diploid chromosome content.

On the webpage the option *Parameter Estimation* implements **Eq. 4** and calculates the required minimal window size. Typical values obtained for the platform used by the authors are $W = 20$ for the detection of single copy gains and $W = 35$ for one copy losses.

2.7. Detection of Genomic Imbalances

The implementation of the topological statistics algorithm *(11)* on our website expects tab-delimited text files in the format id <tab> D1 <tab> ... Dn

generated by the gene sorter. If data from self–self hybridizations are available for bias reduction, the option **Reduce Bias** should be selected and the file containing the data for two or more self–self hybridizations can be uploaded (*see* **Note 9**). As a result the program returns $-\log_2$ of the p-values multiplied by minus one for losses (negative numbers), whereas gains generate positive numbers.

2.8. Recurrent Alterations

One major biological interest in analyzing genomic alterations in cancer is the identification of *recurrent* alterations, particularly the identification of smallest regions of overlap (SRO) which may hint toward the presence of onco-genes in gained regions or tumor-suppressor genes if DNA is lost. It is quite difficult to define an SRO in a mathematically rigorous way if one or more samples do not have an alteration in that region; strictly speaking there is no SRO in this case, whereas heuristics, which exclude the samples without alterations are very prone to false discoveries when the number of samples *increases*. Instead we suggest using the frequency of gains or losses for a clean definition of SRO. For the case when all samples are affected, a region with a frequency of one is equivalent to an SRO. For the case when only part of the samples have a genomic instability, a region of local maximum frequency is an acceptable definition for an SRO.

Our website offers a program that estimates the frequency of alterations directly from the log_2 p-values generated by topological statistics (**Subheading 2.7.**) without the need for a threshold *(11,17)*. On the website select *gain* or *loss* to calculate the frequency of gains or losses, respectively. As a result a file with two columns *id*, and frequency *v* is returned. Note that an approximation based on the average p-value is used to estimate v leading to a continuous distribution of numbers rather than a discrete set one would expect from a small set of samples.

3. Notes

1. The *false-positive* rate is the probability that a statistical test rejects the null hypothesis even though the null hypothesis is true. In this context, the false-positive rate estimates the fraction of tested probes where by chance the statistical test identifies a genomic instability, whereas the true answer is no genomic instability.

2. The *false-negative* rate is the probability that a statistical test accepts the null hypothesis even though the null hypothesis is false. In this context, the false-negative rate estimates the fraction of probes with a genomic instability that remain undetected.

3. In a *self–self* hybridization a DNA sample is split into two groups, labeled with Cy3 and Cy5, respectively, and is then cohybridized on a microarray. At first sight this may seem wasteful because the measurement is apparently uninformative as it is known *a priori* that the measurement should return a ratio of one everywhere.

It may therefore seem that one cannot learn from this type of experiment. In fact, the opposite is true; this setup is one of the few microarray experiments where one has complete knowledge about what the measurement results *should* be. Deviations from this expected behavior provide important information, for example, a *reproducible* pattern points to systematic noise whereas other deviations make it possible to estimate the level of stochastic noise.

4. The self–self hybridizations are used to estimate a potential bias induced by the microarray. It is well known that subtle changes in hybridization conditions may alter the bias pattern. It is therefore advisable to do the self–self hybridizations in parallel with the biological samples in order to capture as similar bias patterns as possible.

5. The probable most frequently used normalization scheme is a global normalization. In this strategy the array-wide median ratio is adjusted to one (or zero if applied to log-transformed data) by dividing each fluorescent ratio by the observed median ratio, or for log-transformed data by subtracting the observed median log ratio. In the print-tip normalization *(15)* this step is done independently for the sets of clones printed with the same needle in order to reduce a potential print-tip bias. LOWESS normalization *(14)* normalizes and reduces intensity-dependent nonlinearities from the data.

6. An implicit assumption of the typical normalization methods (*see* **Note 5**) is that only a small fraction of the probes on the array show a signal ratio different from one (or, somewhat weaker, that gains and losses are symmetric and cancel out on the global scale). For cancer DNA this assumption is very often violated. Genomic alterations of large parts of the genome are not uncommon (*see* **Note 10**). Multimixture models like in **ref.** *9* could be useful to determine the median selectively for the unchanged regions in the presence of larger genomic alterations. However, in our experience, cDNA data is too noisy to be used effectively with this algorithm. Similarly, the copy numbers for the X and Y chromosome are gender dependent and the pseudosignal induced by these chromosomes may distort normalization. If the biological problem allows it may be helpful to repeat normalization after sorting the data in genomic order (**Subheading 2.5.**) and repeat the analysis for the X, Y chromosomes and the autosome separately.

7. The normalization software provided on our website automatically outputs files *with* quality information. For these files it is always necessary to select ***contains quality*** even if the user did not provide quality information in the normalization step. In this case the quality was set to one (perfect) for all data-points.

8. Cell samples from tumors do, in most cases, contain an admixture from normal cells. The normal, diploid genomes of these cells further reduce the expected signal. One way to take this into account is to correct the coefficient Γ accordingly, i.e., if only 50% of the sample are tumor cells, one can use $\Gamma' = 0.5\Gamma$.

9. It is essential that both the data file and the self–self hybridization data contain the *same* number of clones in the same order. Note that the program does not verify if this assumption is met.

10. The normalization schemes remove ploidy information. A change of the global DNA copy number is offset by the adjustment of the median ratio.

References

1. Mittelman, A. (1962) Tumor etiology and chromosome pattern. *Science* **176,** 1340–1341.
2. Forozan, F., Karhu, R., Kononen, J., et al. (1997) Genome screening by comparative genomic hybridization. *Trends Genet.* **13,** 405–409.
3. Pinkel, D., Segraves, R., Sudar, D., et al. (1998) High resolution analysis of DNA copy number variation using comparative genomic hybridization to microarrays. *Nat Genet.* **20,** 207–211.
4. Pollack, J. R., Perou, C. M., Alizadeh, A. A., et al. (1999) Genome-wide analysis of DNA copy-number changes using cDNA microarrays. *Nat. Gen.* **23,** 41–46.
5. Pinkel, D. and Albertson, D. G. (2005) Array comparative genomic hybridization and its applications in cancer. *Nat Genet.* **37 Suppl,** S11–S17.
6. Pollack, J. R., Sorlie, T., Perou, C. M., et al. (2002) Microarray analysis reveals a major direct role of DNA copy number alteration in the transcriptional program of human breast tumors. *Proc Natl Acad Sci USA* **99,** 12,963–12,968.
7. Hyman, E., Kauraniemi, P., Hautaniemi, S., et al. (2002) Impact of DNA amplification on gene expression patterns in breast cancer. *Cancer Res.* **62,** 6240–6245.
8. Beheshti, B., Braude, I., Marrano, P., et al. (2003) Chromosomal localization of DNA amplifications in neuroblastoma tumors using cDNA microarray comparative genomic hybridization. *Neoplasia* **5,** 53–62.
9. Hupe, P., Stransky, N., Thiery, J., et al. (2004) Analysis of array CGH data: from signal ratio to gain and loss of DNA regions, *Bioinformatics* **20,** 3413–3422.
10. Jong, K., Marchiori, E., Meijer, G., et al. (2004) Breakpoint identification and smoothing of array comparative genomic hybridization data, *Bioinformatics* **20,** 3636–3637.
11. Bilke, S., Chen Q. R., Whiteford, C. C., et al. (2005) Detection of low level genomic alterations by comparative genomic hybridization based on cDNA microarrays. *Bioinformatics* **21,** 1138–1145.
12. Sokal, A. (1996) Monte Carlo methods in statistical mechanics: foundations and new algorithms. In: Dewitt-Morette, C., Castier, P., Folacci, A. (eds.) *Lectures at the Cargèse Summer School on "Functional Integration: Basics and Applications."* *Proc. ASI,* Cargèse, France, *p. 431.*
13. Cleveland, W. S. (1979) Robust locally weighted regression and smoothing scatterplots. *J. Amer. Stat. Assoc.* **74,** 829–836.
14. Dudoit, S., Yang, Y. H., Callow, M. J., et al. (2002) Statistical methods for identifying differentially expressed genes in replicated cDNA microarray experiments. *Statist. Sinica.* **12,** 111–139.
15. Yang, Y. H., Dudoit, S., Luu, P., et al. (2002) Normalization for cDNA microarray data: a robust composite method addressing single and multiple slide systematic variation. *Nucl. Acids Res.* **30,** e15.
16. Chen, Q. R., Bilke, S., Wei, J. S., et al. (2004) cDNA Array-CGH profiling identifies genomic alterations specific to stage and *MYCN*-amplification in neuroblastoma. *BMC Genomics* **5,** 70.
17. Bilke, S., Chen, Q. R., Westerman, F., et al. (2005) Inferring a tumor progression model for neuroblastoma from genomic data. *J. Clin. Oncol.* **23,** 7322–7331.

12

Integrated High-Resolution Genome-Wide Analysis of Gene Dosage and Gene Expression in Human Brain Tumors

Dejan Juric, Claudia Bredel, Branimir I. Sikic, and Markus Bredel

Summary

A hallmark genomic feature of human brain tumors is the presence of multiple complex structural and numerical chromosomal aberrations that result in altered gene dosages. These genetic alterations lead to widespread, genome-wide gene expression changes. Both gene expression as well as gene copy number profiles can be assessed on a large scale using microarray methodology. The integration of genetic data with gene expression data provides a particularly effective approach for cancer gene discovery. Utilizing an array of bioinformatics tools, we describe an analysis algorithm that allows for the integration of gene copy number and gene expression profiles as a first-pass means of identifying potential cancer gene targets in human (brain) tumors. This strategy combines circular binary segmentation for the identification of gene copy number alterations, and gene copy number and gene expression data integration with a modification of signal-to-noise ratio computation and random permutation testing. We have evaluated this approach and confirmed its efficacy in the human glioma genome.

Key Words: Array-comparative genomic hybridization; array-CGH; brain tumor; circular binary segmentation; cDNA microarray; gene copy number alteration; gene expression profiling; glioma; permutation testing; signal-to-noise ratio.

1. Introduction

Gene copy number alterations and changes in gene expression patterns are hallmarks of human cancer. Chromosomal instability in particular has been recognized as a major mechanism that confers a selective advantage to tumor cells *(1)*, leading to accelerated inactivation of tumor-suppressor genes, activation of oncogenes, and an increase in proliferation rate because of diminished cell cycle checkpoint controls. Recurrent, nonrandom patterns of genetic alterations have been detected by the systematic cytogenetic exploration in a majority of tumor types *(2)*. Brain tumors demonstrate complex chromosomal aberrations

From: *Methods in Molecular Biology, vol. 377, Microarray Data Analysis: Methods and Applications*
Edited by: M. J. Korenberg © Humana Press Inc., Totowa, NJ

that result in altered gene copy numbers. These aberrations include large regional changes—involving chromosomal fragments, chromosomal arms, or whole chromosomes—that are typically of low amplitude (i.e., gains and losses) and circumscribed alterations of only few neighboring genes (i.e., amplification or deletion), which, on the plus side, can be of high amplitude. Although for some of these altered regions, certain genes have been implicated in gliomagenesis, for others the presumed relevant target genes remain to be identified.

Microarray technology enables the comprehensive high-resolution, genome-wide analysis of gene copy number aberrations in a wide variety of experimental and clinical settings. This technology has revolutionized the systematic exploration of global gene expression and has proved its usefulness in molecular tumor classification, treatment response and survival prediction, and the identification of potential drug targets. However, the molecular processes underlying tumor pathogenesis are highly complex. Comprehensive understanding of the mechanisms and pathways leading to the initiation and the progression of tumors requires the analysis of multiple molecular levels and the integration of data on genetic, epigenetic, transcriptomic, and proteomic determinants of tumor phenotype.

Recent optimization of microarray protocols and the design of advanced bioinformatics tools now allow for the concurrent large-scale profiling of gene expression and gene copy numbers (the latter is commonly referred to as microarray-based comparative genomic hybridization or array-comparative genomic hybridization [CGH]) in a wide variety of biological specimens. This integrated approach provides several advantages compared with the single level analyses. It particularly enables the prioritization of seemingly random gene copy number aberrations in tumors by immediately assessing their effect on the mRNA level. This feature may provide a first-pass means of distinguishing biologically irrelevant bystander genes from potential cancer gene targets. On the other hand, the determination of gene copy number levels adds an additional more consistent dimension to the highly dynamic and fluctuant gene expression profiles of tumors and, thus, facilitates the detection of key transcriptional changes. Finally, such integrated analysis may enhance our understanding of the global influence of genome instability and widespread gene copy number changes on the regulation of gene expression in human tumors.

We here describe the major tools necessary for the integration of microarray gene expression and gene copy number data, and demonstrate their application in brain tumor research using an academic cDNA microarray platform. We focus on data analysis methodologies, in particular on the circular binary segmentation (CBS) algorithm for the identification of gene copy number alterations and on signal-to-noise ratio computations coupled with statistical significance determination by random permutation testing.

2. Materials

2.1. RNA and DNA Isolation From Brain Tumor Specimens

1. RNeasy lipid tissue midi kit (Qiagen, Valencia, CA).
2. DNeasy tissue kit (Qiagen).

2.2. Microarray-Based Comparative Genomic Hybridization

1. *Dpn*II restriction enzyme QIAquick PCR purification kit (Qiagen).
2. Male and female human genomic DNA (Promega, Madison, WI).
3. Bioprime labeling kit (Invitrogen, Carlsbad, CA).
4. 10X dNTP mix: 1.2 mM each of dATP, dGTP, and dTTP, and 0.6 mM of dCTP in TE buffer (pH 8.0).
5. Cy3-dCTP and Cy5-dCTP fluorescent dyes (Amersham Biosciences, Piscataway, NJ).
6. Microcon YM-30 filters (Millipore, Billerica, MA).
7. TE buffer (pH 8.0): 10 mM Tris-HCl, pH 8.0, and 1 mM EDTA.
8. TE buffer (pH 7.4): 10 mM Tris-HCl, pH 7.4, and 1 mM EDTA.
9. Yeast tRNA (Invitrogen).
10. Human Cot-1 DNA (Invitrogen).
11. poly(dA-dT) (Sigma-Aldrich, St. Louis, MO).
12. UltraPure 20X SSC buffer (Invitrogen).
13. 10% SDS.
14. cDNA microarrays and appropriate hybridization and scanning equipment.

2.3. Microarray-Based Gene Expression Profiling

1. 3DNA array 900 Cy3 and Cy5 indirect labeling kits (Genisphere, Hatfield, PA).
2. DyeSaver2 anti-fade coating solution (Genisphere).
3. Universal human reference RNA (Stratagene, La Jolla, CA).
4. cDNA microarrays and appropriate hybridization and scanning equipment.

2.4. Software

Table 1 lists major software packages used in our analysis as well as important alternative tools.

3. Methods

As in any microarray application, the integrated analysis of gene copy number and gene expression data relies on a number of carefully executed and controlled experimental steps, as well as on a data analysis pipeline consisting of raw data acquisition, data normalization and filtering, followed by the identification of gene copy number alterations and significant gene expression changes.

3.1. RNA and DNA Isolation From Brain Tumor Specimens

There is a wide variety of methodologies available for the isolation and purification of total RNA and DNA from tumor samples. We are generally using column-based techniques as supplied by Qiagen for both RNA and DNA

Table 1
Major Software Packages Used by the Authors and Important Alternative Tools

Software	Platform	Source	Reference
General computation			
R[a]	R	www.r-project.org	*(7)*
MATLAB	MATLAB	www.mathworks.com	Commercial
Normalization			
MIDAS*	Java	www.tigr.org/software/tm4/ midas.html	*(4)*
SNOMAD	WWW	pevsnerlab.kennedykrieger.org/ snomad.htm	*(14)*
SMA	R	www.r-project.org	*(15)*
Visualization			
Caryoscope[a]	Java	caryoscope.stanford.edu	*(12)*
Gene copy number aberration identification			
DNAcopy[a]	R	www.bioconductor.com	*(6)*
CGH-Plotter	MATLAB	sigwww.cs.tut.fi/TICSP/CGH-Plotter	*(11)*
CGH-Miner	MS Excel	www-stat.stanford.edu/ ~wp57/CGH-Miner	*(17)*
Multiple hypothesis testing correction			
QVALUE[a]	R	www.bioconductor.org	*(19)*

[a]Software commonly used by authors.

extraction (*see* **Note 1**). We are utilizing 43,000-feature cDNA microarrays manufactured by the Stanford Functional Genomics Facility for both gene expression and gene copy number profiling. Although parallel global assessment of gene expression and gene copy numbers can be performed on multiple platforms, the use of a common (cDNA) platform for both molecular levels reduces the need for downstream data adjustments (*see* **Note 2**).

3.2. Microarray-Based Comparative Genomic Hybridization

1. For microarray-based CGH, we are performing labeling of digested DNA and microarray hybridizations essentially as described by Pollack et al. *(3)*, with slight modifications. For labeling reactions, 6 μg each of normal human reference genomic DNA and tumor DNA are digested separately with *Dpn*II at 37°C for 1.5 h (total volume of 40 μL, 1.5 μL *Dpn*II, and 6 μL *Dpn*II buffer).

2. After *Dpn*II inactivation by heating at 65°C for 20 min, samples are snap-cooled on ice for 2 min. Digests are purified using the QIAquick PCR purification kit. Samples are resuspended in 50 μL of EB buffer (*see* **Note 3**).

3. For microarray hybridization, 2 μg each of digested reference and tumor DNA in a volume of 22.5 μL are separately labeled using the Bioprime labeling kit, with

the kit's dNTP mix substituted with a custom 10X dNTP mix adjusted for dCTP. To each sample, 20 μL of 2.5X random primers are added, the mixture is boiled for 5 min at 100°C and snap-cooled on ice for 5 min. After adding 5 μL of 10X dNTP labeling mix, 3 μL of Cy3-dCTP and Cy5-dCTP fluorescent dye to the paired hybridization samples, and 1 μL of concentrated Klenow enzyme, samples are incubated for 2 h at 37°C.

4. Reactions are stopped by adding 5 μL of stop buffer, placed on ice for 5 min, and centrifuged at 18,000g for 2 min.

5. Labeled products are then purified using Microcon YM-30 filters. Corresponding Cy3- and Cy5-labeled probes are combined to the centrifugal filter unit, 400 μL of 1X TE buffer (pH 7.4) are added, and the mixture is inverted several times and centrifuged at 13,800g for 7 min.

6. After two additional washes with 450 μL of 1X TE (pH 7.4), a mixture of 380 μL of 1X TE (pH 7.4), 20 μL of 5 μg/μL yeast tRNA, 50 μL of 1 μg/μL human Cot-1 DNA, and 2 μL of 10 μg/μL poly(dA-dT) is added to block nonspecific binding, hybridization to repetitive elements, and undesired hybridization to extended poly(A) tails, respectively. The mixture is concentrated to <32 μL by centrifugation at 12,000g for 12 to 14 min. Probes are recovered by inverting filters into a new Microcon tube and centrifugation at 14,000g for 2 min.

7. After adjusting the volume to 32 μL with 1X TE (pH 7.4), 6.8 μL of 20X SSC, and 1.2 μL of 10% SDS are added and the mixture is denatured at 100°C for 2 min. Following a 30-min Cot-1 DNA preannealing step at 37°C, probes are hybridized to cDNA microarrays containing more than 43,000 cDNA sequences (manufactured by the Stanford Functional Genomics Facility) under a 22 × 60-mm glass cover slip and incubated in a hybridization chamber at 65°C for 15–18 h.

8. After overnight hybridization, cover slips are removed by briefly dipping microarrays into a 65°C 2X SSC and 0.03% SDS washing solution. To remove unbound labeled DNA, microarrays are sequentially washed in 2X SSC, 0.03% SDS at 65°C for 5 min, rinsed in 2X SSC at 65°C, followed by shaking washes 5 min each at room temperature in 1X SSC (one wash) and 0.2X SSC (two washes). Microarrays are finally centrifuged dry at 500g for 5 min.

3.3. Microarray-Based Gene Expression Profiling

1. For microarray-based gene expression profiling, we are using total RNA and an indirect labeling approach, utilizing the 3DNA Array 900 labeling system by Genisphere. We are here strictly following the procedural protocol provided by the manufacturer without any modifications. For cDNA synthesis, 3 μg of sample and reference total RNA are separately reverse transcribed using the Cy5- and Cy3-specific primers, respectively (*see* **Notes 4** and **5**).

2. Arrays are hybridized overnight at 65°C (*see* **Note 6**). In our experience, when small amounts of input material are used, the indirect labeling strategy generates robust and reliable gene expression data as compared with traditional direct labeling methods.

3.4. Normalization and Filtering

There are multiple sources of random variation and systematic biases at every step of the microarray experiment. In order to ensure the validity and reliability of the measured gene expression and gene copy number ratios, it is necessary to perform several data normalization and transformation procedures. The basic normalization strategy involves background correction and the application of global mean normalization to the raw array-element intensities, followed by logarithmic ($\log_2$) transformation. Because additional biases are distributed nonuniformly across the microarray surface and across the range of signal intensities, it is important to employ a normalization strategy that takes these factors into account. This is particularly critical in the integrated analysis of gene expression and gene copy number data because local-, spatial-, or intensity-based trends will be erroneously interpreted as regional genomic events.

We are using a local normalization approach that is implemented in the Institute for Genomic Research Microarray Data Analysis System (TIGR MIDAS) function of the freely available Java application-based TM4 microarray software suite *(4)*, which enables the necessary data preprocessing required for subsequent higher level analyses (*see* **Note 7**). After image scanning and data acquisition, using GenePix Pro 5.1 software (Axon Instruments, Union City, CA), raw data ".gpr" files are converted to MIDAS input files using the built-in ExpressConverter. Data are background corrected, filtered using a flag and background filter (1.5-minimal signal-over-background ratio for expression arrays in either channel; 2.5 minimal signal-over-background ratio in the reference channel and regression correlation >0.6 in both channels for array-CGH), and normalized by the LOWESS normalization function in microarray block-by-block mode. Finally, block standard deviation regulation is applied and the normalized $\log_2$-transformed data are exported for downstream analyses. Data normalization and transformation are performed separately for the gene expression and gene copy number datasets. Because we use universal human reference RNA and not RNA derived from the tissue of tumor origin for the reference channel, gene expression values are mean centered. The GoldenPath Human Genome Assembly (http://genome.ucsc.edu) is used to map fluorescent ratios of the arrayed human cDNAs to chromosomal positions.

3.5. Identification of Gene Copy Number Alterations

The method most commonly used for the identification of gene copy number aberrations in array-CGH data applies thresholds to moving average smoothed data *(5)*. These thresholds are usually based on reference self-to-self hybridizations and can be further supported by the hybridization of genomic DNA from cell lines with varying numbers of X chromosomes. However, this approach does not take into account the spatial relationship between the genes along the genome, and therefore alternative methods have been developed that are primarily based on gene position information.

CBS is a novel method for the analysis of array-CGH data developed by Olshen et al. *(6)* and is implemented in the freely available DNAcopy package for R *(7)*. This method is a modification of binary segmentation that translates noisy intensity measurements into regions of equal copy number and that has been successfully applied to the high-resolution characterization of tumor genomes *(8)*.

The use of the DNAcopy package is straightforward and requires minimal knowledge of the R environment. A normalized and quality-filtered gene copy number data matrix, together with genome position index and chromosome assignment vectors, need to be provided in tab-delimited format. The built-in function CNA creates the "copy number array" object used in all subsequent computations. Single point outlier detection and data smoothing are performed by smooth.CNA. The CBS algorithm is executed by segment, which segments gene copy number data into constant level regions that can be visualized by plot.DNAcopy or can be exported for further analyses. Detailed explanations of a number of tuning parameters, which allow modification of the algorithm's sensitivity and computation speed, can be found in the software documentation (*see* **Note 8**). In **Note 9**, we provide some important suggestions for alternative gene copy number aberration identification tools.

Figure 1 exemplary shows the application of the CBS algorithm to the raw array-CGH data derived from a crude brain tumor tissue sample. Low-amplitude gene copy number alterations over large chromosomal regions are apparent and include known cytogenetic changes such as gain of chromosome 7 and losses of chromosomes 10, 17p, and 22. In addition, small high-amplitude changes mirroring gene amplifications (such as epidermal growth factor receptor [EGFR] and cyclin-dependent kinase 4 [CDK4] amplicons) are readily identified by the algorithm.

3.6. Identification of Associations Between Gene Copy Number Level and Gene Expression

Significant associations between gene copy number alterations and gene expression can be detected using signal-to-noise ratio computation and permutation testing. This approach was initially used for the selection of gene markers and class prediction based on gene expression *(9)* and has been successfully applied to the integration of gene expression and gene copy number profiles *(8,10)*.

After initial transformation of the noisy signal intensity measurements for each gene into regions of equal copy number, and assignment of $\log_2$ ratios according to the corresponding chromosomal segment, translated gene copy number values are deemed changed as compared with normal human reference DNA, if they fall beyond the ±3 standard deviations range of distribution of all segmented values of control self-to-self hybridizations. In view of the known

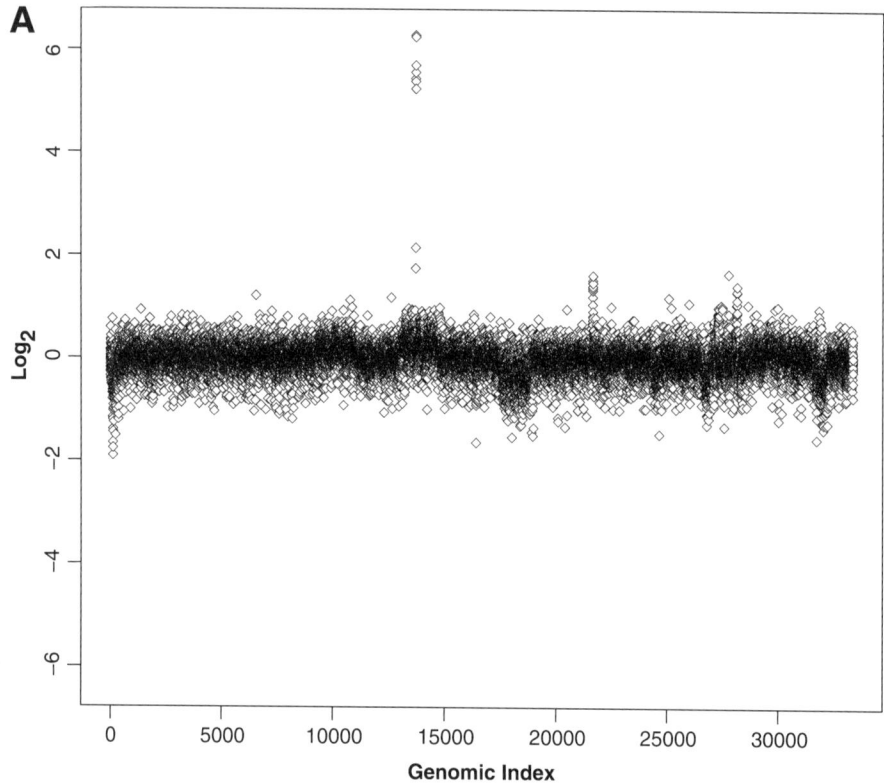

Fig. 1. *(Continued)*

ubiquitous presence and the considerable extent of gene copy number alterations in human tumors, we feel that this is a robust and conservative approach and particularly reasonable for the hypothesis-generating nature of microarray experiments. In our experience, the thresholds that are calculated by this strategy are well in the range of those that are generated by other automated aberrations calling algorithms *(11)*.

The global influence of copy number alteration for each gene on its transcript can then be determined by simple and intuitive computation of a signal-to-noise ratio (*s2n*), as initially described by Hyman et al. *(10)*. The *s2n* is defined by the difference of the means (*m*) of expression levels in the groups of altered (m_1) and unaltered (m_0) samples, divided by the sum of standard deviations (*s*) of expression levels in both groups (s_1 and s_0, respectively).

$$s2n = \frac{m_1 - m_0}{s_1 + s_0}$$

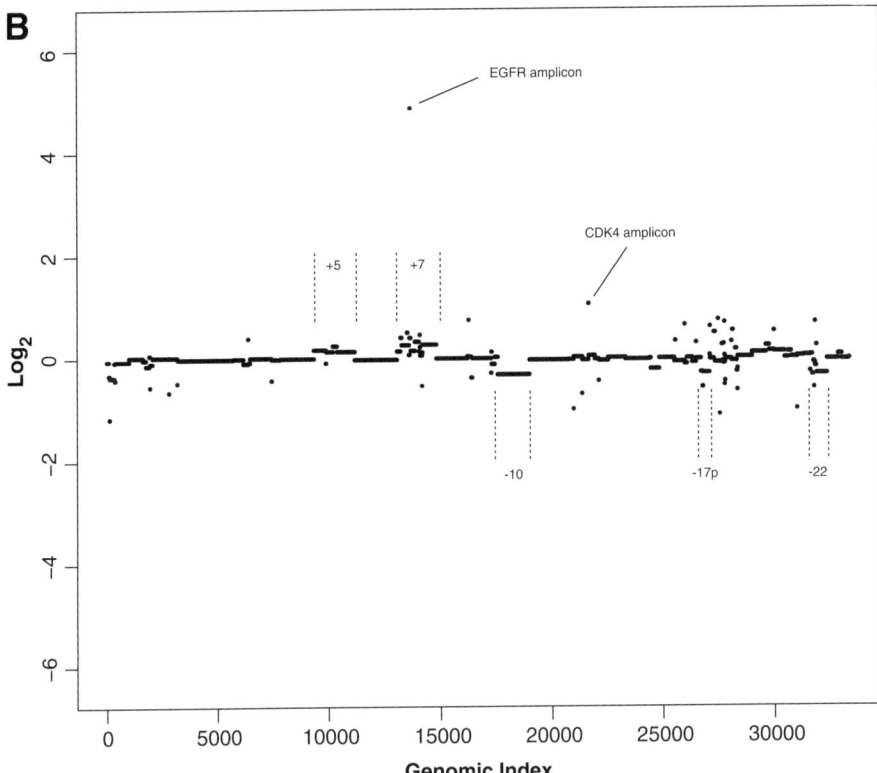

Fig. 1. Circular binary segmentation (CBS) of gene copy number data in a human glioblastoma multiforme. (A) Displays the raw, normalized log$_2$ signal intensity ratios plotted for 38,435 clones in genome order. (B) Shows the result of translating the noisy intensity measurements into regions of equal copy number, using the CBS algorithm. Various characteristic low- and high-amplitude gene copy number alterations have been readily depicted by the algorithm.

Here, the CGH data are first transformed into a binary system and represented by a labeling matrix, in which gene copy number alteration is assigned a value of one and no gene copy number alteration is assigned a value of zero. The significance of all computed ratios can then be assessed by randomly permuting the vector labels multiple times and by applying a probability (p)-value threshold of 0.05 (*see* **Note 10**). These procedures should be performed separately for the genes with gene copy number gain in at least two samples and for those with gene copy number loss in at least two samples. All required computations can be performed in any higher level statistical program. Simple functions s2n and permute displayed next are written in R and execute the signal-to-noise computation and estimation of p-values by permutation testing:

```
# gep, cbs and nperm define input gene expression matrix,
# labeling matrix and the number of required permutations,
# respectively;
# s2nval and pval contain results for s2n and p-values
s2nval <- rep(NA,nrow(gep))
pval <- rep(NA,nrow(gep))
s2n <- function(g,l){
    m1 <- mean(g[which(l==1)])
    m0 <- mean(g[which(l==0)])
    s1 <- sd(g[which(l==1)])
    s0 <- sd(g[which(l==0)])
    return((m1-m0)/(s1+s0))
    }
for(i in 1:nrow(gep)){
  s2nval[i] <- s2n(t(gep[i,]),t(cbs[i,]))
    }
  permute   <- function(g, l, nperm){
      c<-0
      w<-s2n(g,l)
      for(i in 1:nperm){
        wperm <- s2n(g,sample(l))
        if(wperm > w) c <- c+1
        }
      P <- c/nperm
      return(p)
      }
  for(i in 1:nrow(gep)){
      pval[i] <- permute(t(gep[i,]),t(cbs[i,]),nperm)
      }
```

We have evaluated and demonstrated the efficacy of this approach in glial brain tumors. We have revealed a sizable number of genes (8% of genes for which combined gene copy number and gene expression data were available) in the human glioma genome whose expression is significantly influenced by gene copy number alterations.

Because recurrence frequencies of genetic alterations in human tumors provide a natural means for prioritization of detected associations, we have implemented a modification of the previous approach in which we further weigh the computed signal-to-noise ratio for each gene by the relative frequency of alteration of this gene across the whole dataset ($n_1/[n_1 + n_0]$). We have termed this modified ratio recurrence-weighted signal-to-noise ratio ($rs2n$), which is calculated as follows:

$$rs2n = \frac{m_1 - m_0}{s_1 + s_0} \cdot \frac{n_1}{n_1 + n_0}$$

In order to explore the genomic distribution of those genes with significant association between gene copy number and gene expression, i.e., genes whose transcript is genetically regulated, recurrence-weighted signal-to-noise ratios can be visualized in genome order using the Caryoscope software *(12)*. Peak ratios identify candidate genes with top associations between genetic and transcriptional level, weighted for the abundance of the underlying genetic alteration in the sample set. Additionally, such plotting enables the exploration of spatial relationships between genes with and/or without gene copy number-driven gene expression changes. It also maps regions enriched for gene copy number/gene expression associations, that is, regions in which mechanisms of genetic coregulation may be operative (*see* **Note 11**).

The exact and systematic delineation of their boundaries is a challenging problem. Caryoscope provides valuable built-in features that could assist in this task. In particular, moving window computation allows data smoothing based on genomic position of neighboring probes and enables easier detection of possibly important trends in the gene copy number-driven gene expression effects.

As an example, **Fig. 2** shows the result of application of the outlined integrated analysis to the chromosomal region 7p12-p11 in a cohort of human gliomas. Panel A shows the mean gene copy number curves for two subgroups of patients with and without gene copy number alteration in this region, as determined by the CBS algorithm. Although no change in gene copy number is apparent in the group of nonamplified tumors, the amplified tumor group shows increased mean gene copy number across the whole displayed chromosomal segment, which peaks at the EGFR locus known to be amplified in a significant portion of gliomas. Additionally, the mean gene expression level for each gene within the region is indicated for both groups. Increased mean gene expression for a number of genes in the group of tumors with gene copy number alteration is apparent. Panel B reports the calculated symmetric moving average of $rs2n$ ratios for each gene (window size = 11). This curve peaks within the EGFR locus, suggesting the possible existence of a narrow and recurrently altered cluster of genes whose expression is strongly influenced by gene copy number.

4. Notes

1. For the parallel analysis of gene copy number and gene expression profiles, it is critical to isolate genomic DNA and total RNA from the same region of the sample,

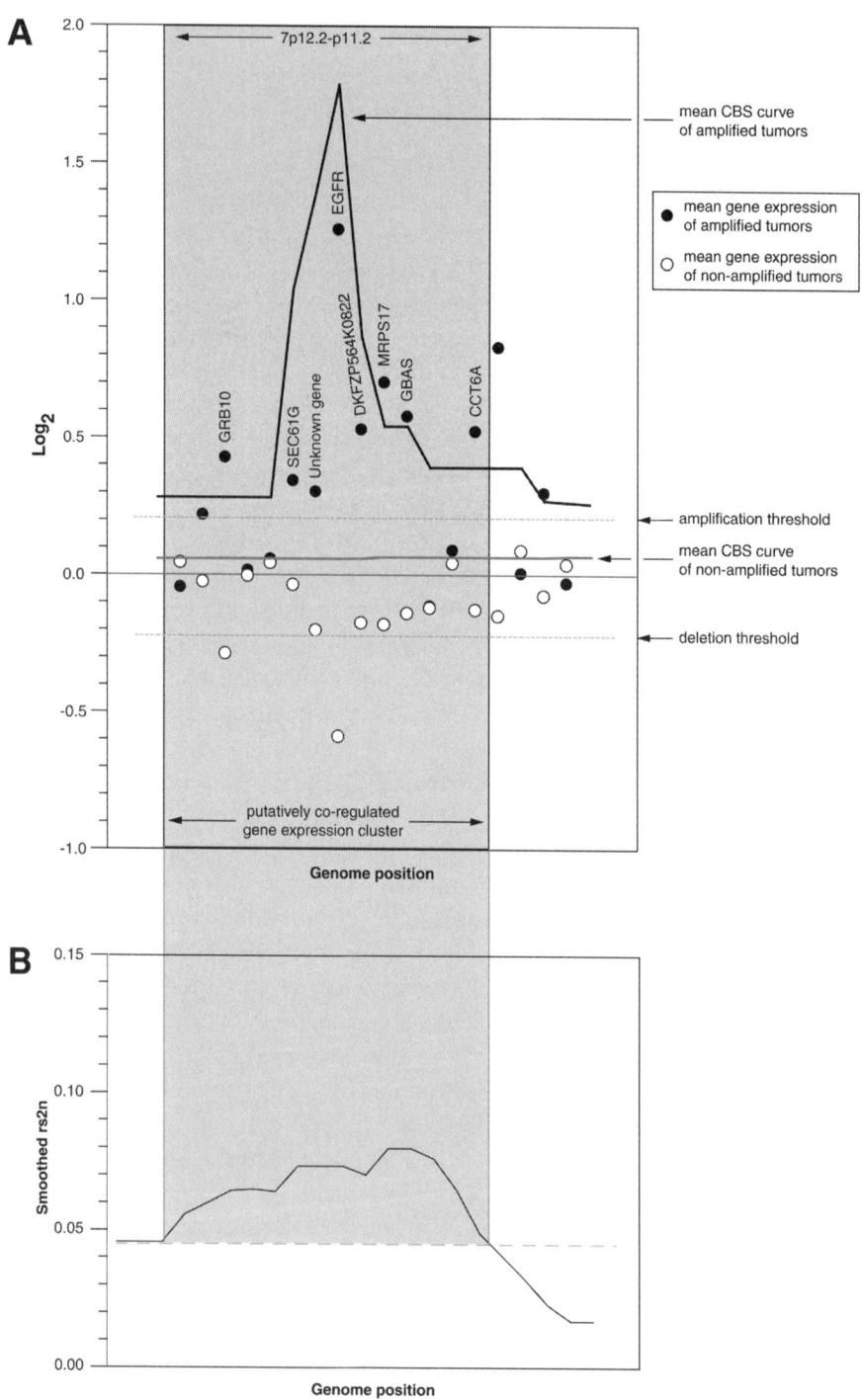

especially if crude tumor samples are analyzed. Genetic heterogeneity present in tumors can hamper the analysis if large tumor tissue samples are dissected in several pieces and the isolation of nucleic acids is not performed on directly neighboring parts. Ideally, protocols that allow concomitant isolation of both DNA and RNA (Qiagen) should be used. However, for RNA isolation from lipid-rich tissues, such as the brain, these kits do not produce optimal results. In brain tumors, the Qiagen RNeasy lipid tissue kit provides an excellent method for RNA recovery. We have noticed that the subsequent extraction of genomic DNA from the organic phase does not meet a quality necessary for array-CGH analysis.

2. Although custom or commercially available cDNA microarrays are a convenient choice for both applications, two different microarray platforms, each optimized for the best results, could also be used. This approach requires array-CGH data interpolation, so that expression measurements can be mapped to their corresponding gene copy number levels. At the same time, it avoids data interpretation difficulties related to possible tight correlations between expression and copy number measurements because of individual probe performance. Although gene expression profiling can be performed on a number of oligonucleotide microarray platforms, the best CGH results have been obtained with cDNA and bacterial artificial chromosome arrays. Use of oligonucleotide arrays for CGH analysis typically requires PCR-based genomic DNA complexity reduction that introduces additional biases. Recently, protocols and bioinformatics analysis tools were developed, which allow high-resolution, genome-wide gene copy number profiling using long oligonucleotide arrays and full-complexity DNA *(13)*.

3. We are routinely quantifying digestion products by spectrophotometry at 260 and 280 λ prior to DNA labeling; because of the considerable non-DNA contamination of even purified genomic DNA from lipid-rich brain tumors, the amount of digest does not properly reflect the amount of input genomic DNA.

4. We have successfully used the scaled-up protocol for the cDNA preparation from our reference RNA described in the Genisphere 3DNA Array 900 manual. This large-scale preparation of reference cDNA is not only highly convenient but also assures

---

Fig. 2. *(Opposite page)* Integration of gene copy number and gene expression data in chromosomal region 7p12-p11 in a cohort of glial brain tumors. (**A**) Shows the mean gene copy number curves for two subgroups of patients with and without gene copy number alteration at this locus, based on calculating the regional gene copy number profile for each tumor using the circular binary segmentation algorithm. The group of tumors with amplification shows increased mean gene copy number across the whole displayed chromosomal segment, which peaks at the epidermal growth factor receptor (EGFR) locus. Mean gene expression levels for all genes within the segment are plotted separately for both subgroups. There is increased mean gene expression for a number of genes in the subgroup of tumors with gene copy number alteration. (**B**) Reports the corresponding smoothed (*see* **Subheading 3.6.**) *rs2n* ratio curve, which peaks at the EGFR locus, indicating that the expression of a cluster of genes around EGFR is primarily genetically regulated and recurrently altered in the tumors.

that a cDNA product as constant as possible over a larger study cohort (which may be hybridized in a number of experiment sets) is used for the reference channel.

5. Because of increased fading of Genisphere Cy5 3DNA Array 900 reagent, we are applying the Genisphere DyeSaver2 anti-fade coating to each microarray immediately after the last wash. We are here sequentially drying the slide by centrifugation at 1000g for 30 s, dipping the microarray into DyeSaver2 for 3 s, and centrifuging the slide for 50 s. This procedure does not add any background to the microarrays but ensures that there is enough time (several hours) to scan the microarrays.

6. Using the Genisphere labeling protocol, bovine serum albumin prehybridization for background reduction is not necessary on our microarrays. We are only performing a postprocessing procedure (immediately before hybridization) in which the microarrays are sequentially ultraviolet cross-linked with 60 mJ, agitated in isopropanol for 15 min, placed into boiling nuclease-free water (95°C) for 2 min, and dried by centrifugation at 400g for 1 min.

7. In addition to the MIDAS software, other normalization tools are readily available. Particularly easy to use is the standardization and normalization of microarray data (SNOMAD) tool *(14)*, an internet-accessible interactive application that is excellent for the normalization and preprocessing of smaller sample sets. An alternative for the R environment is the SMA package developed by Yang and Dudoit *(15)*.

8. In order to increase the number of change points obtained by CBS, we routinely set the `alpha` parameter of the `segment` function to 0.05. For the purpose of parallel analysis of gene copy number and gene expression, we do not use the available "undo" option that eliminates change points, which are not at least three standard deviations apart. An independent measurement by the gold standard that is usually needed to remove "unnecessary" change points is rarely available for all the genes in the dataset.

9. The CBS algorithm is also implemented in the more user-friendly CGHPRO data analysis tool offering interactive graphical interface *(16)*. Several alternative gene copy number aberration identification tools are available. CGH-Plotter is a freely available MATLAB-toolbox for array CGH data analysis *(11)*. It enables a quick analysis of large datasets and includes a highly customizable graphical output. Similar to the CBS algorithm, actual gain/loss calling depends on user defined thresholds. CGH-Miner *(17)* uses a new "Cluster along chromosomes" algorithm for the identification of chromosomal regions with different gene copy number levels. It provides an automated gain/loss calling function with false discovery rate estimation based on normal–normal array hybridizations. It also generates "consensus curves" that reflect the recurrence of gene copy number alterations in a study set. This program is written in R and available as Excel add-in. A comprehensive comparative analysis of various tools and algorithms for CGH data analysis is also available and provides valuable insights into their performance characteristics *(18)*.

10. These *p*-values are determined in the context of multiple hypothesis testing. Appropriate procedures have to be used in order to control the number of falsely positive results. The use of the permutation-based *q*-value, which measures statistical significance in terms of false discovery rate, offers a sensible balance between the number of true and false positives and provides an automatically calibrated and

easily interpreted approach for the estimation of statistical significance in genome-wide studies *(19)*.

11. Proper interpretation of the observed regions requires careful handling of probe redundancy because a major source of focal effects detected in *rs2n* plots could be because of the presence of multiple probes per gene. Combination of multiple $\log_2$ ratios or, alternatively, *rs2n* values into one estimate per gene avoids this difficulty.

References

1. Albertson, D. G., Collins, C., McCormick, F., and Gray, J. W. (2003) Chromosome aberrations in solid tumors. *Nat. Genet.* **34**, 369–376.

2. Mertens, F., Johansson, B., Hoglund, M., and Mitelman, F. (1997) Chromosomal imbalance maps of malignant solid tumors: a cytogenetic survey of 3185 neoplasms. *Cancer Res.* **57**, 2765–2780.

3. Pollack, J. R., Perou, C. M., Alizadeh, A. A., et al. (1999) Genome-wide analysis of DNA copy-number changes using cDNA microarrays. In: *Nat. Genet.*, Vol. 23, pp. 41–46.

4. Saeed, A. I., Sharov, V., White, J., et al. (2003) TM4: a free, open-source system for microarray data management and analysis. *Biotechniques* **34**, 374–378.

5. Pollack, J. R., Sorlie, T., Perou, C. M., et al. (2002) Microarray analysis reveals a major direct role of DNA copy number alteration in the transcriptional program of human breast tumors. *Proc. Natl. Acad. Sci. USA* **99**, 12,963–12,968.

6. Olshen, A. B., Venkatraman, E. S., Lucito, R., and Wigler, M. (2004) Circular binary segmentation for the analysis of array-based DNA copy number data. *Biostatistics* **5**, 557–572.

7. Ihaka, R. and Gentleman, R. (1996) R: a language for data analysis and graphics. *J. Comput. Graph. Stat.* **5**, 299–314.

8. Aguirre, A. J., Brennan, C., Bailey, G., et al. (2004) High-resolution characterization of the pancreatic adenocarcinoma genome. *Proc. Natl. Acad. Sci. USA* **101**, 9067–9072.

9. Golub, T. R., Slonim, D. K., Tamayo, P., et al. (1999) Molecular classification of cancer: class discovery and class prediction by gene expression monitoring. *Science* **286**, 531–537.

10. Hyman, E., Kauraniemi, P., Hautaniemi, S., et al. (2002) Impact of DNA amplification on gene expression patterns in breast cancer. *Cancer Res.* **62**, 6240–6245.

11. Autio, R., Hautaniemi, S., Kauraniemi, P., et al. (2003) CGH-Plotter: MATLAB toolbox for CGH-data analysis. *Bioinformatics* **19**, 1714–1715.

12. Awad, I. A., Rees, C. A., Hernandez-Boussard, T., Ball, C. A., and Sherlock, G. (2004) Caryoscope: an Open Source Java application for viewing microarray data in a genomic context. *BMC Bioinformatics* **5**, 151.

13. Brennan, C., Zhang, Y., Leo, C., et al. (2004) High-resolution global profiling of genomic alterations with long oligonucleotide microarray. *Cancer Res.* **64**, 4744–4748.

14. Colantuoni, C., Henry, G., Zeger, S., and Pevsner, J. (2002) SNOMAD (Standardization and NOrmalization of MicroArray Data): web-accessible gene expression data analysis. *Bioinformatics* **18**, 1540–1541.

15. Yang, Y. H., Dudoit, S., Luu, P., et al. (2002) Normalization for cDNA microarray data: a robust composite method addressing single and multiple slide systematic variation. *Nucleic Acids Res.* **30**, e15.

16. Chen, W., Erdogan, F., Ropers, H. H., Lenzner, S., and Ullmann, R. (2005) CGH-PRO—a comprehensive data analysis tool for array CGH. *BMC Bioinformatics* **6**, 85.

17. Wang, P., Kim, Y., Pollack, J., Narasimhan, B., and Tibshirani, R. (2005) A method for calling gains and losses in array CGH data. *Biostatistics* **6**, 45–58.

18. Lai, W. R., Johnson, M. D., Kucherlapati, R., and Park, P. J. (2005) Comparative analysis of algorithms for identifying amplifications and deletions in array CGH data. *Bioinformatics* **21**, 3763–3770.

19. Storey, J. D. and Tibshirani, R. (2003) Statistical significance for genomewide studies. *Proc. Natl. Acad. Sci. USA* **100**, 9440–9445.

13

Progression-Associated Genes in Astrocytoma Identified by Novel Microarray Gene Expression Data Reanalysis

Tobey J. MacDonald, Ian F. Pollack, Hideho Okada,
Soumyaroop Bhattacharya, and James Lyons-Weiler

Summary

Astrocytoma is graded as pilocytic (WHO grade I), diffuse (WHO grade II), anaplastic (WHO grade III), and glioblastoma multiforme (WHO grade IV). The progression from low- to high-grade astrocytoma is associated with distinct molecular changes that vary with patient age, yet the prognosis of high-grade tumors in children and adults is equally dismal. Whether specific gene expression changes are consistently associated with all high-grade astrocytomas, independent of patient age, is not known. To address this question, we reanalyzed the microarray datasets comprising astrocytomas from children and adults, respectively. We identified nine genes consistently dysregulated in high-grade tumors, using four novel tests for identifying differentially expressed genes. Four genes encoding ribosomal proteins (*RPS2, RPS8, RPS18, RPL37A*) were upregulated, and five genes (*APOD, SORL1, SPOCK2, PRSS11, ID3*) were downregulated in high-grade by all tests. Expression results were validated using a third astrocytoma dataset. *APOD*, the most differentially expressed gene, has been shown to inhibit tumor cell and vascular smooth muscle cell proliferation. This suggests that dysregulation of *APOD* may be critical for malignant astrocytoma formation, and thus a possible novel universal target for therapeutic intervention. Further investigation is needed to evaluate the role of *APOD*, as well as the other genes identified, in malignant astrocytoma development.

Key Words: Astrocytoma; tumor progression; gene expression; microarray.

1. Introduction

Astrocytoma is the most common brain tumor in children and adults. Although adult and childhood astrocytomas can be distinguished by distinct clinical and genetic characteristics, the malignant forms are histologically identical and share a dismal prognosis, regardless of patient age *(1–4)*. The World Health Organization (WHO) grades astrocytomas based on histopathological characteristics as pilocytic (WHO grade I), diffuse (WHO grade II), which often progresses to high-grade

From: *Methods in Molecular Biology, vol. 377, Microarray Data Analysis: Methods and Applications*
Edited by: M. J. Korenberg © Humana Press Inc., Totowa, NJ

astrocytoma, anaplastic (WHO grade III), and glioblastoma multiforme (WHO grade IV) *(1)*. Pilocytic astrocytomas, the most common brain tumor in children, rarely exhibit malignant progression, and are considered to be a biologically distinct entity from nonpilocytic astrocytomas *(1)*. Because these are well circumscribed and rarely infiltrative, a complete surgical resection and cure is expected in the majority of patients. In contrast, nonpilocytic astrocytomas, which account for the vast majority of astrocytomas in adults and a sizeable subgroup of astrocytomas in children, are diffusely infiltrative and are often not amenable to complete resection *(5)*. Upon recurrence, grade II diffuse astrocytomas have a tendency for malignant progression to anaplastic astrocytoma and, ultimately, glioblastoma multiforme. There is increasing evidence that the progression from grade II to higher grade astrocytoma is the result of a sequence of genetic alterations that are acquired during the process of transformation *(5–7)*. Glioblastomas evolving from a previous lower grade astrocytoma are defined as secondary (ScGBM), while those arising without any evidence of a previous low-grade precursor are termed primary (PrGBM) *(1)*. Although PrGBM and ScGBM are histologically indistinguishable, the two types exhibit distinct molecular alterations *(8–11)*. PrGBM usually occur in older patients and are characterized by amplification and overexpression of *EGFR*, *PTEN* mutations, and loss of *INK4a* *(8–11)*. ScGBM tend to occur in younger adults and are associated with *TP53* mutations and overexpression of *PDGFA* and its receptor *(8–12)*. High-grade astrocytomas of childhood clinicopathologically resemble PrGBM of adulthood, yet these tumors rarely demonstrate *EGFR* amplification *(13)*. Childhood HGA also rarely exhibit *TP53* mutations. However, overexpression of *EGFR* and the *TP53* gene and protein is common *(14–16)*.

Determining whether there are common underlying molecular changes associated with malignant astrocytoma, independent of patient age, and demonstrating a critical role for these changes in the formation of malignant astrocytoma may ultimately lead to the development of novel and universal cancer therapies targeting these alterations. Microarray gene expression analysis has been an invaluable tool with which to unveil unforeseen patterns in the molecular alterations of cancers with indistinguishable phenotypes and histological characteristics *(17–21)*. Cancer progression from benign-to-malignant grade, in which tumor cells acquire the ability to migrate away from the primary tumor, invade through the surrounding microenvironment, initiate angiogenesis, and establish a distant colony is a highly complex process that is dependent on critical genetic changes. In principle, the identification of significantly differentially expressed genes between benign and malignant astrocytomas from patients of all ages should provide insight into the underlying genetic regulation of this process in this disease.

In this study, we sought to identify and validate gene expression patterns that universally differentiate higher from lower grades of astrocytomas. To this end, we reanalyzed two previously published microarray datasets of expression intensities of astrocytomas, comprised of childhood and adult astrocytomas, respectively *(14,22)*. We applied a series of four novel supervised statistical analyses to identify differentially expressed genes, followed by unsupervised clustering of the samples using leave-one-out validation and cross-dataset predictions to assess classification error. Each gene set was used to cluster the tumor samples in both the datasets. We performed iterative cross-validation on the union of the genes found to be significant in both datasets by all tests. A list of marker genes was created that was comprised of genes found to be significant under all tests in both datasets. We then validated our derived gene list using a third published dataset *(23)* and found the same genes differentially expressed using the same tests.

2. Materials

2.1. The K dataset

This study analyzed the expression of 12,625 probe sets (Affymetrix U95Av2 oligonucleotide array) in 13 childhood astrocytoma samples of two classes *(14)*. Out of 13, 6 samples were low-grade astrocytomas (LGAs) whereas 7 were of high grades (HGAs). The aim of their study was to determine an overlap of astrocytoma progression markers with a preselected gene list of angiogenesis markers. They used expression profiling of 133 angiogenesis-related genes and found a list of 44 differentially expressed genes (17 overexpressed and 27 underexpressed), which were also present in their list of angiogenesis markers. They used hierarchical clustering and principal components analysis and succeeded in classifying HGAs from LGAs using all genes as well as 133 angiogenesis-related genes. These data were downloaded from http://microarray.cnmcresearch. org/resources.htm.

2.2. The V Dataset

This study compared the expression profiles of 7,129 probe sets (Affymetrix HUGFL oligonucleotide array) in 16 astrocytoma samples (HGAs and LGAs) *(22)*. Of 16 samples, 8 were of primary and 8 were of recurrent high-grade astrocytomas. They identified 66 genes that exhibited twofold or greater difference in expression between primary and higher grade tumors. They further validated 12 of those genes by further analysis. These data were downloaded from http://dot.ped.med.umich.edu:2000/pub/astrocytoma/index.html

Both datasets are also "on-tap" for ease of reanalysis in our online open source gene expression analysis web application (http://bioinformatics.upmc.edu/GE2/ GEDA.html).

2.3. The Validation Set

This study used the Atlas Human Cancer 1.2 Array (Clontech), comprised of 1185 genes, to profile 21 newly diagnosed glioblastoma, 8 high-grade recurrent tumors (comprising two astrocytoma WHO grade III and 6 grade IV), and 24 LGA *(23)*. Data was obtained from supplementary data section from http://cancerres.aacrjournals.org.

3. Methods

3.1. Data Quality Measures

Data quality was checked for both the datasets by calculating the global correlation of group means (all genes). If the number of strongly differentially expressed genes between sample groups is low, correlation among the group means in a clean dataset should be around or greater than 95%. We also calculated the between-array coefficient of variation, which ideally should be as low as possible (<0.3 is generally acceptable). To detect undesirable and unanticipated structure or associations among the samples that cannot be accounted by blocking in the experimental design, we calculated a measure called the confounding index (CI). It is the ratio of sum of mean array-wide Pearson correlations of group A and group B over two times the correlation between the group means **(Eq. 1)**. The ideal CI value is 1.0; values of CI up to 1.1 are acceptable.

$$CI = \frac{\bar{r}_A + \bar{r}_B}{2\bar{r}_{AB}} \tag{1}$$

3.2. Preprocessing of Expression Data

The data obtained from both the research groups were already preprocessed by Affymetrix software MicroArray Suite. According to the published descriptions, the datasets were background subtracted, normalized, and log-transformed. Given this preprocessing of the data, we assumed the data quality far refined and verified the same by observing the box and whisker plots for both the datasets. We therefore did not apply any preprocessing algorithm on the data. We also analyzed some of the data under other preprocessing strategies to evaluate the robustness of our results.

3.3. Selection of Differentially Expressed Genes

The expression data from both datasets were analyzed using the Gene Expression Data Analyzer (http://bioinformatics.upmc.edu/GE2/GEDA.html). We applied multiple tests for identification of differentially expressed genes. These included permutation versions of the pooled variance *t*-test, the J5 test, the permutation percentile separability test (PPST), and the ABA test *(24–26)*.

Because we know that all genes do not exhibit the same distribution, even within sample groups, it does not make sense to apply a single threshold of significance for all genes. Instead, we randomized the sample labels 1000 times to determine a null distribution of the test statistic(s) for each gene. All permutation tests were performed at $\alpha = 0.05$.

3.4. Pooled Variance t-Test

The difference in gene expression for each gene is determined by comparing the average expression value within each group using a studentized test statistic, t, which employs the pooled variance error term (**Eq. 2**). This form of the test statistic is more appropriate as it takes into account the difference in number of samples in the two groups.

$$t = \frac{\mu_1 - \mu_2}{\sqrt{\left[\dfrac{(n_1 - 1)s_1^2 + (n_2 - 1)s_2^2}{n_1 + n_2 - 2}\right]\left(\dfrac{1}{n_1} + \dfrac{1}{n_2}\right)}} \tag{2}$$

3.5. J5 Test

The J5 test gives a statistic based on the magnitude of the difference between the means (**Eq. 3**). It essentially compares the difference of means in any gene to the average difference in means over the whole array. This test appears to be most useful when the number of samples is low.

$$J_i = \frac{\bar{a}_i - \bar{b}_i}{\dfrac{1}{n}\displaystyle\sum_{k=1}^{n}|\bar{a}_k - \bar{b}_k|} \tag{3}$$

3.6. Significance Analysis of Microarrays

Significance analysis of microarrays (SAM) determines genes to be statistically significant based on changes in their expression determined by gene-specific modified t-tests (27,28). An individual score is assigned to each gene based on the change in their expression relative to the sum of standard deviation and a fudge factor for repeated measurements for that gene. The score is in fact a t-statistic with an added fudge factor in the denominator. The purpose of the fudge factor is to prevent a large test statistic for genes with low variance. Genes carrying a score over a set threshold are identified as significant. The set of genes called significant is large or small depending on the threshold. SAM uses permutations to construct a null distribution for the t-values and estimating the proportion of significant genes identified by chance, which is termed as the false discovery rate (FDR). FDR is estimated by counting and averaging the number of false-positives over the permutations of the measurements. SAM is incorporated in caGEDA; a detailed description of SAM can be found at http://bioinformatics.upmc.edu/Help/SAM/SAMINFO.htm.

3.7. PPST Test

PPST is a test for detecting genes that exhibit a significant *number of samples* of one group that exhibit expression intensities that are beyond a certain percentile of the observed intensities of the samples in the other group *(24–26)*. Differentially expressed genes are generally reported as being either overexpressed or underexpressed in case or control samples. The PPST is capable of identifying genes that are differentially expressed in only a subset of samples in one group, which may have been missed by tests that compare population-level differences (means). In general, the search for differentially expressed genes should include the search for genes that are differentially expressed in a subset of patients to foster the transition toward individualized medicine.

For each gene, the number of samples in group A (e.g., HGA) was counted that had intensities above the 95th percentile of the intensities of group B (i.e., LGAs). This number is s_1. To this number is added the number of samples in group B that exhibit expression intensities below the 5th percentile of group A (s_2). These scores are calculated for all 1000 permutations and a null distribution for each gene is generated. Genes with $s_1 + s_2$ values beyond the sum $s_1 + s_2$ associated with a 5% type I error risk (gleaned from the null distribution resulting from permutation) are classified as overexpressed in HGAs. Similar scores are calculated for the opposite pattern (underexpression in HGA; $s_3 + s_4$) and compared with the $s_3 + s_4$ null distribution.

3.8. ABA Test

The ABA test identifies genes with two significant subsets with opposite expression differentials *(24–26)*. Genes that exhibit an unusual expression (ABA or BAB) patterns are likely to be missed by the tests that seek population-level biomarkers. Genes that have significant $s_1 + s_2$ or $s_3 + s_4$ scores are either over- or underexpressed in HGAs. Some genes can have both significant $s_1 + s_2$ and $s_3 + s_4$ scores, and are said to exhibit ABA (A > B > A) pattern. The PPST test is slightly reformulated to determine ABA patterns so the *number of occurrences* of ABA-type patterns becomes the statistic of interest. Under the ABA test, a gene is significant if and only if it is differentially expressed in both directions (i.e., simultaneously overexpressed and underexpressed in a significant number of samples).

3.9. Distance and Clustering

Once a set of genes has been identified as significant the samples are clustered in a "semi-supervised" mode because the user first identifies significant genes (feature selection) in a supervised manner, and the samples are classified using the retained genes as features *(29)*. The clustering algorithm does not use the sample label to enforce the clustering. We performed a variety of clustering algorithms to assess the importance of the known and unknown assumptions implied by each clustering method. Classification trees for each dataset were generated using distances measured by a variety of distance metrics to assess the robustness of the various gene lists to known and unknown assumptions implied by each distance metric.

3.10. Computational Validation

The true validation we have performed is a result of the discovery of genes differentially expressed in two separate populations using data originating in two separate laboratories. Within-dataset computational validation of the results using leave-one-out validation was also performed. In leave-one-out validation, samples are removed, one at a time, and the feature set is determined using a test applied to the remaining $n - 1$ samples. These features are then applied to make a prediction on the placement of the sample left out. The procedure is repeated for all samples, and a score (usually the proportion of correct predictions) is tallied. Leave-one-out validation uses $n - 1$ samples as a training set, predicts on the sample left out, and the score of $1 - P$ (correct inference) leads to a classification error rate *(30)*.

3.11. Validation

The previously mentioned tests were applied identically to the Godard dataset *(23)* for comparison in order to validate the derived marker gene list from the Khatua *(14)* and van den Boom *(22)* studies.

4. Results

4.1. Visualization of Data Quality

The quality of datasets was judged by observing the box and whisker plots for both datasets **(Fig. 1)**. The datasets exhibit similar distributions and therefore no significant variability among the samples was detected. The data quality parameters are within acceptable limits **(Table 1)**. Overall, the two datasets exhibit high among sample all-gene distributions, and appeared to require no further normalization.

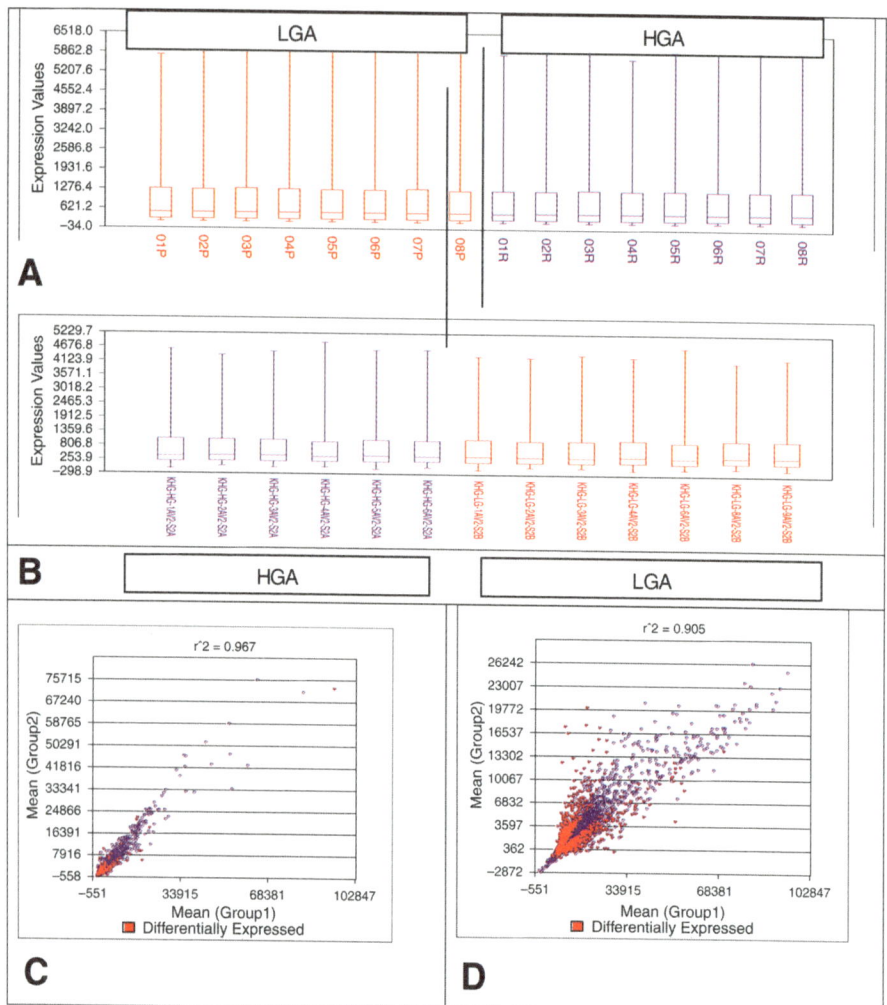

Fig. 1. Box and Whisker plots for the three datasets. Plots for van den Boom et al. *(22)* (**A** and **C**) and Khatua et al. *(14)* (**B** and **D**) present the quality of the data. The *x*-axis represents the samples and *y*-axis represents the expression intensities. Samples of group one are high grade (blue) and those of group two are low-grade (red) astrocytomas. In an ideal experiment, the median (or mean), first standard deviation, upper and lower second medians, and the 95th percentile should be approximately the same across all arrays. HGA, high-grade astrocytoma; LGA, low-grade astrocytoma.

4.2. Differentially Expressed Genes

The K and V datasets were reanalyzed using pooled variance *t*-test and J5 test (both in conjunction with jackknife), and the PPST and ABA tests. The numbers of genes found to be significant at the 5% type 1 error rate are summarized in

Table 1
Data Quality Metric

Parameters	K dataset	V dataset	Permissible limit
Between-mean group correlation (r^2)	0.974	0.967	>0.95
Among array coefficient of variation	0.018	0.014	<0.30
Confounding index	1.017	1.018	<1.1

Table 2. A total of nine genes were identified as differentially expressed between LGAs and HGAs in three datasets; of these, four genes were consistently overexpressed by HGAs, and five genes were downregulated in HGAs (**Table 3**). The overexpressed genes are *RPL37A*, *RPS18*, *RPS2*, and *RPS8* all encoding ribosomal proteins, and the five downregulated genes are *SORL1*, *APOD*, *SPOCK2*, *PRSS11*, and *ID3*. These nine genes were differentially expressed in the validation dataset *(23)* under all tests.

4.3. Classification of Tumor Samples

For all tests examined, the LGA and HGA samples of the K dataset clustered onto separate branches of the classification, indicating that the gene expression patterns of the selected genes are more alike within one tumor class than between tumor classes (**Fig. 2**). Leave-one-out cross-validation classification error was low (0) in the K dataset. In the V dataset, a correct classification was obtained under the pooled variance *t*-test at 40 genes, but leave-one-out cross-validation classification error ranged from 25 to 60%. This may suggest that other genes in addition to the 8 or 18 we have focused on may also be clinically significant, perhaps in unique ways for each patient.

4.4. Chromosomal Location of Dysregulated Genes

The chromosomal locations of the differentially expressed genes are shown in **Fig. 3**. The gene expression results lead us to speculate that there may be dysregulation of the genes because of cytogenetic alterations that have been previously described at loci for chromosomes 1 and 10.

5. Discussion

Our reanalysis of astrocytoma expression profiles from three independent datasets using novel bioinformatics tools reveals new and inherently distinct patterns of gene expression commonly shared among HGAs and LGAs, regardless of patient age. This is the first report of astrocytoma progression-associated marker genes found to be consistently differentially expressed in separate microarray studies of astrocytoma- spanning tumors from early childhood to

Table 2
Genes Found to be Significant in Three Datasets Using the Five Tests

Test	Threshold	K dataset	V dataset
Pooled *t*-test	$\alpha = 0.01$	217	39
SAM	$\Delta = 0.6$	874	53
J5 test	T = 4.0	331	847
PPST	$\alpha = 0.05$	1281	2304
ABA	$\alpha = 0.05$	51	69

Note: In pooled *t*-test with Jackknife, there were six overlapping genes (three overexpressed and three underexpressed in HGAs), whereas J5 with jackknife results showed five overlapping genes (two overexpressed and three underexpressed in HGAs). In the PPST test we observed five overlapping genes (four overexpressed and one underexpressed in HGAs) and ABA test gave out five overlaps in the lists of differentially expressed genes in K and V datasets.

Table 3
Genes Identified as Differentially Expressed in Three Datasets by the Four Tests

Symbol	Gene name	GenBank accession	Chromosomal location	Protein function
Overexpressed in high-grade astrocytomas				
RPS8	Ribosomal protein S8	X67247	1p34.1	Unknown
RPS2	Ribosomal protein S2	AB007147	16p13.3	Unknown
RPS18	Ribosomal protein S18	X69150	6p21.3	Unknown
RPL37A	Ribosomal protein L37A	L11567	5p	Unknown
Underexpressed in high-grade astrocytomas				
APOD	Apolipoprotein D	J02611	3q26.2	Lipid metabolism & transport
PRSS11	Protease, serine, 11	AF157623	10q25.3	Cell growth regulation
ID3	Inhibitor of DNA binding 3	X73428	1p36.13	Transcription corepressor
SORL1	Sortilin-related receptor 1	Y08110	11q23.2	Lipid metabolism & transport
KIAA0275	SPOCK2	D87465	10q21	Cell differentiation

Data set	Dendrogram
Khatua et.al., 2003, 'K data set'	
VanDen Boom et al., 2003, 'V data set'	

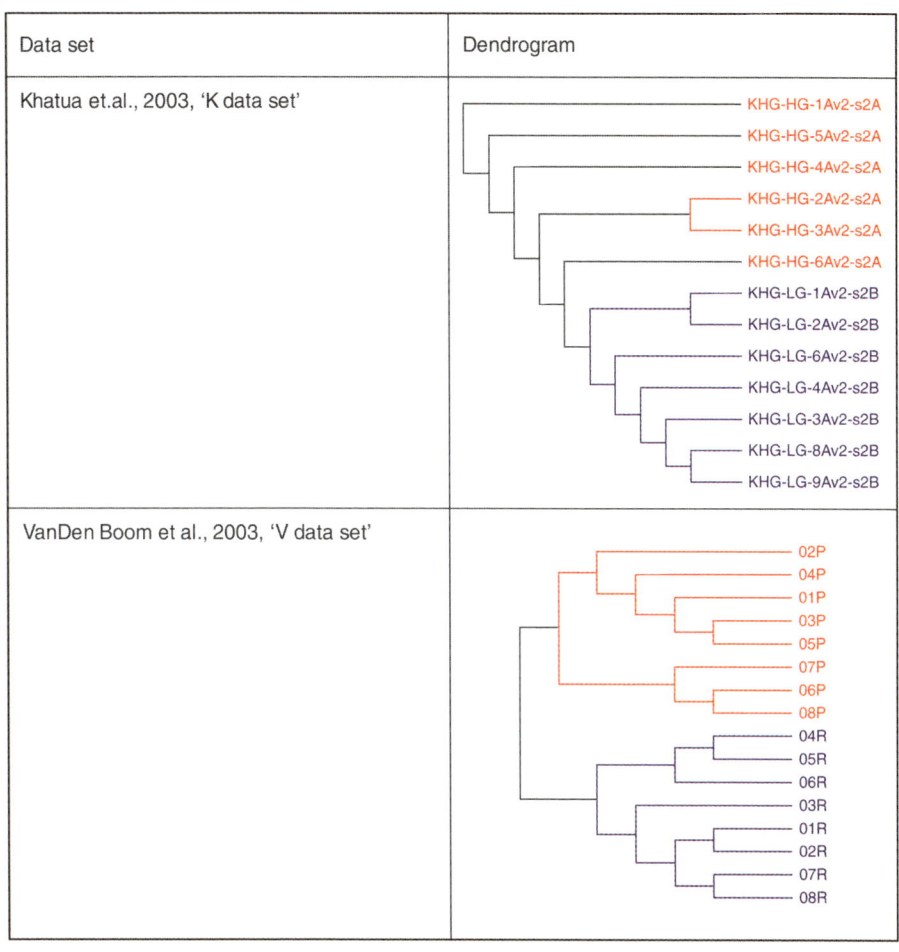

Fig. 2. Dendrograms of samples from the two datasets *(14,22)* (van den Boom et al. and Khatua et al.). High-grade astrocytomas in both groups are shown in red, low-grade astrocytomas are shown in blue.

late adulthood. This may have profound implications because common patterns of gene expression changes across these datasets suggests that there exist universal markers of malignant astrocytoma development and progression, which may indeed be novel candidates for therapeutic intervention in both children and adults. Further independent investigation is needed to confirm whether these marker genes are functionally relevant, rather than merely associative, and whether they represent a set of genes core to the malignant progression of all astrocytomas, regardless of the patient's age at diagnosis.

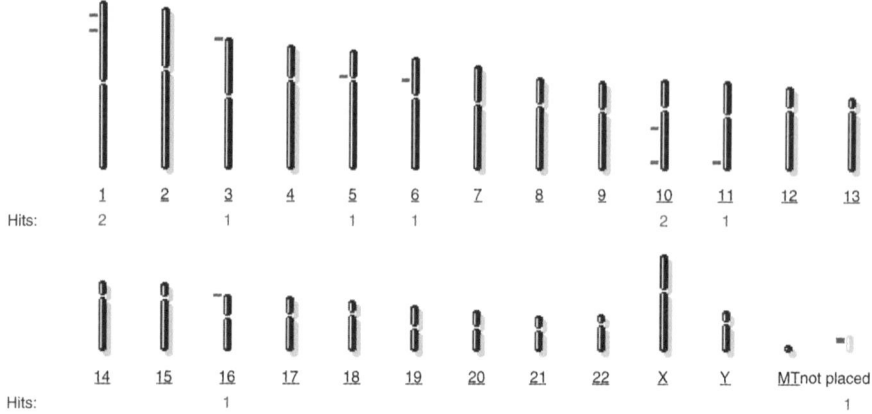

Fig. 3. Chromosomal locations of the differentially expressed genes. Chromosome map from NCBI Map Viewer (http://www.ncbi.nlm.nih.gov/mapview) shows the cytogenetic locations of nine differentially expressed genes. The red bars indicate locations of differentially expressed genes. High-grade astrocytomas showed overexpression of *RPS8* (1p), *RPL37A* (5p), *RPS18* (6p) and *RPS2* (16p) and downregulation of *ID3* (1p), *APOD* (3q), *PRSS11 SPOCK2* (10q), and *SORL1* (11q).

Interestingly, all of the genes upregulated in HGAs are highly conserved genes that encode for ribosomal proteins. The mammalian functions of these proteins are largely unknown, except for RPS2, which in a very recent report appears to act as a substrate for arginine methyltransferase 3, which catalyzes the formation of dimethylarginine *(31)*. Increased expression of RPS2 has been reported in murine liver tumors and is associated with hepatocyte proliferation *(32)*.

The consistently downregulated genes by HGA have more evidence in the literature for their functional roles and possible interrelatedness to malignant astroctyoma progression (**Fig. 4**). These include *Id3, PRSS11, SPOCK2, SORL1,* and *APOD. Id* genes encode proteins that interfere with transcriptional activation and are required to maintain neuronal differentiation and invasiveness of the vasculature for angiogenesis *(33)*. Id3 protein has been previously demonstrated in endothelial cells of astrocytic tumor blood vessels and its expression correlates with tumor vascularity *(34)*. Downregulation of *Id3* by HGA in this study suggests a more potent role of Id3 in promoting dedifferentiation from LGA to HGA rather than maintaining angiogenesis.

Fig. 4. *(Opposite page)*   Schema of functional relatedness of the differentially expressed genes. Functional gene ontologies and construction of schema showing inter-relatedness of gene functions was generated using GeneInfo Viz: Constructing and Visualizing Gene Relation Networks (http://genenet1.utmem.edu/geneinfoviz/search.php).

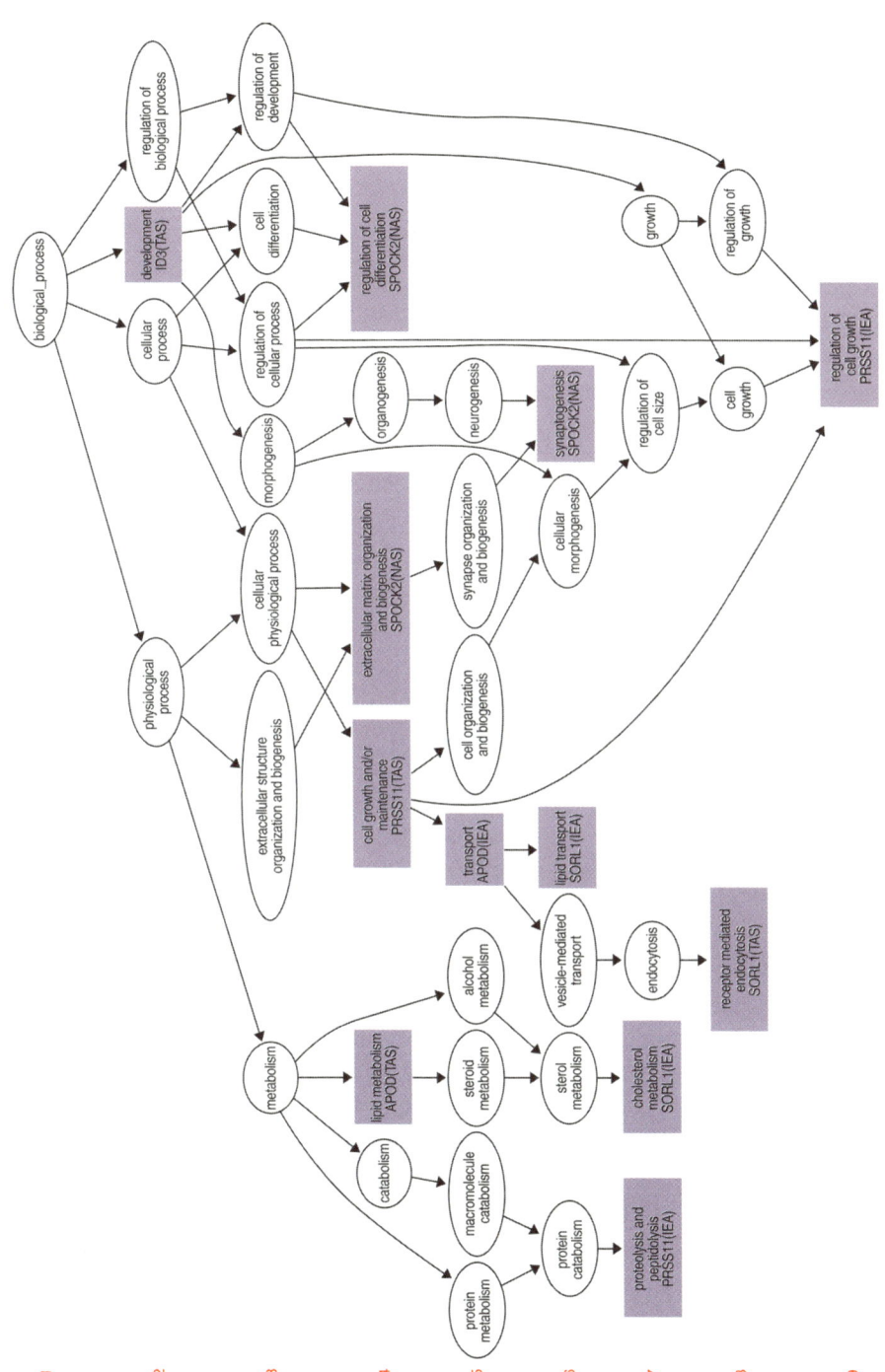

Fig. 4.

215

The former concept is supported by studies showing that forced expression of Id3 in erythroid progenitor cells promotes erythroid differentiation and down-regulation of Id3 by retinoic acid-induced differentiation of neuroblastoma cells *(35,36)*. Similarly, treatments with phorbol ester, another inducer of neuroblastoma cell differentiation, also resulted in coordinated downregulation of *Id3* gene expression, underscoring the significant role of *Id* genes in differentiation *(36)*. Finally, downregulation of *Id3* has also been observed in primary ovarian tumors and was detected in only about 30% cases *(37)*.

PRSS11 encodes the serine protease HtrA1, a candidate tumor suppressor implicated in protease-induced cell death *(38)*. Downregulation of *PRSS11* has been observed in ovarian cancer as well as during melanoma progression *(39,40)*. Furthermore, microarray analysis of metastatic melanoma cells identified downregulation of *PRSS11* compared with nonmetastatic melanoma cells and overexpression of *PRSS11* resulted in the inhibition of melanoma growth *(41)*. Differential expression of *PRSS11* has also been observed between highly migratory U373MG glioma cells compared with slower moving primary glioblastoma cells *(42)*. Taken together, these findings implicate *PRSS11* as a potential tumor suppressor gene in astrocytoma as well.

KIAA0275 (SPOCK2, testican 2) encodes a calcium-binding proteoglycan primarily expressed in the brain, but to date there is very little information regarding its function. SPOCK2 has been recently shown to remove inhibition of MT1- or MT3-MMP-mediated pro-MMP-2 activation by other testican family members *(43)*. This would appear to be counterproductive to promoting protease-mediated invasion. However, as in our study, expression levels of all testican family members in astrocytomas were found to decrease as tumor grade increases *(43)*. These findings would appear to indicate an alternative, and as yet unknown, functional role for SPOCK2.

SORL1 (SorLA/LR11) encodes a recently identified member of the LDL receptor superfamily, which is broadly expressed in the nervous system and functions as a neuronal apolipoprotein E receptor. Unlike the other downregulated genes, a connection between *SORL1* and tumor progression has yet to be demonstrated. Of note is that significant and consistent loss of the LR11 protein in histologically normal-appearing neurons has been observed in Alzheimer patients *(44)*. LR11 has also been shown to interact with the plasminogen-activating system and PDGF-BB signaling, which has potential implications for astrocytoma progression *(45)*.

APOD encodes a human plasma protein, apolipoprotein D, which belongs to the lipocalin superfamily. Our results showing downregulation of *APOD* across all age groups with malignant astrocytoma is further supported by other reports that have showed *APOD* as a marker for low-grade, noninfiltrating astrocytomas *(46,47)*. Moreover, in human breast cancer cells, increased expression of *APOD* was accompanied by an inhibition of cell proliferation and a progression through a more differentiated phenotype *(48)*. Likewise, apo-D secretion was inversely correlated to cell proliferation and cell density in human prostate cancer cells *(49)*.

The most frequent cytogenetic changes observed in astrocytomas have been losses of loci on 9p, 10, and 22 along with gains on 7, 19, and 20. The chromosomal locations of the differentially expressed genes identified by our approach reside on chromosomes 1, 3, 6, 7, 10, 11, and 16; however, only downregulation of *PRSS11*, located on chromosome 10, would possibly follow previously described changes. Thus, dysregulation of these genes would appear to be secondary to factors other than cytogenetic alterations.

6. Conclusion

The genes we have found to be differentially expressed are robust to the test used in two different datasets generated by two separate laboratories. The genes can distinguish between low-grade and high-grade astrocytomas, independent of age of the patient at diagnosis. These results imply that these genes may indeed be universal targets and hence most appropriate for therapeutic intervention in all malignant astrocytomas. However, the functional roles of these genes in astrocytomas need confirmation, and further studies are needed to characterize their roles in the regulatory pathways. Larger studies are also warranted to ensure that the associated genes maintain their patterns of expression observed in this study.

Acknowledgments

We are grateful to the participants in the Khatua study and Dr. Hanash and other researchers involved in the van den Boom study for making their respective datasets available to the public. This research was funded by a grant from the Claude Worthington Benedum Foundation to Dr. Michael Becich and Dr. Ronald Herberman. These funds helped establish the Benedum Oncology Informatics Center and the Oncology Informatics program at UPCI. The specific funds used for this project were from Dr. Lyons-Weiler's faculty recruitment package.

All the tests for differentially expressed genes and cluster analysis can be applied to any dataset on the Cancer Gene Expression Data Analysis (caGEDA) web application at http://bioinformatics.upmc.edu/GE2/GEDA.html. It is part of the University of Pittsburgh Bioinformatics Web Application Collection (http://bioinformatics.upmc.edu), which runs on a JAVA web server and is available for public usage. Researchers can upload their data and run a wide range of analyses options available in caGEDA, or download and install caGEDA locally.

References

1. Kleihues, P. and Cavenee, W. K. (ed.) (2000) *Pathology & Genetics. Tumours of the Nervous System.* IARC Lyon, France.
2. Batchelor, T. T., Betensky, R. A., Esposito, J. M., et al. (2004) Age-dependent prognostic effects of genetic alterations in glioblastoma. *Clin. Cancer Res.* **10,** 228–233.
3. Simmons, M. L., Lamborn, K. R., Takahashi, M., et al. (2001) Analysis of complex relationships between age, p53, epidermal growth factor receptor, and survival in glioblastoma patients. *Cancer Res.* **61,** 1122–1128.
4. Sung, T., Miller, D. C., Hayes, R. L., Alonso, M., Yee, H., and Newcomb, E. W. (2000) Preferential inactivation of the p53 tumor suppressor pathway and lack of EGFR amplification distinguish de novo high grade pediatric astrocytomas from de novo adult astrocytomas. *Brain Pathol.* **10,** 249–259.
5. Wessels, P. H., Weber, W. E., Raven, G., Ramaekers, F. C., Hopman, A. H., and Twijnstra, A. (2003) Supratentorial grade II astrocytoma: biological features and clinical course. *Lancet Neurol.* **2,** 395–403.
6. Shapiro, J. R. (2002) Genetic alterations associated with adult diffuse astrocytic tumors. *Am. J. Med. Genet.* **115,** 94–201.
7. Fan, X., Munoz, J., Sanko, S. G., and Castresana, J. S. (2002) PTEN, DMBT1, and p16 alterations in diffusely infiltrating astrocytomas. *Int. J. Oncol.* **21,** 667–674.
8. Biernat, W., Tohma, Y., Yonekawa, Y., Kleihues, P., and Ohgaki, H. (1997) Alterations of cell cycle regulatory genes in primary (de novo) and secondary glioblastomas. *Acta. Neuropathol. (Berl).* **94,** 303–309.
9. Watanabe, K., Tachibana, O., Sato, K., Yonekawa, Y., Kleihues, P., and Ohgaki, H. (1996) Overexpression of the EGF receptor and *p53* mutations are mutually exclusive in the evolution of primary and secondary glioblastomas. *Brain Pathol.* **6,** 217–224.
10. Kleihues, P. and Ohgaki, H. (1999) Primary and secondary glioblastomas: from concept to clinical diagnosis. *Neuro-oncol.* **1,** 44–51.
11. Hegi, M. E., zur Hausen, A., Rüedi, D., Malin, G., and Kleihues, P. (1997) Hemizygous or homozygous deletion of the chromosomal region containing the p16INK4a gene is associated with amplification of the EGF receptor gene in glioblastomas. *Int. J. Cancer* **73,** 57–63.
12. Hermanson, M., Funa, K., Koopmann, J., et al. (1996) Association of loss of heterozygosity on chromosome 17p with high platelet-derived growth factor alpha receptor expression in human malignant gliomas. *Cancer Res.* **56,** 164–171.
13. Kraus, J. A., Felsberg, J., Tonn, J. C., Reifenberger, G., and Pietsch, T. (2002) Molecular genetic analysis of the TP53, PTEN, CDKN2A, EGFR, CDK4 and MDM2 tumour-associated genes in supratentorial primitive neuroectodermal tumours and glioblastomas of childhood. *Neuropathol. Appl. Neurobiol.* **28,** 325–333.
14. Khatua, S., Peterson, K. M., Brown, K. M., et al. (2003) Overexpression of the EGFR/FKBIP/HIF-2a patahway identified in chilhood astrocytomas by angiogenesis gene profiling. *Cancer Res.* **63,** 1865–1870.

15. Bredel, M., Pollack, I. F., Hamilton, R. L., and James, C. D. (1999) Epidermal growth factor receptor expression and gene amplification in high-grade non-brainstem gliomas of childhood. *Clin. Cancer Res.* **5,** 1786–1792.

16. Pollack, I. F., Finkelstein, S. D., Woods, J., et al. (2002) Expression of p53 and prognosis in children with malignant gliomas. *N. Engl. J. Med.* **346,** 420–427.

17. Golub, T. R., Slonim, D. K., Tamayo, P., et al. (1999) Molecular classification of cancer: class discovery and class prediction by gene expression monitoring. *Science* **286,** 531–537.

18. Perou, C. M., Sorlie, T., Eisen, M. B., et al. (2000) Molecular portraits of human breast tumours. *Nature* **406,** 747–752.

19. Bittner, M., Meltzer, P., Chen, Y., et al. (2000) Molecular classification of cutaneous malignant melanoma by gene expression profiling. *Nature* **406,** 536–540.

20. Alon, U., Barkai, N., Notterman, D. A., et al. (1999) Broad patterns of gene expression revealed by clustering analysis of tumor and normal colon tissues probed by oligonucleotide arrays. *Proc. Natl. Acad. Sci. USA* **96,** 6745–6750.

21. Alizadeh, A., Eisen, M., Davis, R. E., et al. (2000) Distinct types of diffuse large B-cell lymphoma identified by gene expression profiling. *Nature* **403,** 503–511.

22. van den Boom, J., Wolter, M., Kuik, R., et al. (2003) Characterization of gene expression profiles associated with astrocytoma progression using ologinucleotide-based microarray analysis and real-time reverse transcription-polymerase chain reaction. *Am. J. Path.* **163,** 1033–1043.

23. Godard, S., Gatz, G., Delorenzi, M., et al. (2003) Classification of human astrocytic astrocytomas on the basis of gene expression: A correlated group of genes with angiogenic activity emerges as a strong predictor of subtypes. *Cancer Res.* **63,** 6613–6625.

24. Lyons-Weiler, J., Patel, S., Becich, M. J., and Godfrey, T. E. (2004) Tests for finding complex patterns of differential expression in cancers: towards individualized medicine. *BMC Bioinformatics* **12,** 110.

25. Bhattacharya, S., Long, D., and Lyons-Weiler, J. (2003) Overcoming confounded controls in the analysis of gene expression data from microarray experiments. *Appl. Bioinformatics* **2,** 197–208.

26. Lyons-Weiler, J., Patel, S., and Bhattacharya, S. A. (2003) Classification-based machine learning approach for the analysis of genome-wide expression data. *Genome Res.* **13,** 503–512.

27. Pan, W. (2002) A comparative review of statistical methods for discovering differentially expressed genes in replicated microarray experiments. *Bioinformatics* **18,** 546–554.

28. Tusher, V. G., Tibshirani, R., and Chu, G. (2001) Significance analysis of microarrays applied to the ionizing radiation response. *Proc. Natl. Acad. Sci. USA* **98,** 5116–5121.

29. Shannon, W., Culverhouse, R., and Duncan, J. (2003) Analyzing microarray data using cluster analysis. *Pharmacogenomics* **4,** 41–52.

30. Hastie, T., Tibshirani, R., Eisen, M. B., et al. (2000) 'Gene shaving' as a method for identifying distinct sets of genes with similar expression patterns. *Genome Biol.* **1,** RESEARCH0003.

220   *MacDonald et al.*

31. Swiercz, R., Person, M. D., and Bedford, M. T. (2005) Ribosomal protein S2 is a substrate for mammalian PRMT3 (protein arginine methyltransferase 3). *Biochem. J.* **386,** 85–91.
32. Kowalczyk, P., Woszczynski, M., and Ostrowski, J. (2002) Increased expression of ribosomal protein S2 in liver tumors, posthepactomized livers, and proliferating hepatocytes in vitro. *Acta. Biochim. Pol.* **49,** 615–624.
33. Lyden, D., Young, A. Z., Zagzag, D., et al. (1999) Id1 and Id3 are required for neurogenesis, angiogenesis and vascularization of tumour xenografts. *Nature* **401,** 670–677.
34. Vandeputte, D. A., Troost, D., Leenstra, S., et al. (2002) Expression and distribution of id helix-loop-helix proteins in human astrocytic tumors. *Glia* **38,** 329–338.
35. Deed, R. W., Jasiok, M., and Norton, J. D. (1998) Lymphoid-specific expression of the Id3 gene in hematopoietic cells. Selective antagonism of E2A basic helix-loop-helix protein associated with Id3-induced differentiation of erythroleukemia cells. *J. Biol. Chem.* **273,** 8278–8286.
36. Lopez-Carballo, G., Moreno, L., Masia, S., Perez, P., and Barettino, D. (2002) Activation of the phosphatidylinositol 3-kinase/Akt signaling pathway by retinoic acid is required for neural differentiation of SH-SY5Y human neuroblastoma cells. *J. Biol. Chem.* **277,** 25,297–25,304.
37. Arnold, J. M., Mok, S. C., Purdie, D., and Chenevix-Trench, G. (2001) Decreased expression of the Id3 gene at 1p36.1 in ovarian adenocarcinomas. *Br. J. Cancer* **84,** 352–359.
38. Gray, C. W., Ward, R. V., Karran, E., et al. (2000) Characterization of human HtrA2, a novel serine protease involved in the mammalian cellular stress response. *Eur. J. Biochem.* **267,** 5699–5710.
39. Shridhar, V., Sen, A., Chien, J., et al. (2002) Identification of underexpressed genes in early- and late-stage primary ovarian tumors by suppression subtraction hybridization. *Cancer Res.* **62,** 262–270.
40. Baldi, A., De Luca, A., Morini, M., et al. (2002) The HtrA1 serine protease is down-regulated during human melanoma progression and represses growth of metastatic melanoma cells. *Oncogene* **21,** 6684–6688.
41. Baldi, A., Battista, T., De Luca, A., et al. (2003) Identification of genes down-regulated during melanoma progression: a cDNA array study. *Exp. Dermatol.* **12,** 213–218.
42. Tatenhorst, L., Senner, V., Puttmann, S., and Paulus, W. (2004) Regulators of G-protein signaling 3 and 4 (RGS3, RGS4) are associated with glioma cell motility. *J. Neuropathol. Exp. Neurol.* **63,** 210–222.
43. Nakada, M., Miyamori, H., Yamashita, J., and Sato, H. (2003) Testican 2 abrogates inhibition of membrane-type matrix metalloproteinases by other testican family proteins. *Cancer Res.* **63,** 3364–3369.
44. Scherzer, C. R., Offe, K., Gearing, M., et al. (2004) Loss of apolipoprotein E receptor LR11 in Alzheimer disease. *Arch. Neurol.* **61,** 1200–1205.
45. Gliemann, J., Hermey, G., Nykjaer, A., Petersen, C. M., Jacobsen, C., and Andreasen, P. A. (2004) The mosaic receptor sorLA/LR11 binds components of the

plasminogen-activating system and platelet-derived growth factor-BB similarly to LRP1 (low-density lipoprotein receptor-related protein), but mediates slow internalization of bound ligand. *Biochem. J.* **381**, 203–212.

46. Gutmann, D. H., James, C. D., Poyhonen, M., et al. (2003) Molecular analysis of astrocytomas presenting after age 10 in individuals with NF1. *Neurology* **61**, 1397–1400.

47. Hunter, S., Young, A., Olson, J., et al. (2002) Differential expression between pilocytic and anaplastic astrocytomas: identification of apolipoprotein D as a marker for low-grade, non-infiltrating primary CNS neoplasms. *J. Neuropathol. Exp. Neurol.* **61**, 275–281.

48. Lopez-Boado, Y. S., Tolivia, J., and Lopez-Otin, C. (1994) Apolipoprotein D gene induction by retinoic acid is concomitant with growth arrest and cell differentiation in human breast cancer cells. *J. Biol. Chem.* **269**, 26,871–26,878.

49. Sugimoto, K., Simard, J., Haagensen, D. E., and Labrie, F. (1994) Inverse relationships between cell proliferation and basal or androgen-stimulated apolipoprotein D secretion in LNCaP human prostate cancer cells. *J. Steroid Biochem. Mol. Biol.* **51**, 167–174.

14

Interpreting Microarray Results With Gene Ontology and MeSH

John D. Osborne, Lihua (Julie) Zhu, Simon M. Lin, and Warren A. Kibbe

Summary

Methods are described to take a list of genes generated from a microarray experiment and interpret these results using various tools and ontologies. A workflow is described that details how to convert gene identifiers with SOURCE and MatchMiner and then use these converted gene lists to search the gene ontology (GO) and the medical subject headings (MeSH) ontology. Examples of searching GO with DAVID, EASE, and GOMiner are provided along with an interpretation of results. The mining of MeSH using high-density array pattern interpreter with a set of gene identifiers is also described.

Key Words: Microarray; GO; MeSH; protocol; DAVID; HAPI; SOURCE; MatchMiner; Interpret.

1. Introduction

After identifying a list of differentially expressed genes, researchers often ask, "what is known about the biological function of these genes? What biochemical properties are known for the encoded proteins? What functional categories/pathways/networks do these genes belong to? What diseases are these genes associated with?" Answers to these questions can be directly addressed or inferred by looking at names of the gene, inspecting their database entries, or reading related literature. These time-consuming and error-prone steps can be facilitated by a formal computational approach using ontologies. An ontology is a controlled vocabulary. It has a formal structure that relates the concepts represented by each term in the ontology with other terms in the same ontology *(1)*. Each concept, such as "induction of apoptosis," is coded with an identifier. Further, each relationship, such as "apoptosis" (a kind of "cell death"), is also coded so that database and computational inference can be done with them.

Gene ontology (GO; *[2]*) is one of the vocabularies of open biomedical ontologies and it is designed to describe knowledge of the biological process, the

From: *Methods in Molecular Biology, vol. 377, Microarray Data Analysis: Methods and Applications*
Edited by: M. J. Korenberg © Humana Press Inc., Totowa, NJ

molecular function of gene products, and the localization/compartmentaliza-tion/aggregation of gene products, much as medical subject headings (MeSH) ontology is designed to describe medical findings and implications. GO can be used to *annotate* the biological knowledge of a gene or gene product, just as MeSH can be used to annotate medical literature (*see* **Note 1**).

Formally, GO is comprised of three separate "knowledge trees" describing biological process, biochemical function, and cellular location/compartmental-ization. Each tree is a directed acyclic graph, with the property that the path from any node (term) to the root term (e.g., biological process) must be true. These properties are key to many of the computationally important uses of GO in knowledge discovery.

MeSH is a controlled vocabulary developed by National Library of Medicine for indexing, cataloging, and retrieving medical literatures. MeSH contains about 22,568 descriptors and their relationships among each other. MeSH descriptors are organized in 15 categories and each category is further divided into subcategories. Within each subcategory, descriptors are organized as a tree structure with the most general descriptors on the top and the most specific descriptors as leaves. Each MeSH descriptor appears in one or more branches in the trees. The disease category in MeSH complements GO biological processes and molecular function for describing a gene. Database for annotation, visuali-zation, and integrated discovery (DAVID), Gene Ontology Miner (GOMiner), and Expression Analysis Systematic Explorer (EASE) allows one to group genes according to GO biological process and molecular functions, whereas high-density array pattern interpreter (HAPI) *(3)* allows one to relate genes according to disease-related MeSH descriptors.

The general workflow in this chapter is described as follows (*see* **Note 2**). Retrieving the GO annotation for each gene identified in a microarray experi-ment is facilitated through database lookups using appropriate gene identifiers, such as Genbank, LocusLink, or Unigene. Converting the identifiers to a stan-dard, interoperable identifier is a prerequisite to use many ontology analysis programs and is sometimes the first step of the ontology analysis illustrated in **Fig. 1**. The starting list of genes (identified by any identifier) should be con-verted by a program such as SOURCE *(4)* or MatchMiner *(5)* to a set of iden-tifiers used by GOMiner or DAVID if necessary. The resulting list of identifiers can then processed by DAVID or GOMiner to retrieve annotation lookups for each gene, and the results are clustered according to GO. Clusters with lower p-values may indicate biologically important areas of functionality or biologi-cal process for the gene list. The same gene list can also be analyzed through HAPI to search for conceptual clusters according to MeSH.

2. Materials

Software tools and databases are listed in **Table 1** (*see* **Note 3**).

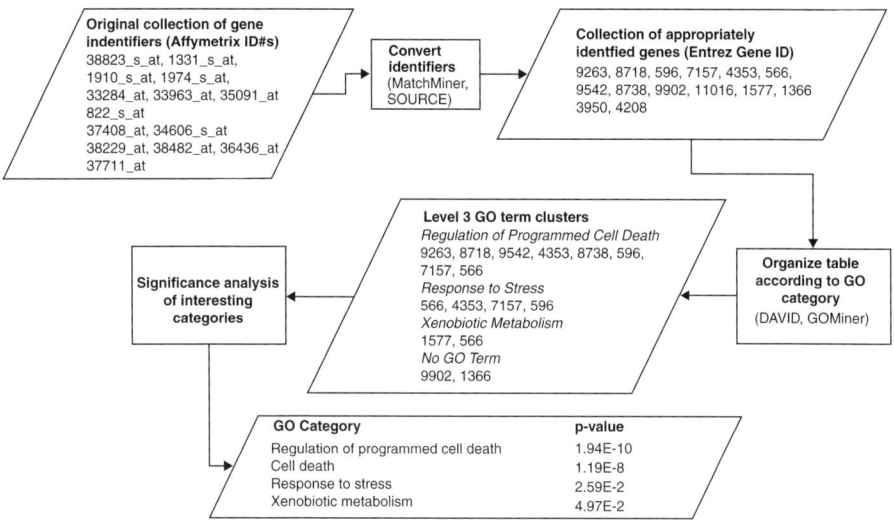

Fig. 1. Using functional annotation with GO to interpret a list of genes.

3. Methods

3.1. Conversion of Identifiers With SOURCE

Because different microarray platforms and different public genomic databases might use different gene product identifiers for a given gene, there is a need to map or translate between major gene product identifiers. For example, Affymetrix oligonucleotide microarray uses Affymetrix Probe ID and one frequently needs to translate that identifier to a GenBank accession number, Unigene name, and symbol. Customized cDNA array usually uses IMAGE cloneID and one needs to translate it to GenBank accession, Unigene name, and symbol. Both SOURCE and MatchMiner can be used to convert from a gene product identifier in one database to a gene product identifier in a different source database including those for human, and mice and rat gene and gene products. **Figure 2** is a screen shot of SOURCE interface that has translated a dbEST Clone ID into a Unigene cluster ID, Unigene Name, Unigene Symbol, LocusLink ID, and UniProt ID.

Table 2 contains the results from running SOURCE with cloneID of 1568950, 4524419, and 1240116 as input; Unigene cluster ID, Unigene Name, Unigene Symbol, LocusLink ID, and UniProt ID representative as output selection.

MatchMiner can be used to translate Affymetrix probe ID to the previously mentioned gene identifiers. One of the common uses of MatchMiner is to convert Affymetrix probe ID to a gene symbol that is one of the accepted gene identifiers for GoMiner to perform GO classification and over-representation analysis.

Table 1
Websites for GO and MeSH Analysis

Website	URL	Description
Gene Ontology Consortium	http://www.geneontology.org	Consortium for maintaining GO and annotating genomes with GO
AmiGO	http://www.godatabase.org/cgi-bin/amioO/go.cgi	Browsing and searching GO
OBO	http://obo.sourceforge.net/	Open biological ontologies
MeSH	http://www.nlm.nih.gov/mesh/meshhome.html	NLM's biomedical terminology thesaurus
SOURCE	http://source.stanford.edu/cgi-bin/source/sourceSearch	Batch conversion of gene identifiers
MatchMiner	http://discover.nci.nih.gov/matchminer/html/index.jsp	Batch conversion of gene identifiers
NCBI Entrez	http://www.ncbi.nlm.nih.gov/	An integrated search and retrieval system at the NCBI for major genomic databases and literature
DAVID/EASE	http://david.niaid.nih.gov/david/ease.htm	Batch extraction of GO annotations, conversion of gene identifiers, and statistical analysis of significant GO terms
GoMiner	http://discover.nci.nih.gov/gominer/	Interprets conceptual similarities of a group of genes with GO
HAPI	http://array.ucsd.edu/hapi/	Interprets conceptual similarities of a group of genes with MeSH
Bioconductor	http://www.bioconductor.org/	Statistical analysis of microarray results using R programming language (see **Note 3**)

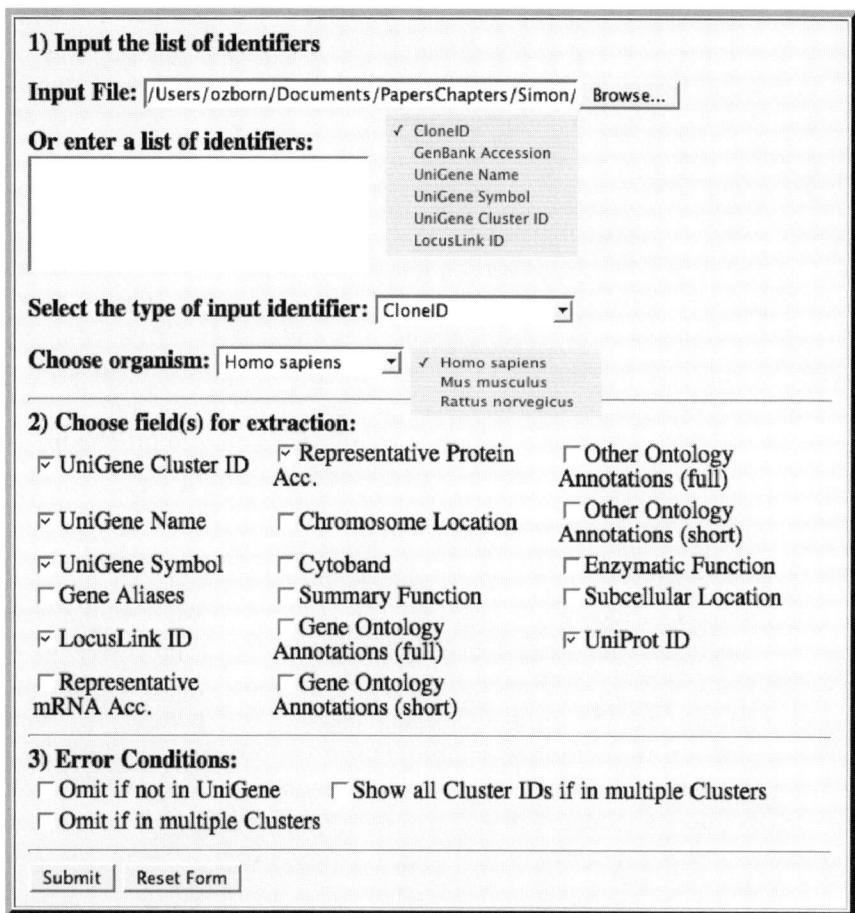

Fig. 2. Interface for conversion of identifiers by SOURCE.

Table 2
Conversion of Identifiers With SOURCE

CloneID	UGCluster	Name	Symbol	LLID	LLRepProtAcc	UniProt
1568950	Hs.513915	Claudin 7	CLDN7	1366	NP_001298	O95471
4524419	Hs.408312	Tumor protein p53	TP53	7157	NP_000537	P04637
1240116	Hs.408515	Neuregulin 2	NRG2	9542	NP_053588	O14511

SOURCE is currently using UniGene as the central database to which all other databases are linked. Therefore, a gene of interest must be in UniGene in order for data to be available for it. To exclude the identifiers not in any UniGene cluster, one can check the box "Omit if not in UniGene" in the "Error

Conditions" section of the form. Similarly, to exclude the identifiers in multiple UniGene clusters, one can check the box "Omit if in multiple Clusters." To include the identifiers in multiple UniGene clusters, one can check the box "Show all Cluster IDs if in multiple Clusters."

3.2. Browsing GO

An ontology is usually a hierarchical structure similar to the "table of contents" in a book. As mentioned earlier, GO is organized into a structure known as a directed acyclic graph (DAG). Each ontology term is encoded with a unique identifier to precisely specify the concept and prevent it from being confused with similar terms. For instance, the GO term "apoptosis" is assigned the identifier GO:0006915, which is a special case of its parent term of "programmed cell death." The AMIGO browser is an easy way to browse the GO ontology, and a screenshot is shown in **Fig. 4** of **Subheading 3.4**.

3.3. Retrieving GO Annotations of a Gene

GO is an international standard for annotating the biological function of genes and gene products, including cellular components (where—location of the event), molecular functions (what—physical activity), and biological processes (why— biological goals). Instead of using free text to describe the function(s) of a gene product, GO can be used to annotate very precisely the published literature describing the function of the gene product *(6)*. For example, the function of the *bax* gene can be either described verbally **(Fig. 3A)** or using GO **(Fig. 3B)**. The use of ontology renders further analysis easier because well-defined concepts can be parsed and linked to associated terms easily. As we discuss later, the use of a standard vocabulary, and even better, the use of GO enables comparative studies between experiments and enables multiple labs working on similar processes in different systems to compare results. For example, apoptosis can be found as a significant process of a certain type of cancer in both clinical samples and mouse models studied by different labs.

GO annotations can be retrieved as needed from a variety of different programs including the NCBI Entrez database (that is useful when browsing for a single gene), third party software like DAVID/EASE (useful for retrieving annotations from a list), or directly from the GO consortium. The GO database is easy to set up and use locally.

3.4. Retrieving Genes Associated With a GO Annotation

The AmiGO website can be used to retrieve a list of genes associated with a GO annotation. These associations are made through "evidence codes" that couple the annotation, the gene, and the experimental evidence behind the assignment of each annotation to a gene or gene product **(Table 3)**.

A

Summary: The protein encoded by this gene belongs to the BCL2 protein family. BCL2 family members from hetero-or homodimers and act as anti- or pro-apoptotic regulators that are involved in a wide variety of cellular activities. This protein forms a heterodimer with BCL2, and functions as an apoptotic activator. This protein is reported to interact with, and increase the opening of, the mitochondrial voltage-dependent anion channel (VDAC), which leads to the loss in membrane potential and the release of cytochrome c. The expression of this gene is regulated by the tumor suppressor P53 and has been shown to be involved in P53-mediated apoptosis. Six alternatively spliced transcript variants, which encode different isoform, have been reported for this gene.

B

Function
molecular function unknown

Process
apoptosis
apoptotic mitochondrial changes
germ cell development
induction of apoptosis
induction of apoptosis by extracellular signals
negative regulation of cell cycle
negative regulation of survival gene product activity
regulation of apoptosis

Component
integral to membrance

Fig. 3. (**A**) GO text annotation of the *bax* gene. (**B**) GO annotation of the *bax* gene.

For instance, to perform a basic search in AmiGO enter a single GO term using the name "apoptosis" or identifier "0006915" (**Fig. 4A**). If a term name is used to search, the precise term name (apoptosis) must be selected from the retrieved list of other terms containing the search term. In an advanced search a list of terms may be used to search AmiGO. The results may also be filtered, in the example shown in **Fig. 4B**, and only human genes are queried by setting the gene product filter to *H. sapiens*.

3.5. Annotating a Gene List Using DAVID

To facilitate the biological interpretation of gene lists derived from the analysis of microarray and proteomic experiments, gene lists can be grouped into different GO categories. DAVID can be used to classify the gene lists according to a GO term or branch in the GO hierarchy. Using DAVID requires a list of genes to annotate in tab-delimited format. Only the identifiers need be present but additional columns can be present in the file.

The process is intuitive and is as follows (**Fig. 5**):

1. Select the identifier to query with from the drop-down box. This can be a Genbank accession #, Affymetrix ID, or any of the other identifiers listed.

A
apoptosis

Accession: GO:0006915
Aspect: biological_process
Synonyms: type I programmed cell death
Definition:

A form of programmed cell death induced by external or internal signals that trigger the activity of proteolytic caspases, whose actions dismantle the cell and result in cell death. Apoptosis begins internally with condensation and subsequent fragmentation of the cell nucleus (blebbing) while the plasma membrane remains intact. Other characteristics of apoptosis include DNA fragmentation and the exposition of phosphatidylserine.

B

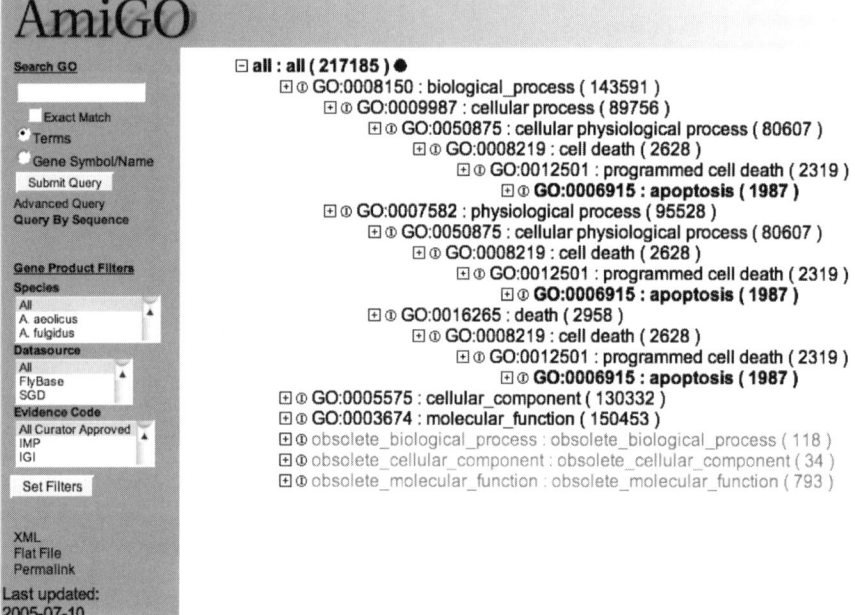

Fig. 4. Browsing GO annotations with AMIGO. "Apoptosis" is defined and assigned with an identifier (**A**). The tree structure (**B**) of GO can also be represented graphically as a directed acrylic graph (DAG) (**C,** facing page).

2. Enter the list of genes either by browsing and selecting a file prepared earlier or by copying the list into the text box. Click on submit.
3. After submission, the next step requires the selection of annotation types to annotate your gene list with. Using DAVID 2.0 the default annotation results will

C

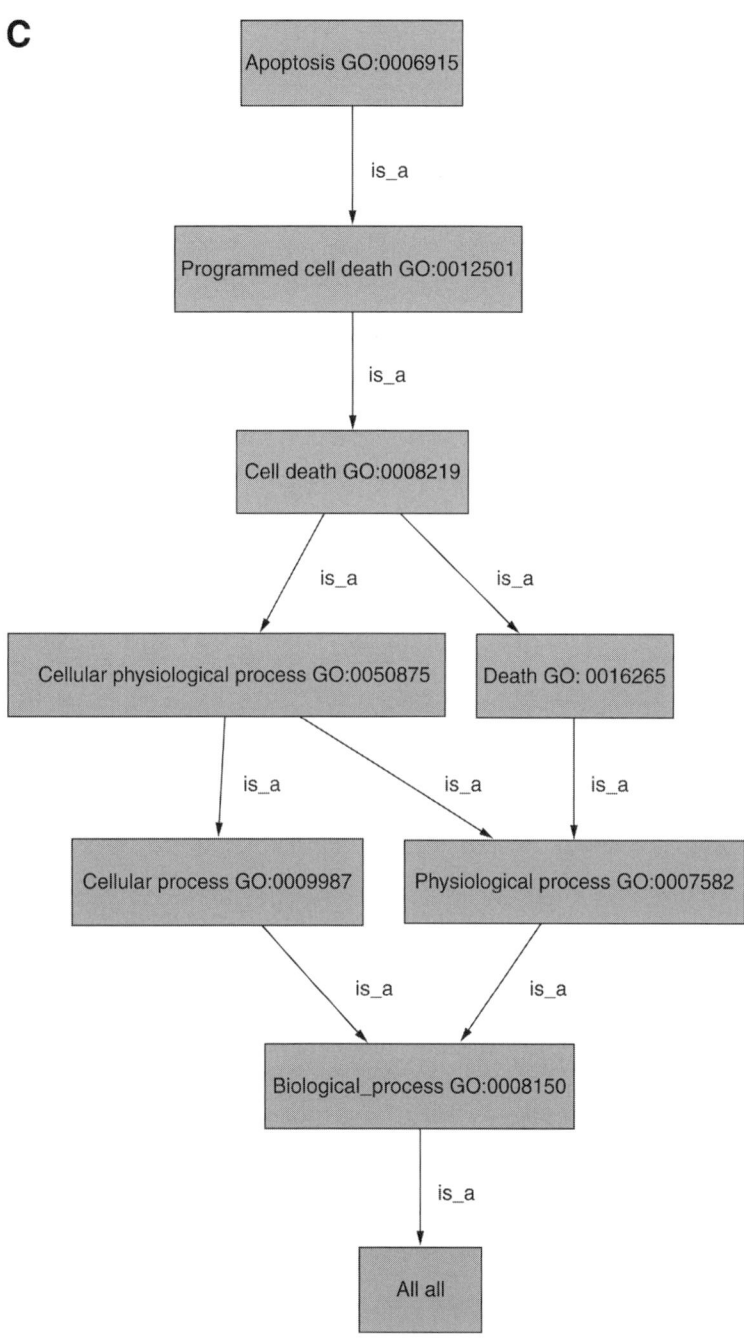

Fig. 4 *(continued).*

Table 3
Human Genes Associated With GO Annotation of Apoptosis

Gene Symbol	Datasource	Evidence	Full Name
A4_HUMAN ATGCC/GOst	UniProt	TAS	Amyloid beta A4 protein precursor
AA2AR_HUMAN ATGCC/GOst	UniProt	TAS	Adenosine A2a receptor
ABS_HUMAN ATGCC/GOst	UniProt	TAS	DEAD-box protein abstrakt homolog
ADA1A_HUMAN ATGCC/GOst	UniProt	TAS	Alpha-1A adrenergic receptor
AG22_HUMAN ATGCC/GOst	UniProt	TAS	Type-2 angiotensin II receptor
AHR_HUMAN ATGCC/GOst	UniProt	TAS	Aryl hydrocarbon receptor precursor
APAF_HUMAN ATGCC/GOst	UniProt	TAS	Apoptotic protease activating factor 1
APGB_HUMAN ATGCC/GOst	UniProt	TAS	Autophagy protein 12-like
ARHG6_HUMAN ATGCC/GOst	UniProt	TAS	Rho guanine nucleotide exchange factor 6
B2L10_HUMAN ATGCC/GOst	UniProt	TAS	Apoptosis regulator Bcl-B
BCLX_HUMAN ATGCC/GOst	UniProt	TAS	Apoptosis regulator Bcl-X

GO contains a number of "Evidence Codes" to validate an annotation. The TAS evidence code indicates a Traceable Author Statement, generally from a review paper or a book.

include the Entrez gene ID#, Uniprot ID#, GO biological process and molecular function, KEG and Biocarta pathway information, and Swiss Prot PIR keywords (SP_PIR_KEYWORDS). In our example we also selected GOTERM, Biological Process with a "Level" value of 3 (*see* **Note 4**). Click on "Get Annotation."

4. The genes are now annotated. To display the annotations of a particular type in a chart format, click on the "Chart" button in the Data Source Summary. The entire gene list, complete with all annotations, can be viewed by checking all the annotations on the Data Source Summary section and then selecting "Create Table."

5. To export the annotated gene list, select the format to export in (html, txt, or xls) and click on all pages. Exporting in either html or txt format will make it easy to use the list again.

Figure 6 is a screen shot after running DAVID on the sample input list, selecting for "Biological Process" classifications at level 3. There are 32 (19.5%)

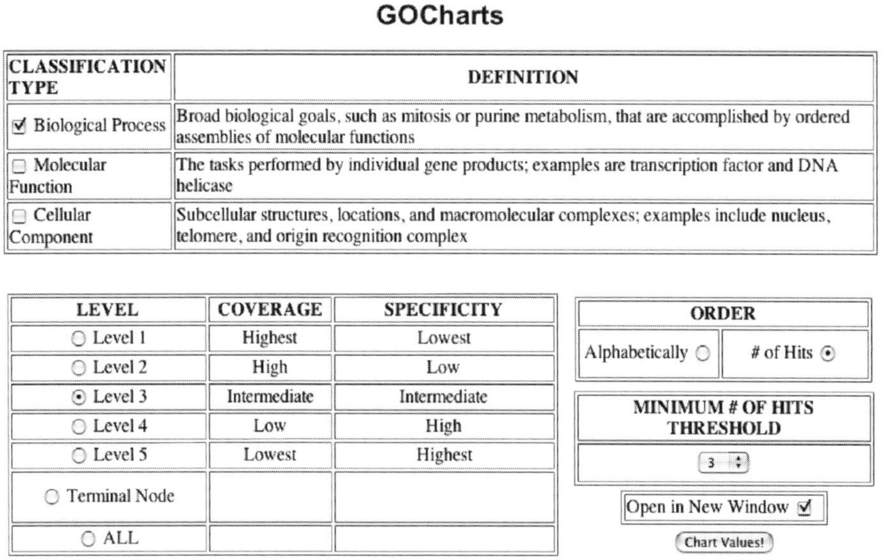

GOCharts

CLASSIFICATION TYPE	DEFINITION
☑ Biological Process	Broad biological goals, such as mitosis or purine metabolism, that are accomplished by ordered assemblies of molecular functions
☐ Molecular Function	The tasks performed by individual gene products; examples are transcription factor and DNA helicase
☐ Cellular Component	Subcellular structures, locations, and macromolecular complexes; examples include nucleus, telomere, and origin recognition complex

LEVEL	COVERAGE	SPECIFICITY
○ Level 1	Highest	Lowest
○ Level 2	High	Low
◉ Level 3	Intermediate	Intermediate
○ Level 4	Low	High
○ Level 5	Lowest	Highest
○ Terminal Node		
○ ALL		

ORDER	
Alphabetically ○	# of Hits ◉

MINIMUM # OF HITS THRESHOLD
3 ⇕

Open in New Window ☑

(Chart Values!)

Fig. 5. Running DAVID for GO classifications.

genes falling into the cell growth and/or maintenance category, 32 (19.5%) genes falling into the cell transduction category, 26 (15.9%) genes falling into the response to external stimulus category, 18 (11%) genes falling into the cell-to-cell signaling category, 17 (10.4%) genes falling in the response to stress category, and a significant portion of unannotated genes (*see* **Note 5**).

DAVID can also be used to inspect the "Molecular Function" graph (**Fig. 7**). There are 12 (7.3%) genes falling into the DNA binding category, 10 (6.1%) genes falling into the transmembrane receptor category, 9 (5.5%) genes falling into the purine nucleotide binding category, and 7 (4.3%) genes falling into the cytokine activity category. In both cases, the number of categories found is small and maps to a small region in the total GO graph.

3.6. Statistical Analysis of Significant GO Categories

Having said that, the annotation is clustered in a small region of the GO graph and should therefore be "biologically meaningful." How can we demonstrate that statistically (*see* **Note 6**)? The GO classification from DAVID gives the number and percentage of genes falling into each GO category for a given level. However, the classification alone does not tell us whether the number of genes falling into a specific category is because of random chance or treatment effects. To address the statistical significance of the number of genes in each GO category, one can use EASEonline, EASE, or GOMiner, all of which use Fisher's exact test (*see* **Note 7**).

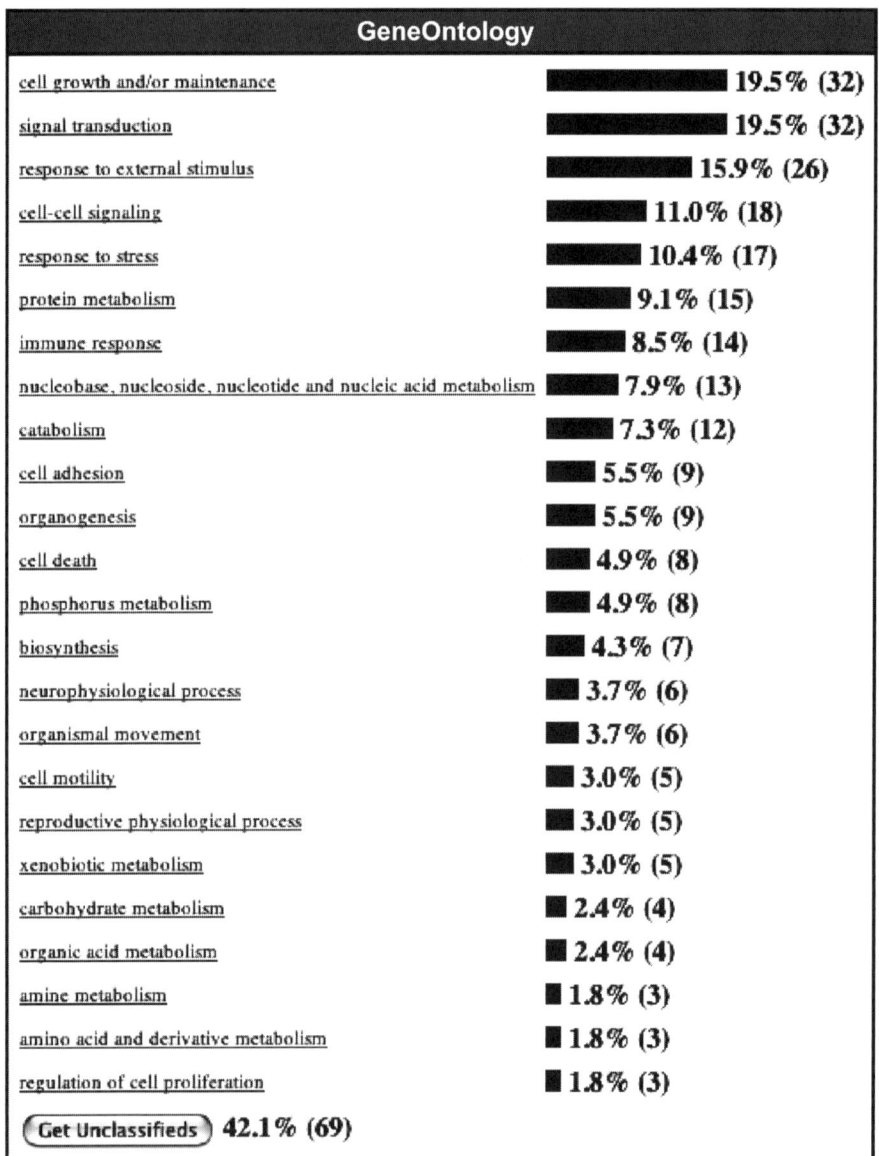

Fig. 6. GO biological process classification results analyzed at level 3 using the demo gene list 1 at http://david.niaid.nih.gov/david/upload.asp.

Figure 8 is a screen shot from running EASEonline with the sample gene list in **Subheading 3.5**. The system column contains the system of categorizing genes that can be any of the three structured GO graphs, i.e., biological process, molecular function, or cellular component. The category column contains the

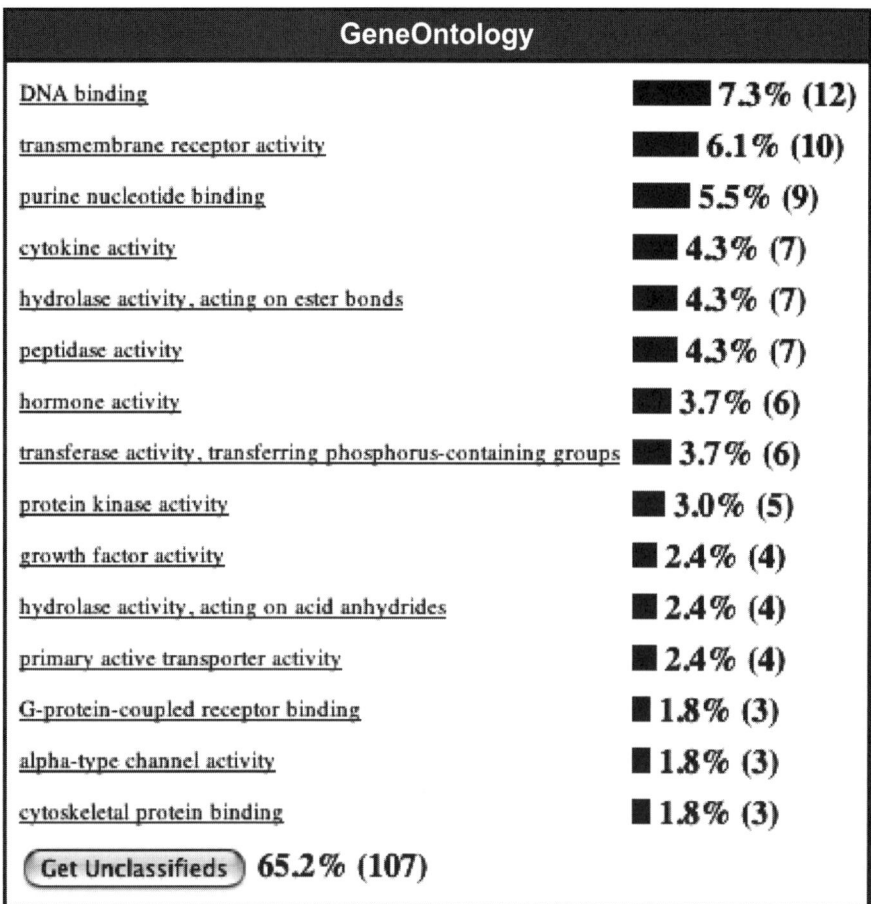

Fig. 7. GO molecular function classification analyzed at level 3 using the demo gene list 1 at http://david.niaid.nih.gov/david/upload.asp.

specific category of terms within a specific system such as extracellular and cytosol within the cellular component system, response to chemical substance and cell-cell signaling within the biological process system, and receptor binding and hormone activity within the molecular function system. The "List Hits" column contains the number of genes in the input gene list that belong to the specific category. In the above table, 29 genes belong to the extracellular category within the cellular component system, 13 genes belong to the response to chemical substance category within the biological process system, and 11 genes belong to the receptor binding category within the molecular function system (*see* **Note 8**). The List "Total" column contains the total number of genes in the input gene list that are annotated with the specific system. In the above table,

System	Category	LH	LT	PH	PT	EASE Score	Fisher Exact
cellular_component	extracellular	29	92	604	5501	0.000000203	0.0000000677
biological_process	response to chemical substance	13	98	143	6079	0.00000232	0.000000378
biological_process	response to abiotic stimulus	13	98	231	6079	0.000277	0.0000747
biological_process	xenobiotic metabolism	5	98	24	6079	0.000507	0.0000328
biological_process	response to xenobiotic stimulus	5	98	26	6079	0.000696	0.0000494
biological_process	defense response to fungi	3	98	5	6079	0.00244	0.0000397
biological_process	cell-cell signaling	15	98	390	6079	0.00331	0.00132
biological_process	response to external stimulus	25	98	856	6079	0.0034	0.0018
biological_process	response to fungi	3	98	6	6079	0.00363	0.0000785
biological_process	response to stimulus	28	98	1027	6079	0.00449	0.00255
biological_process	response to pest/pathogen/parasite	13	98	324	6079	0.00513	0.00194
cellular_component	extracellular space	11	92	239	5501	0.00577	0.00197
biological_process	chemotaxis	6	98	79	6079	0.00829	0.00162
biological_process	taxis	6	98	79	6079	0.00829	0.00162
biological_process	organismal physiological process	26	98	1003	6079	0.0127	0.00752
biological_process	response to stress	17	98	555	6079	0.0138	0.0069
biological_process	defense response	17	98	557	6079	0.0143	0.00715
biological_process	response to biotic stimulus	18	98	608	6079	0.0148	0.00766
cellular_component	cytosol	8	92	176	5501	0.0258	0.00884
molecular_function	receptor binding	11	91	341	6169	0.026	0.0111
biological_process	sodium ion transport	4	98	48	6079	0.0406	0.00718
biological_process	inflammatory response	6	98	124	6079	0.0475	0.0146
biological_process	innate immune response	6	98	128	6079	0.0531	0.0168
biological_process	regulation of apoptosis	7	98	170	6079	0.0536	0.0194
biological_process	induction of apoptosis by extracellular signals	3	98	25	6079	0.0596	0.00722
molecular_function	hormone activity	4	91	63	6169	0.0638	0.0136
biological_process	pregnancy	3	98	30	6079	0.0822	0.012
molecular_function	G-protein-coupled receptor binding	3	91	35	6169	0.0919	0.0145
molecular_function	chemokine activity	3	91	35	6169	0.0919	0.0145

Fig. 8. EASE Online using the demo gene list 1 at http://david.niaid.nih.gov/david/ upload.asp.

92 genes in the input gene list have cellular component annotation, 98 genes in the input gene list have biological process annotation, and 91 genes in the input list have molecular function annotation. The "Population Hits" column contains number of genes assayed that fall into the specific category. In **Fig. 8**, 604 genes assayed fall into the extracellular category within the cellular component system, 143 genes fall into the response to chemical substance category within the biological process system, and 341 genes fall into the receptor binding category within the molecular function system. The "Population Total" column contains the number of genes assayed and annotated within the specific system. For example, 5501 genes assayed have cellular component annotation, 6079 genes assayed have biological process annotation, and 6169 genes assayed have molecular function annotation. The Fisher Exact column contains the Fisher exact probability of observing the number of "List Hits" in the "List

Total" given the frequency of "Population Hits" in the "Population Total." The EASE score column contains the adjusted Fisher exact probability using the Jackknife Fisher exact test that strongly penalizes the significance of categories supported by few genes and negligibly penalizes categories supported by many genes. It therefore yields more robust results and the EASE score is the default metric used by EASE to rank categories of genes by over-representation.

Based on the EASE score and using 0.05 as a cut-off value in **Fig. 8**, we will conclude that the input gene list is over-represented by genes whose products are likely to be found extracellularly, in the extracellular space or in the cytosol. The input gene list is also over-represented in more than 20 biological function categories ranging from chemical substance response to inflammatory response. Note that all of the categories in the above table including regulation of apoptosis and induction of apoptosis by extracellular signals would have been considered as over-represented in the input list if Fisher Exact probability had been used instead of EASE score (*see* **Note 9**).

3.7. Search and Browsing MeSH

Similar to GO, MeSH is an ontology to describe concepts and relationships in medical research and practice *(2)*. MeSH has been used to index medical literature. Annotating genes with a controlled vocabulary of MeSH terms provides the disease context for understanding the gene list of interest.

As an example, MeSH defines the term "apoptosis" **(Fig. 9)**, provides synonyms and spelling variations, and relates this term to other terms **(Fig. 10)**. The development of MeSH is independent from GO. We can see some overlaps between the MeSH ontology and the GO ontology in some areas. A unifying open biological ontology is under active development to describe all biomedical phenomena *(7)*.

3.8. Interpreting a Gene List Using MeSH Terms

Clusters of genes that have been identified through DAVID or other software can be analyzed through HAPI to search for similarities in MeSH category and descriptors among the genes in the cluster. HAPI takes a tab-delimited text file with the first column identified by GenBank accession numbers, Affymetrix probeset identifiers, or UniGene identifiers, and outputs the number of matches in each MeSH category and the number of matches for individual MeSH descriptors in each MeSH category. **Figure 10** shows the most significant MeSH descriptor matches in the disease category of MeSH from running HAPI using the sample gene list. There are 35 genes associated with neoplasms and 5 of those genes are related to leukemia. One can view the detailed gene information

A

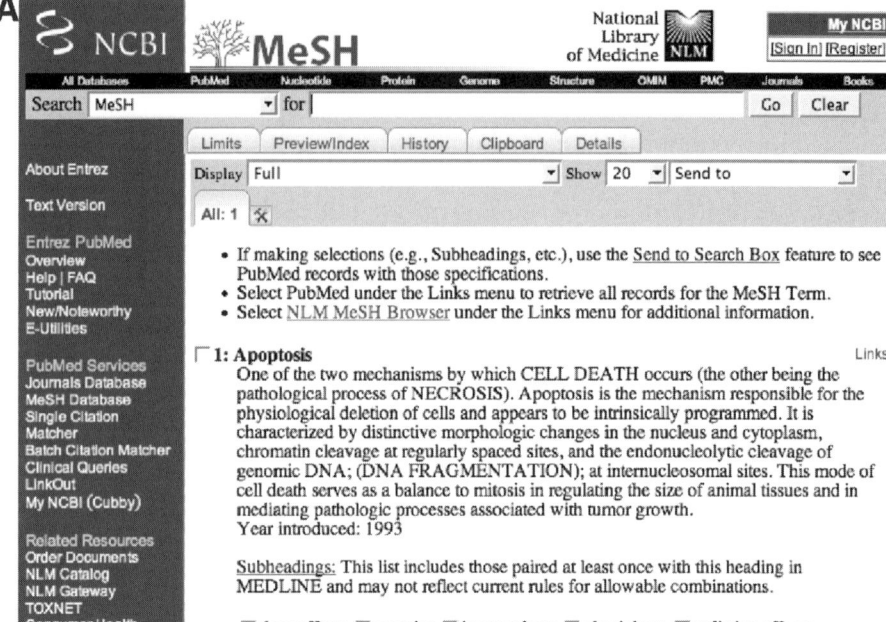

1: Apoptosis
One of the two mechanisms by which CELL DEATH occurs (the other being the pathological process of NECROSIS). Apoptosis is the mechanism responsible for the physiological deletion of cells and appears to be intrinsically programmed. It is characterized by distinctive morphologic changes in the nucleus and cytoplasm, chromatin cleavage at regularly spaced sites, and the endonucleolytic cleavage of genomic DNA; (DNA FRAGMENTATION); at internucleosomal sites. This mode of cell death serves as a balance to mitosis in regulating the size of animal tissues and in mediating pathologic processes associated with tumor growth.
Year introduced: 1993

B Entry Terms:

- Apoptoses
- Cell Death, Programmed
- Cell Deaths, Programmed
- Death, Programmed Cell
- Deaths, Programmed Cell
- Programmed Cell Death
- Programmed Cell Deaths

Previous Indexing:

- Cell Survival (1972-1992)

See Also:

- Necrosis
- Caspase 1
- In Situ Nick-End Labeling
- Cellular Apoptosis Susceptibility Protein
- Genes, Transgenic, Suicide

All MeSH Categories
  Biological Sciences Category
    Biological Phenomena, Cell Phenomena, and Immunity
      Cell Physiology
        Cell Death
          Apoptosis
            Anoikis
            DNA Fragmentation

Fig. 9. (**A**) MsSH definition of "apoptosis." (**B**) – MeSH browsing details of the term "apoptosis."

```
┌─────────────────────────────────────────────────────────────────┐
│                           Diseases                                │
╞═══════════════════════════════════════════════════════════════════╡
│ Neoplasms (35) {<.001}                                            │
│    Cysts (1) {<.001}                                              │
│       Bone Cysts (1) {<.001}                                      │
│          Jaw Cysts (1) {<.001}                                    │
│             Odontogenic Cysts (1) {<.001}                         │
│                Basal Cell Nevus Syndrome (1) {<.001}              │
│    Neoplasms by Histologic Type (16) {<.001}                      │
│       Leukemia (5) {<.001}                                        │
│          Leukemia, Lymphocytic (2) {<.001}                        │
│             Leukemia, Lymphocytic, Acute (1) {<.001}              │
│          Leukemia, Myeloid (3) {<.001}                            │
│             Leukemia, Myeloid, Chronic (1) {>.6}                  │
│             Leukemia, Nonlymphocytic, Acute (2) {<.001}           │
│                Leukemia, Promyelocytic, Acute (2) {<.005}         │
└─────────────────────────────────────────────────────────────────┘
```

Fig. 10. Output from running HAPI using the demo gene list 1 at http://david.niaid. nih.gov/david/upload.asp.

by clicking the Pubmed ID, GenBank accession number supplied in the number links besides the MeSH descriptors (*see* **Note 10**).

4. Notes

1. Assigning ontology terms to each gene is called annotation. The annotation process is achieved by a combination of human curation of literature and computer inference from sequence similarity. For individual genes, the gene ontology annotation in current databases can be neither complete nor accurate. However, the collective GO evidence from a list of many genes can be statistically meaningful. Thus, GO analysis can extract relevant biological information despite its limitations in the annotation process.

2. Because the utility of ontology in interpreting gene lists was demonstrated *(8)*, a vast number of tools, either commercial or free, have been designed *(9–11)*. Traditional GO analysis of microarray results, as discussed in this chapter, starts from a list of differentially expressed genes to retrieve ontology annotations, and then infers the statistical significance of each ontology term. Alternatively, we can start from retrieving all the gene expression data associated with a particular ontology term first, and then assess their probability of differential expression as a group. The latter strategy is recently established and is supposed to identify more subtle changes of differential expression *(12)*.

3. We only discuss point-and-click software tools in this chapter. Readers can use Bioconductor to customize the analysis with more programming control.

4. The "Level" in the gene otology tree will affect the output of the classification results. There are 5 levels of choice in DAVID, that is 1, 2, 3, 4, and 5. Level 1 has

Table 4
Fischer Test Notes Example

	List	Population	Totals
Hits	29	604	633
NonHits	92	5501	5593
Totals	121	6105	6226

the highest coverage of terms but lowest specificity, whereas level 5 has the lowest coverage of terms but highest specificity (**Fig. 5**). Level 3 is recommended for immediate coverage and specificity.

5. Many of the genes do not have assigned GO annotation that might result in biased classification results. In the sample input list, there are about 42% accessions that do not have annotation within biological process (**Fig. 2**) and about 65% accessions that do not have annotation within the molecular function system (**Fig. 3**). Therefore, before our knowledge of gene and GO association becomes nearly complete, we must be cautious not to use GO classification as the only evidence to draw conclusions about over-representation of genes in terms of the GO category.

6. There are several assumptions of the significance analysis of GO. (1) The function annotation by GO is complete and accurate. As discussed in **Notes 1** and **5**, we know this assumption is problematic. (2) A GO category is "statistically significant" if there are a disproportionably large number of genes in this category differentially expressed. However, the sheer number does not necessary reflect the biology. Sometimes, a very small number of genes in the category change, they can be biologically important, although not reaching the level of statistical significance.

7. Usually this problem is modeled as a hypothesis test on the equality of the two proportions. A 2×2 contingency table is constructed for each GO category (**Table 4**). A Fisher's exact test can be performed with this contingency table. The p-value from the test indicates whether the proportion of a certain category in the gene list could have resulted from a random drawing of the genes in the population.

8. DAVID reports the number of accessions supplied in the input list that belong to a given category despite the possibility that there might be multiple accessions representing one gene. To avoid genes with multiple accessions receiving more than one "vote" in the over-representation analysis, EASE converts all accessions to LocusLink ID before reporting counts although LocusLink ID might not be the best way to uniquely identify a gene either.

9. As a large number of GO terms are tested in parallel for their significance, we run into a multiple testing problem. Furthermore, these hypotheses are not independent of each other because of the nest structure of ontologies. Thus, the interpretation of the p-values shall be with caution, depending on whether it is reported as uncorrected p-values or corrected values using various methods, such as false discovery rate or Holm correction *(11)*.

10. HAPI annotates genes with a rudimentary process of extracting information from Medline records. This process is error prone. Thus, any statistically significant categories should be manually checked by clicking the links from HAPI results.

References

1. Bard, J. B. and Rhee, S. Y. (2004) Ontologies in biology: design, applications and future challenges. *Nat. Rev. Genet.* **5,** 213–222.
2. Gene Ontology Consortium (2006). The Gene Ontology project in 2006. *Nucleic Acids Res.* **34** (database issue), D322–D326.
3. Lowe, H. J. and Barnett, G. O. (1994) Understanding and using the medical subject headings (MeSH) vocabulary to perform literature searches. *JAMA* **271,** 1103–1108.
4. Diehn, M., Sherlock, G., Binkley, G., et al. (2003) SOURCE: a unified genomic resource of functional annotations, ontologies, and gene expression data. *Nucleic Acids Res.* **31,** 219–223.
5. Bussey, K. J., Kane, D., Sunshine, M., et al. (2003) MatchMiner: a tool for batch navigation among gene and gene product identifiers. *Genome Biol.* **4,** R27.
6. Ashburner, M., Ball, C. A., Blake, J. A., et al. (2000) Gene ontology: tool for the unification of biology. The Gene Ontology Consortium. *Nat. Genet.* **25,** 25–29.
7. Blake, J. (2004) Bio-ontologies-fast and furious. *Nat. Biotechnol.* **22,** 773–774.
8. Masys, D. R., Welsh, J. B., Fink, J. L., et al. (2001) Use of keyword hierarchies to interpret gene expression patterns. *Bioinformatics* **17,** 319–326.
9. Khatri, P., Draghici, S., Ostermeier, C., and Krawetz, S. (2002) Profiling gene expression using onto-express. *Genomics* **79,** 266–270.
10. Al-Shahrour, F., Diaz-Uriarte, R., and Dopazo, J. (2004) FatiGO: a web tool for finding significant associations of Gene Ontology terms with groups of genes. *Bioinformatics* **20,** 578–580.
11. Beissbarth, T. and Speed, T. P. (2004) GOstat: find statistically overrepresented Gene Ontologies within a group of genes. *Bioinformatics* **20,** 1464–1465.
12. Ben-Shaul, Y., Bergman, H., and Soreq, H. (2005) Identifying subtle interrelated changes in functional gene categories using continuous measures of gene expression. *Bioinformatics* **21,** 1129–1137.

15

Incorporation of Gene Ontology Annotations to Enhance Microarray Data Analysis

Michael F. Ochs, Aidan J. Peterson, Andrew Kossenkov, and Ghislain Bidaut

Summary

Typical microarray or GeneChip™ experiments now provide genome-wide measurements on gene expression across many conditions. Analysis often focuses on only a few of the genes, looking for those that are "differentially expressed" between conditions or groups of conditions. However, the large number of measurements both present statistical problems to such single gene approaches and offers a tremendous amount of information for methods focused on biological processes rather than individual genes. Here we provide a method to utilize biological annotations in the form of gene ontologies to interpret the results of individual or multiple pattern recognition analyses of a microarray experiment.

Key Words: Microarray; gene ontology; biological process; pattern recognition; clustering.

1. Introduction

Microarrays and GeneChips™ have become standard tools in molecular biology, providing researchers with the ability to probe the levels of thousands of gene transcripts routinely. GeneChips are high-density oligonucleotide arrays providing multiple probes per gene, with these individual probe measurements being combined to estimate the expression level of a gene *(1)*. Microarrays typically are coated microscope slides with spots placed on the surface either through robotic spotting of liquid containing cDNA or oligonucleotide *(2)* or through *in situ* growth of individual oligonucleotides using modified inkjet technology *(3)*. These platforms followed the initial use of arrays using older technologies *(4,5)*.

Microarrays provide insight into cellular processes on an unprecedented scale. The ability to query the transcript level for essentially every known gene, as well as most predicted genes in a single hybridization, allows researchers to ask questions on a global scale. However, perhaps because microarrays originally grew

From: *Methods in Molecular Biology, vol. 377, Microarray Data Analysis: Methods and Applications*
Edited by: M. J. Korenberg © Humana Press Inc., Totowa, NJ

out of the concept of Northern blots *(6)*, the analytical approach often searches for individual genes that show differences in transcript levels between conditions. This approach requires very careful statistical analysis because the number of measurements being made is far larger than in a Northern blot *(7–10)*.

Another approach to analysis more fitting to the global scale of the measurement being made is to focus on biological processes that involve regulation of sets of genes. In general, the biological transcriptional response of an organism is not the differential expression of a single gene, but instead the initiation of a complex response involving changes in the transcription of many genes in addition to other processes (such as regulation of transport). By focusing on processes rather than individual genes, a number of problems related to the large number of simultaneous measurements can be avoided. This approach does require additional information to allow the genes to be queried as a group.

The most natural way to identify genes that are likely to be coregulated is through transcription factors. However, our information on the links between transcription factors and regulated genes is still small, as reflected in the limited information available in transcription factor databases *(11,12)*. Although this information is growing, especially through the use of ChIP-on-chip approaches *(13)*, transcription factors alone may not link to biological processes, unless they can be directly linked to known signaling pathways or have other detailed information. More information is available, however, from the growing gene ontology databases *(14)*. Gene ontology comprises a set of three parallel annotations, biological process, molecular function, and cellular location *(15)*. The biological process annotation is of particular interest because upregulation of a set of genes with the same process annotation provides evidence that a specific cellular process has been activated. This approach can even link an expression signature on a microarray-to-signaling pathway activity *(16)*.

2. Materials

1. The TIGR Multiexperiment Viewer (MeV or TMEV) from The Institute for Genomics Research (Rockville, MD) is described in **ref. *17*** and is downloadable from http://www.tigr.org/software/tm4/mev.html.
2. The dataset used in this chapter is described in **ref. *18*** and a preprocessed version is available from http://bioinformatics.fccc.edu/papers/methods/.
3. A new version of the automated sequence annotation pipeline (ASAP II) is available at http://bioinformatics.fccc.edu/software/OpenSource/ASAP/ASAP.shtml. The original version is described in **ref. *19***.
4. The ClutrFree visualization and gene ontology analysis tool is available from http://bioinformatics.fccc.edu/software/OpenSource/ClutrFree/clutrfree.shtml and is described in **ref. *20***.
5. The Go Tree Machine web analysis system can be found at http://genereg.ornl.gov/gotm/ and is described in **ref. *21***.

6. The SOURCE tool for conversion between various gene identifiers can be found at http://source.stanford.edu.
7. The Bayesian Decomposition tool and a description of the advanced analysis using it and ASAP are given at http://bioinformatics.fccc.edu/methods/BD, as the detailed description was too long for this chapter.

Nomenclature for this chapter includes *italic* for on-screen text, SMALL CAPS for buttons, and `courier font` for files and folders.

3. Methods

The methods outlined next describe the analysis of microarray data using gene ontology. It is assumed that the reader can perform standard procedures including preprocessing to correct for background hybridization and to normalize the data, as well as create a tab-delimited file summarizing an experiment. The tools listed in **Subheading 2.** will all work with a tab-delimited file with the first row being a header and the first column being Gene IDs. This data file should have a file extension .txt. The header row labels the conditions, one column for each condition. If available, an auxiliary file with the extension .unc containing uncertainties will be used by Bayesian Decomposition. In addition, Bayesian Decomposition has the ability to run in a supervised learning mode, with assignment of conditions to groups *(22)*, which is accomplished by providing details on the number of conditions per group in a file with the extension `.cls`. This chapter will focus on application of K-means clustering using the TMEV tool and of Bayesian Decomposition to the Project Normal data, collection of annotations using SOURCE or ASAP, and interpretation of the results using ClutrFree and GOTreeMachine.

3.1. Simple Clustering and GoTree Machine

3.1.1. Applying K-Means Clustering With TMEV

The downloaded dataset comprises six files with extensions `.txt`, `.unc`, and `.cls`. The following steps focus on using the `ProjNormSmall` files, as these have only 827 genes and are more useful for rapid analysis. The `ProjNormLarge` files may be used in the same way and contain 3024 genes chosen for having gene ontology annotations. Only the `ProjNormSmall.txt` file is needed for this step, as K-means clustering does not use uncertainty or classification information, which is contained in the `.unc` and `.cls` files, respectively. Start the `TMEV` by double-clicking on the TMEV.bat file on Windows or on the `MEV_3_0_Mac_OSX` file on Macintosh (OS X required). Use the following steps to load the data and generate K-means clusters on the log-transformed data.

1. Choose *"New Multiple Array Viewer"* from the *"File"* menu.
2. Choose *"Load Data"* from the *"File"* menu on the new window.

3. Choose *"Stanford Files (*.txt)"* from the pop-up menu at the top of the page.
4. Navigate to the folder with the downloaded data and choose the `ProjNormSmall.txt` file.
5. Click on the first data-point (1) in the second column, first row of white boxes.
6. Click on LOAD button.
7. Choose *"Log2 Transform"* from the *"Adjust Data"* menu (note that the image will not change, but cluster images will reflect the log transformation).
8. Click on the KMC button near the top of the screen. In the pop-up menu enter 4 for the number of clusters, then press on the OK button.
9. Click on the expand icon next to *"KMC – genes" (1)* on the left window, then click on the expand icon next to *"Expression Images"* and then click on *"Cluster 1."*
10. Right-click (CTRL + click on the Mac) next to the cluster image on the right and choose *"Save All Clusters…"* Save the clusters with the name Kmeans4.

3.1.2. Using GO Tree Machine to Interpret the Clusters

In this step, a cluster for further analysis will be chosen from the four clusters generated in **Subheading 3.1.1.**, and GO Tree Machine will be used to look for significant enhancement of gene ontology terms relative to their representation on the array used. The inclusion of the full gene list available is vital to correct calculation of significance of gene ontology enhancement because any array is biased by the genes included. Owing to limitations of most bioinformatics tools, a number of steps are needed to move from a gene list with accession numbers to a gene ontology measurement. Here SOURCE will be used for the conversion, whereas in **Subheading 3.2.2.** ASAP will be used (*see* **Note 1**). Note that many tools also do not handle certain characters in files (Excel will insert commas for instance in certain cases), resulting in the need for more cutting and pasting than is desirable. First, a cluster should be chosen. So, look at each cluster using TMEV and pick the cluster showing higher expression in the last four conditions (related to testis tissue). This will appear as in **Fig. 1**. Then, open the `Kmeans4-N.txt` file with Excel or another spreadsheet program, where N is the number of the cluster you have picked (clustering algorithms generally proceed from a random starting point so N may vary on repeat of this method).

1. Highlight the column with accession numbers and copy it into a new spreadsheet. Remove the header cell and save this file in tab-delimited format as `ClusterGeneList.txt`.
2. Open the `ProjNormSmall.txt` data file and repeat **step 1**, creating a new file `FullGeneList.txt` in a tab-delimited format.
3. Go to the SOURCE website and click on the link for *"Batch SOURCE."* Open the two files created in **steps 1** and **2**. For each file, cut and paste the gene accession numbers into the box under *"Or enter a list of identifiers:"* (note, SOURCE does not correctly parse files from all types of computers, but this method always works).

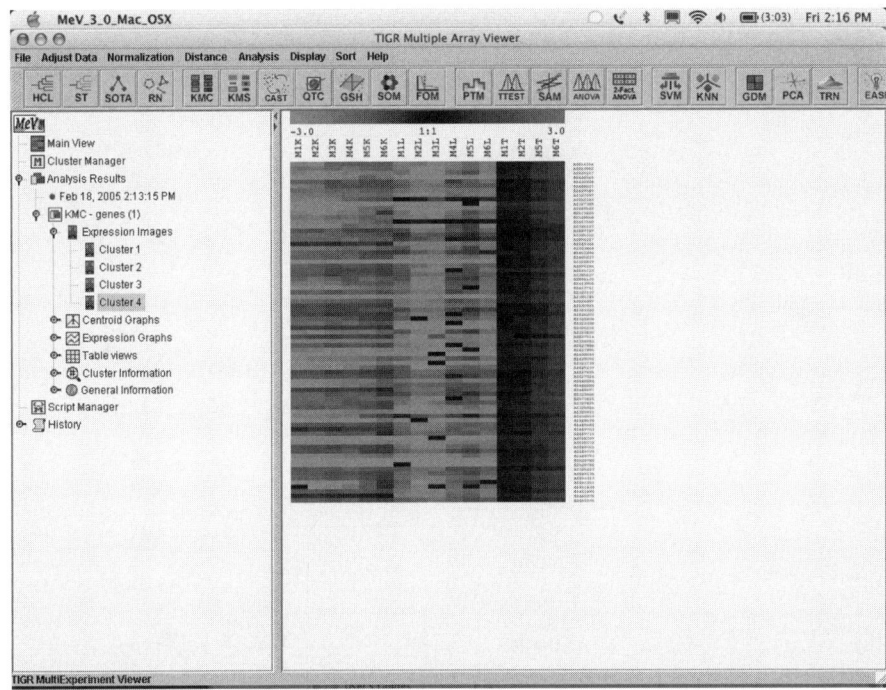

Fig. 1. The screen image of the desired cluster for analysis in the TMEV tool. The cluster here shows higher expression (red on screen, gray here) in the last four conditions.

4. Choose *"GenBank Accession"* as the type of input identifier and *"Mus musculus"* as the organism. Then in section 2 check the box next to *"UniGene Cluster ID"* and then click the SUBMIT button. Do **steps 3** and **4** separately for each file, saving the downloaded files as `ClusterUGList.txt` and `FullUGList.txt`, respectively.

5. Open the files created in **step 4** and remove the accession number column and header row, leaving only the UniGene Cluster IDs (i.e., Mm.NNNNN labels).

6. Go to the GO Tree Machine website and log in (you will need to register the first time, however this is free). After the Welcome page, you will see a Make New Tree page.

7. Enter a name for the analysis in section 1, choose the *"UNIGENE ID"* option in section 2, and the *"interesting gene list vs. reference gene list"* option in section 3.

8. In section 4, click on the BROWSE (or CHOOSE FILE) button and choose the `ClusterUGList.txt` file. In section 5b, click on the BROWSE button and choose the `FullUGList.txt` file.

9. Click on the MAKE TREE button. A new screen will open showing the progress in uploading files and building the gene ontology tree. When completed, click on the CHECK GO TREE button.

10. The tree can be navigated by clicking on the + icons opening up the tree structures. To see the enhanced GO categories directly, click on the number link at the top (*NN*): *"Gene Numbers in NN GO Categories were relatively enhanced."*

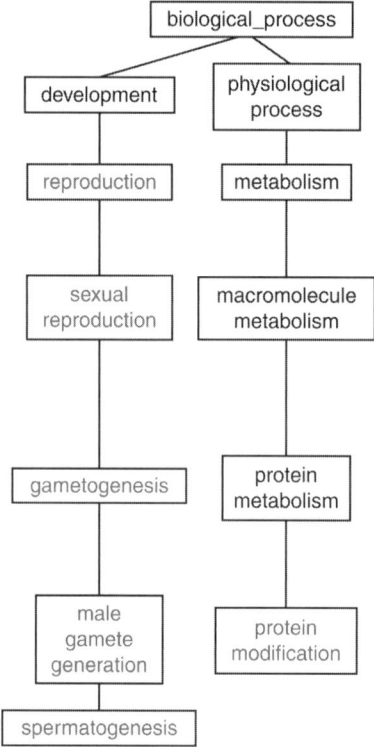

Fig. 2. The directed acyclic graph output option from GO Tree Machine. The gene ontology categories with significant enhancement will be highlighted in red on the web-page (here in gray). On the webpage, all three gene ontology categories are shown, however here we have focused only on biological process ontology.

11. In this example, the testis tissue-related cluster was chosen, and the biological processes related to testis show as significantly enhanced.
12. A bar chart and a directed acyclic graph (DAG) can also be created (*see* **Fig. 2**), by clicking on the BAR CHART or DAG VIEW buttons, respectively, at the top of the page. The level of the gene ontology for the bar chart is given by the pop-up menu under the button. The tree can also be exported as a text file by clicking on the EXPORTGOTREE button.

3.2. Multiple Clustering and ClutrFree

3.2.1. Applying K-Means Clustering With TMEV

Start TMEV as in **Subheading 3.1.1**. Use the following steps to load the data and generate multiple K-means clustering results on the log transform data.

1. Choose *"New Multiple Array Viewer"* from the *"File"* menu.
2. Choose *"Load Data"* from the *"File"* menu on the new window.

3. Choose "*Stanford Files (*.txt)*" from the pop-up menu at the top of the page.
4. Navigate to the folder downloaded with the data and choose the `ProjNormSmall.txt` file.
5. Click on the first data-point (1) in the second column, first row of white boxes.
6. Click on LOAD button.
7. Choose "*Log2 Transform*" from the "*Adjust Data*" menu (note that the image will not change, but clusters will reflect the log transformation).
8. Click on the KMC button near the top of the screen. In the pop-up menu enter 4 for the number of clusters, then press on the OK button.
9. Repeat **step 8** for 5, 6, and 7 clusters.
10. Click on the expand icon next to "*KMC – genes*" *(1)* on the left window, then click on the expand icon next to "*Expression Images*" and then click on "*Cluster 1.*"
11. Right-click (CTRL + click on the Mac) next to the cluster image on the right and choose "*Save All Clusters...*" Save the clusters with the name `Kmeans4`.
12. Repeat **steps 10** and **11** for 5 through 7 cluster results, naming each `KmeansN` where N is the number of clusters.

3.2.2. Obtaining Gene Ontology Information With ASAP (Optional)

ASAP permits users to generate custom queries that link to multiple local and web-based resources. Included with the download is a preset query to retrieve gene ontology information for a list of genes. ASAP will retrieve the ontology data as a tab-delimited file with a format compatible to ClutrFree. Installation of ASAP requires knowledge of the MySQL open source database and Apache open source web server. Details are provided in the installation guide, but this optional section is recommended only for individuals with advanced computer skills.

1. Go to the ASAP web page as established during the installation. Log in and choose the "*Query*" link at the top of the page.
2. Click on the UniGene annotation plan link (*db/UniGeneAnnotation*).
3. Click on the BROWSE (CHOOSE FILE on some computers) button in the "*INPUT*" section of the page. Use the browser to choose the `PNSmallforAnnot.txt` file. This file contains the accession numbers for the `ProjNormSmall` data set in the format created in **Subheading 3.1.2**.
4. Click on the "*Get only organism specific*" radio button and the "*ClutrFree format*" radio button.
5. Click on the button QUERY at the bottom that will initialize the queries. The queries involve locally cached databases only so will take only a few minutes.
6. Retrieve the annotations from the web server by going to the "*status*" page and clicking on the name or number of your annotation run. The key annotations for this work are biological process, so click on the "*Download*" link on the "*Biological Process*" ClutrFree format file line.
7. Save this file as `ontology.txt` in your experiment folder (*see* **Note 2**).

3.2.3. Interpreting the Multiple Results With ClutrFree

ClutrFree is a visualization tool for linking the results of multiple analyses, either performed with the same algorithm as here or performed with different algorithms. It also presents visualization of gene ontology or other annotations if associated files are present. These files have been created using the ASAP system, as described in **Subheading 3.3.2.**, and included with the downloaded datasets for users who skip **Subheading 3.2.2.** Start the ClutrFree tool by double-clicking on the `ClutrFree.jar` file (*see* **Note 3**).

1. Place the results of the K-means clustering into a file structure such that there is a single parent folder named experiment (the name is unimportant). In this folder, create a series of folders `analysis1`, `analysis2`, `analysis3`, and `analysis4` (any series of names can be used). Into these folders place the output files from the TMEV with one set of clusters in each folder (e.g., `Kmeans4-1.txt` ... `Kmeans4-4.txt` into the folder `analysis1`). Also place the `ontology.txt` file generated in **Subheading 3.2.2.** in the experiment folder.

2. If **Subheading 3.2.2.** was skipped, place the `ProjNormSmall.ann` in the experiment folder and rename it to `ontology.txt`. Place the `ProjNormSmall.gnm` file in the `experiment` folder and rename it to `annot.txt`. Place the `ProjNormSmall.exp` file in the `experiment` folder and rename it to `exp-names.txt`. These files provide ClutrFree with gene ontology data, gene ID's, and condition names, respectively. Descriptions of their formats can be found in the ClutrFree user guide.

3. Choose the "*Import data...*" option from the "*File*" menu in ClutrFree and navigate to the folder containing the `experiment` folder you created in **step 1**. Highlight the `experiment` folder icon and click on the CHOOSE button. ClutrFree will load the data and bring up a window for viewing the cluster shapes and a tree relating the clusters to each other for each analysis (*see* **Fig. 3**). The >> button allows the user to view the individual cluster shapes (or patterns). Click on this arrow until the pattern looks like **Fig. 3** (the number at the top may differ as clustering causes the labeling of groups to be random from run to run). The key is that the bars are down in the first 12 conditions (kidney and liver) and up in the last 4 (testis). Note the number before the : at the top of the screen (call it N). This is the cluster that shows genes with high expression in testis.

4. There are many options for exploring the data using ClutrFree. Here we will focus on two features that utilize gene ontology and the ability of ClutrFree to determine if genes are consistently assigned in clusters. Press on the GENE TABLE button to open two new windows (*see* **Fig. 4**).

5. The gene table contains a listing of all the genes in the experiment file, a measure of the strength of assignment of each gene to a pattern (here since clustering is used this is binary), and a measure of the persistence of the gene in the cluster (i.e., a measure of how many levels of the tree linked to the node contain the gene). Full options are described in the User's Guide downloaded with ClutrFree.

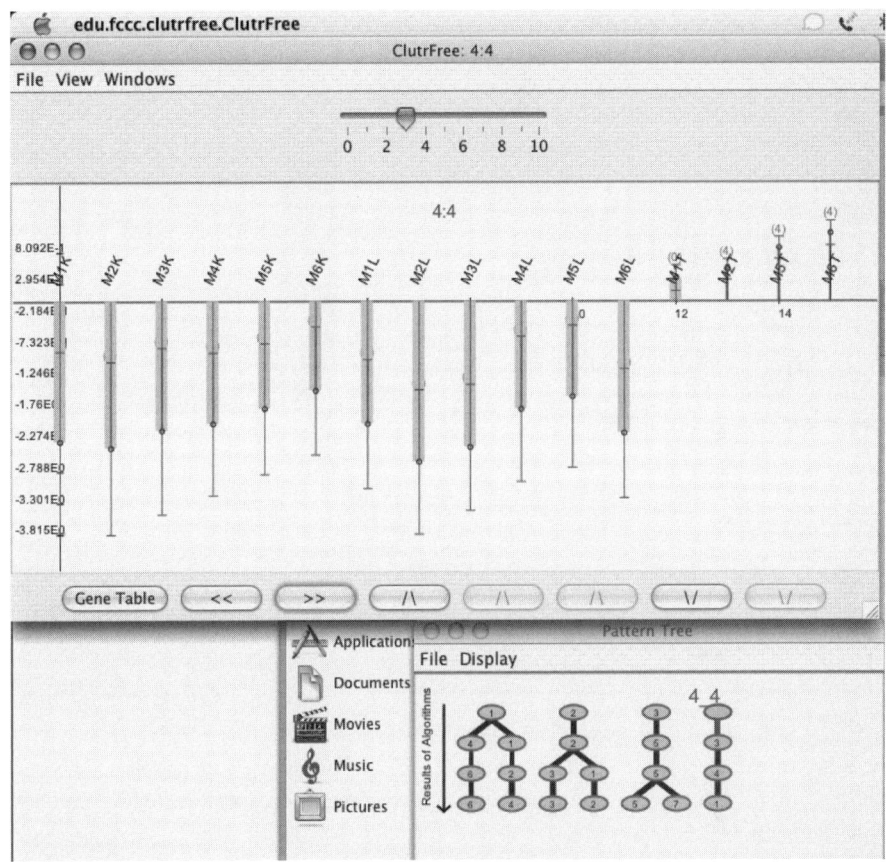

Fig. 3. ClutrFree pattern page. The tree window shows the relationship between the patterns identified by different methods, whereas the main window shows the pattern behavior across conditions. The pattern shown (found by using the >> button) is the pattern that has higher expression in the last four conditions related to the testis tissue.

6. Each header cell sorts the contents from highest to lowest. For this work, click twice on the second occurrence of N from **step 3** in the header at the top of the 4 Patterns window. This will order the rows from highest to lowest. The number in blue is the persistence of the gene in the pattern (i.e., the number of occurrences in the tree shown in **Fig. 4**).

7. Move to the bottom window and adjust the two sliders. For the slider on the left, move it to *150*, which will require an enhancement of 1.5 to highlight a cell. For the slider on the right, move it to *5* to eliminate GO terms with less than 5 occurrences. Then click twice on E(N). It shows that *spermatogenesis, male gamete generation, etc.* are enhanced in this pattern, as expected for a pattern related to testis tissue.

8. A shift click on any column will invert the order, which is useful for using the *p*-values from the hypergeometric test (i.e., *P-v(N)* column heading).

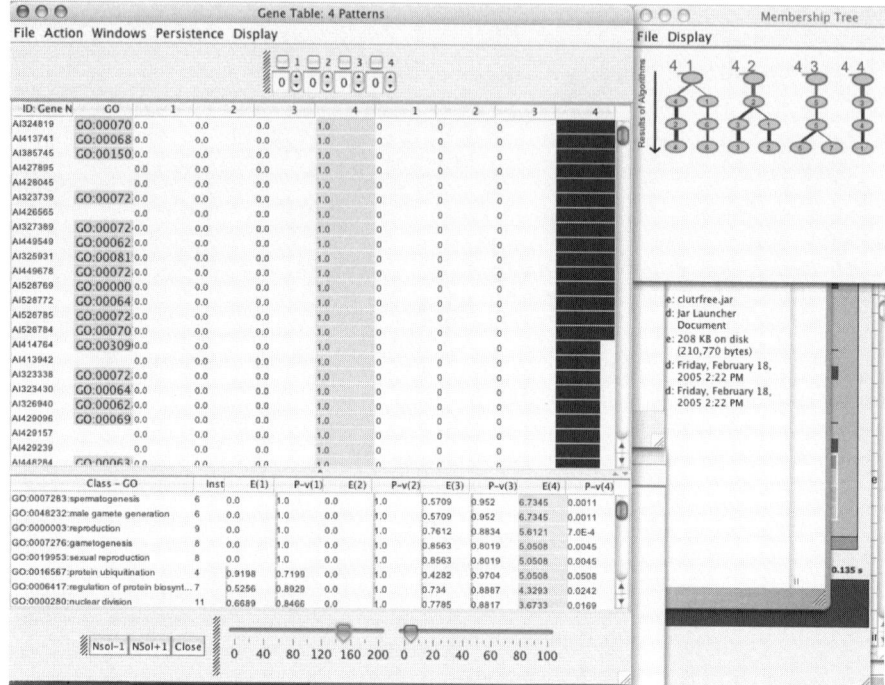

Fig. 4. ClutrFree gene page. The gene tree window shows the relative reliability of the assignment of genes to patterns as you increase the number of patterns. The gene window provides a view of the persistence of genes in patterns and the enhancement of gene ontology.

3.3. Bayesian Decomposition With ClutrFree Visualization

Bayesian Decomposition is a more complex pattern recognition algorithm that creates overlapping patterns of behavior with genes assigned in varying strengths to these patterns. Although the algorithm itself and the interpretation are more complex, it more closely models the complexity of biological systems. Genes generally provide multiple, overlapping functions within cells, and Bayesian Decomposition specifically handles separation of such functional units. The detailed description of the use of this method is too long for this text but can be found at http://bioinformatics.fccc.edu/methods/bd/.

4. Notes

1. An ongoing problem in bioinformatics is the instability of input and output formats as well as URLs on web resources. It is therefore possible that between the final editing of this chapter and the publication, certain web resources may change format requirements or links. The site http://bioinformatics.fccc.edu/methods will

maintain a web request form to answer problems that users have applying the techniques described here. We will also maintain a list that updates any needed changes discovered in response to submitted requests.

2. The annotation methods used by SOURCE and ASAP differ slightly, so there will be minor differences in annotations on some accession numbers. This is typical for annotations as different databases have different ways of linking accession numbers to UniGene clusters.

3. The annotation information available on genes changes rapidly. For instance, the original publication on the type of analysis presented here was done by hand calculation after retrieval of data using the original annotation pipeline *(22)*. In the intervening 3 yr, numerous tools have appeared to aid such calculations and the number of mouse genes with gene ontology annotation has roughly quadrupled. In addition, assignment of accession numbers (and therefore nucleotide sequences) to UniGene clusters changes routinely. As such, application of **Subheading 3.2.** with newly downloaded annotations may not yield exactly the same results as shown previously, although the broad results will remain consistent, demonstrating the value of using a global method in a domain where individual measurements are subject to error. In addition, some genes on the microarray may no longer be linked to gene ontology information, however this will not cause problems for the tools described previously.

References

1. Lockhart, D. J., Dong, H., Byrne, M. C., et al. (1996) Expression monitoring by hybridization to high-density oligonucleotide arrays. *Nat. Biotechnol.* **14,** 1675–1680.

2. Schena, M., Shalon, D., Davis, R. W., and Brown, P. O. (1995) Quantitative monitoring of gene expression patterns with a complementary DNA microarray. *Science* **270,** 467–470.

3. Hughes, T. R., Mao, M., Jones, A. R., et al. (2001) Expression profiling using microarrays fabricated by an ink-jet oligonucleotide synthesizer. *Nat. Biotechnol.* **19,** 342–347.

4. Augenlicht, L. H., Wahrman, M. Z., Halsey, H., Anderson, L., Taylor, J., and Lipkin, M. (1987) Expression of cloned sequences in biopsies of human colonic tissue and in colonic carcinoma cells induced to differentiate in vitro. *Cancer Res.* **47,** 6017–6021.

5. Augenlicht, L. H. and Kobrin, D. (1982) Cloning and screening of sequences expressed in a mouse colon tumor. *Cancer Res.* **42,** 1088–1093.

6. Rohde, W. and Sanger, H. L. (1981) Detection of complementary RNA intermediates of viroid replication by Northern blot hybridization. *Biosci. Rep.* **1,** 327–336.

7. Claverie, J. M. (1999) Computational methods for the identification of differential and coordinated gene expression. *Hum. Mol. Genet.* **8,** 1821–1832.

8. Ideker, T., Thorsson, V., Siegel, A. F., and Hood, L. E. (2000) Testing for differentially-expressed genes by maximum-likelihood analysis of microarray data. *J. Comput. Biol.* **7,** 805–817.

9. Newton, M. A., Kendziorski, C. M., Richmond, C. S., Blattner, F. R., and Tsui, K. W. (2001) On differential variability of expression ratios: improving statistical

inference about gene expression changes from microarray data. *J. Comput. Biol.* **8,** 37–52.

10. Tusher, V. G., Tibshirani, R., and Chu, G. (2001) Significance analysis of microarrays applied to the ionizing radiation response. *Proc. Natl. Acad. Sci. USA* **98,** 5116–5121.

11. Matys, V., Fricke, E., Geffers, R., et al. (2003) TRANSFAC: transcriptional regulation, from patterns to profiles. *Nucleic Acids Res.* **31,** 374–378.

12. Ghosh, D. (1992) TFD: the transcription factors database. *Nucleic Acids Res.* **20 Suppl,** 2091–2093.

13. Ren, B., Robert, F., Wyrick, J. J., et al. (2000) Genome-wide location and function of DNA binding proteins. *Science* **290,** 2306–2309.

14. Consortium, T. G. O. (2001) Creating the gene ontology resource: design and implementation. *Genome Res.* **11,** 1425–1433.

15. Ashburner, M., Ball, C. A., Blake, J. A., et al. (2000) Gene ontology: tool for the unification of biology. The Gene Ontology Consortium. *Nat. Genet.* **25,** 25–29.

16. Bidaut, G., Moloshok, T. D., Grant, J. D., Manion, F. J., and Ochs, M. F. (2002) Bayesian Decomposition analysis of gene expression in yeast deletion mutants. In: *Methods of Microarray Data Analysis II* (Johnson, K. and Lin, S., eds.), Kluwer Academic, Boston, pp. 105–122.

17. Saeed, A. I., Sharov, V., White, J., et al. (2003) TM4: a free, open-source system for microarray data management and analysis. *Biotechniques* **34,** 374–378.

18. Pritchard, C. C., Hsu, L., Delrow, J., and Nelson, P. S. (2001) Project normal: defining normal variance in mouse gene expression. *Proc. Natl. Acad. Sci. USA* **98,** 13,266–13,271.

19. Kossenkov, A., Manion, F. J., Korotkov, E., Moloshok, T. D., and Ochs, M. F. (2003) ASAP: automated sequence annotation pipeline for web-based updating of sequence information with a local dynamic database. *Bioinformatics* **19,** 675–676.

20. Bidaut, G. and Ochs, M. F. (2004) ClutrFree: cluster tree visualization and interpretation. *Bioinformatics* **20,** 2869–2871.

21. Zhang, B., Schmoyer, D., Kirov, S., and Snoddy, J. (2004) GOTree Machine (GOTM): a web-based platform for interpreting sets of interesting genes using Gene Ontology hierarchies. *BMC Bioinformatics* **5,** 16.

22. Moloshok, T. D., Datta, D., Kossenkov, A. V., and Ochs, M. F. (2003) Bayesian Decomposition classification of the Project Normal data set. In: *Methods of Microarray Data Analysis III* (Johnson, K. F. and Lin, S. M., eds.), Kluwer Academic, Boston, pp. 211–232.

16

Predicting Survival in Follicular Lymphoma Using Tissue Microarrays

Michael J. Korenberg, Pedro Farinha, and Randy D. Gascoyne

Summary

A tissue microarray (TMA) containing diagnostic biopsies was used to develop predictors of outcome in a group of 105 patients having advanced-stage follicular lymphoma (FL). The patients were staged and uniformly treated, and the usable cases had been randomly divided into a subgroup of 50 patients with outcomes identified, and a reserved subgroup of 43 patients whose outcomes were masked for blind testing of the predictors. Using training-input data from some patients with known outcomes, parallel cascade identification developed two predictors of overall survival based on a number of biomarkers. Both predictors had statistically significant performance over the remaining patients with known outcomes. The first predictor had been identified with model architectural settings and encoding scheme chosen, for the particular training input used, to enhance classification accuracy over remaining patients in the known subgroup. The second predictor was obtained without changing the settings and encoding scheme, but from an entirely different training input corresponding to novel cases from the TMA. Not surprisingly, the first predictor showed much higher accuracy over the known subgroup, but when tested over the reserved subgroup of 43 patients, averaged about 58% correct and did not reach statistical significance. The other predictor performed very similarly over the known and the reserved subgroups, with prediction on the reserved subgroup highly significant at $p = 0.0056$ in Kaplan–Meier survival analysis. We conclude that a predictor based on a number of biomarkers obtainable at diagnosis has the potential to improve prediction of overall survival in FL.

Key Words: Overall survival; clinical outcome; treatment response; biomarkers; tissue microarrays; follicular lymphoma.

1. Introduction

Follicular lymphoma (FL) frequently exhibits a long clinical course, with median survival time of 8–10 yr *(1,2)*. A follicular lymphoma international prognostic index, based on five clinical variables, has been used to predict clinical outcome *(3,4)*. Recent success in building a gene expression-based predictor of outcome has demonstrated that molecular characteristics present in tumor samples at time of diagnosis of FL are important for determining survival *(1)*.

From: *Methods in Molecular Biology, vol. 377, Microarray Data Analysis: Methods and Applications*
Edited by: M. J. Korenberg © Humana Press Inc., Totowa, NJ

The latter study found both a favorable pattern or signature of gene expression associated with good prognosis, and an unfavorable signature predicting decreased survival. Both signatures were mostly derived from the nonmalignant cells of a tumor microenvironment. The favorable signature was enriched with genes characteristic of T-cells and the unfavorable one with genes expressed on macrophages and dendritic cells.

Farinha et al. *(4)* recognized that the genes involved in this unfavorable signature suggested the importance of macrophages in influencing FL survival. They built a tissue microarray (TMA) with diagnostic biopsies from 105 patients with advanced-stage FL uniformly treated at the British Columbia Cancer Agency with a BP-VACOP protocol consisting of chemotherapy (bleomycin, cisplatin, etoposide, doxorubicin, cyclophosphamide, vincristine, and prednisone) followed by radiotherapy of the involved sites. The protein expression of different markers in both malignant and nonmalignant cells was studied using immunohistochemistry and scored in terms of cell content as well as morphological patterns. Fourteen biomarkers were defined. Of these biomarkers, they found that a lymphoma-associated macrophage (LAM) score predicted overall survival independently of the clinical prognostic index *(4)*. In particular, a LAM score of more than 15 cells per high-power field predicted a poor outcome (12 patients). Their results revealed the importance of macrophages in the biology of FL. None of the markers other than the LAM score appeared to be predictive of outcome *(4)*.

In this chapter, essentially the same TMA is used to build a predictor of overall survival, this time based on a number of biomarkers to see whether it leads to increased accuracy. The present work has two main objectives. The first is to develop a predictor whose accuracy is verified over a reserved subgroup of patients where the outcomes have been masked. The second objective is to investigate whether the predictor can discriminate over the low-macrophage subgroup of patients (81 usable cases), all of whom would be predicted to survive based on the LAM score.

2. Materials and Patient Samples

2.1. Tissue Microarray

1. The data are the same as in **ref. 4**. In particular, the TMA was constructed using duplicate 1-mm cores from biopsy material in paraffin blocks (Beecher Instruments, Silver Spring, MD).
2. Hematoxylin and eosin staining was used on the TMA; further details of histology and immunohistochemistry are presented in **ref. 4**. Although 14 biomarkers were analyzed there *(4)*, some were scored by multiple measures, such as for both architectural pattern and number of positively stained cells. CD20 was performed to ensure tumor cell content in all cores.

3. Counting each of the measures separately, we used 20 biomarkers (**Table 1**), a number of which were scored qualitatively, such as CD10(F), CD10(IF), CD21, BCL2, and BCL-XL.

2.2. Patient Samples

1. All patients had advanced-stage indolent follicular lymphoma, and had been uniformly treated at the British Columbia Cancer Agency between July of 1987 and May 1993 *(4)*.
2. Informed consent was obtained. The University of British Columbia–British Columbia Cancer Agency provided approval to review, analyze, and publish the data.
3. In total, 93 FL cases were available where all 20 biomarkers had been assessed, and survival status was known. Of these, 50 (29 alive/21 dead) were randomly selected and the clinical outcome indicated for each. The remaining 43 had outcome masked and were for validation.

3. Methods

The following approach to building parallel cascade identification (PCI) predictors of treatment response and clinical outcome has previously been used with gene expression data *(5,6)*, and was also briefly reviewed in **ref. 7**.

3.1. Numerically Encoding Biomarkers

1. Because some of the biomarkers were qualitatively assessed, they had to be assigned numerical values for analysis (**Table 1**, right column). As examples, for BCL-XL, BCL2, CD10(F), CD10(IF), and TIA1(10%), "negative" was scored as 1 and "positive" as –1. For CD68(cells), the number N of cells per high-power field was converted to $-N/5$. This helped to keep the magnitude similar to that for other biomarkers, so that the resulting predictor did not overemphasize one measure.
2. The scoring system tended to give lower values to features believed to be unfavorable to outcome, such as a higher MIB1 proliferation rate. However, there are some inconsistencies in this pattern, such as oppositely scoring CD3(int) and CD7(int), although higher values of both these biomarkers are believed to be favorable. However, a training input that is consistently lower for failed outcomes does not typically have the variability useful for system identification (*see* **Note 1**), and changes in scoring had been introduced into **Table 1** to increase the effectiveness of the resulting first predictor over the subgroup of patients with known outcomes.

3.2. Forming a Training Input and Output

1. Building an outcome predictor began with forming a training input from a selected number of cases from the TMA associated with failed and successful outcomes. For the first predictor, the training input used biomarkers from the first three cases of the TMA for patients with failed (F) outcome, denoted F1–F3, and the first three cases for survivors (S), denoted S1–S3 (*see* **Note 2**). In particular, the average value of each biomarker for the three failed outcomes was compared with that for the

Table 1
Biomarkers Used in the Study

Biomarker	Description	Scoring system
BCL-XL	Antiapoptotic factor BCL2 related (POS or NEG).	POS→ –1, NEG→ 1
MIB1	Proliferation rate graded in 1, 2, 3 (<10%, <50% and >50%, respectively).	Grades 1, 2, 3 → –1, –2, –3, respectively
BCL2	Antiapoptotic gene (POS or NEG). Its over-expression is a hallmark of FL.	POS→ –1, NEG→ 1
BCL6	Presence of BCL6+ cells. Scored as 0–2, where NEG(0), POS(1 and 2), or NEG(0,1) and POS (2).	Grades 0, 1, 2→ 0, –1, –2, respectively
CD10(F)	POS/NEG for tumor cells.	POS→ –1, NEG→ 1
CD10(IF)	Presence of positive neoplastic CD10+ cells outside the follicles (POS or NEG).	POS→ –1, NEG→ 1
CD68	Intensity of the infiltrate of macrophages within the tumor (0-weak/1-strong).	Grades 0, 1→ 0, –1, respectively
CD68cells	Number of CD68+ cells per high power field (cut-off = 15 cells HPF, but three groups can be defined: 0–10 cells; 10–20 cells and >20 cells)	Number of cells $N\rightarrow$ $-N/5$
CD3(arch)	Architectural pattern of T-cells (reactive cells responsible for the immune response against the tumor)—Perifollicular or diffuse	Perifollicular → 1, Diffuse → –1
CD3(int)	Intensity of the infiltrate of CD3 T-cells (0-weak/1-strong)	Grades 0, 1→ 0, –1, respectively
CD7(arch)	Architectural pattern of CD7 T-cells—perifollicular or diffuse	Perifollicular → 1, Diffuse → –1
CD7(int)	Intensity of the infiltrate of CD7 T-cells (0-weak/1-strong)	Grades 0, 1→ 0, 1, respectively
TIA1(10%)	Samples POS/NEG in >10% of the TIA1 positive T-cells	OS→ –1, NEG→ 1P
CD21	Architectural pattern of the neoplastic follicles based on FDC cells (follicular dendritic cells), the meshwork of follicles—follicular or expanded.	Follicular → 1, Expanded → –1

(Continued)

Table 1 *(Continued)*

Biomarker	Description	Scoring system
CD4	Intensity of the infiltrate of CD4 T-cells (0-weak/1-strong)	Grades 0, 1→ 0, 1, respectively
CD4/8	Predominance of type of T-cells—CD4, CD8, or mixed (CD4 and CD8).	Grades 4, 4/8, 8→ −1, 0, 1, respectively
CD4/8(arch)	Architectural pattern of all T-cells scored simultaneously—perifollicular or follicular	Follicular → 1, Perifollicular → −1
CD8	Intensity of the infiltrate of CD8 T-cells (0-weak/1-strong)	Grades 0, 1→ 0, 1, respectively
CD57(int)	Intensity of the infiltrate of CD57 T-cells (0-absent/1-weak/ 2-moderate/ 3-strong)	Grades 0,...,3 → 0,...,3, respectively
CD57(arch)	Architectural pattern of the CD57 T-cells subset—perifollicular or follicular	Follicular → −1, Perifollicular → 1

three successful outcomes, and only 13 biomarkers were found to differ between outcomes. The unused biomarkers were BCL-XL, BCL2, BCL6, CD10(F), CD21, CD4, and CD57(int).

2. The remaining 13 biomarkers were numerically encoded and the values appended, in the same order as in **Table 1**, to form an F segment corresponding to the case for the first failed outcome. Similar segments were prepared for the remaining five cases, and then all the segments were concatenated to form a 78-point training input (**Fig. 1**, dotted line).

3. The corresponding training output (**Fig. 1**, solid line) was defined as −1 over each of the three F segments and as 1 over the three S segments of the training input.

4. The nonlinear system having this input/output relation can be viewed as an ideal classifier. In particular, the model identified from the training record is expected to have negative output corresponding to a case for a failed outcome, and positive output for a successful outcome.

3.3. Identifying a Classifier Model

The parallel cascade model used in this work is shown in **Fig. 2**. Each *L*-block denotes a linear element that is dynamic, i.e., has memory. This means that each model output value (and hence ultimately its prediction of outcome) depends on more than one biomarker, and the number of biomarkers involved depends on the memory length. Each *N*-block denotes a static nonlinearity, in the form of a polynomial. If the polynomial degree exceeds one, then the model output would depend upon nonlinear interactions (products) of biomarkers. Previously Palm *(8)*, to uniformly approximate discrete-time nonlinear Volterra

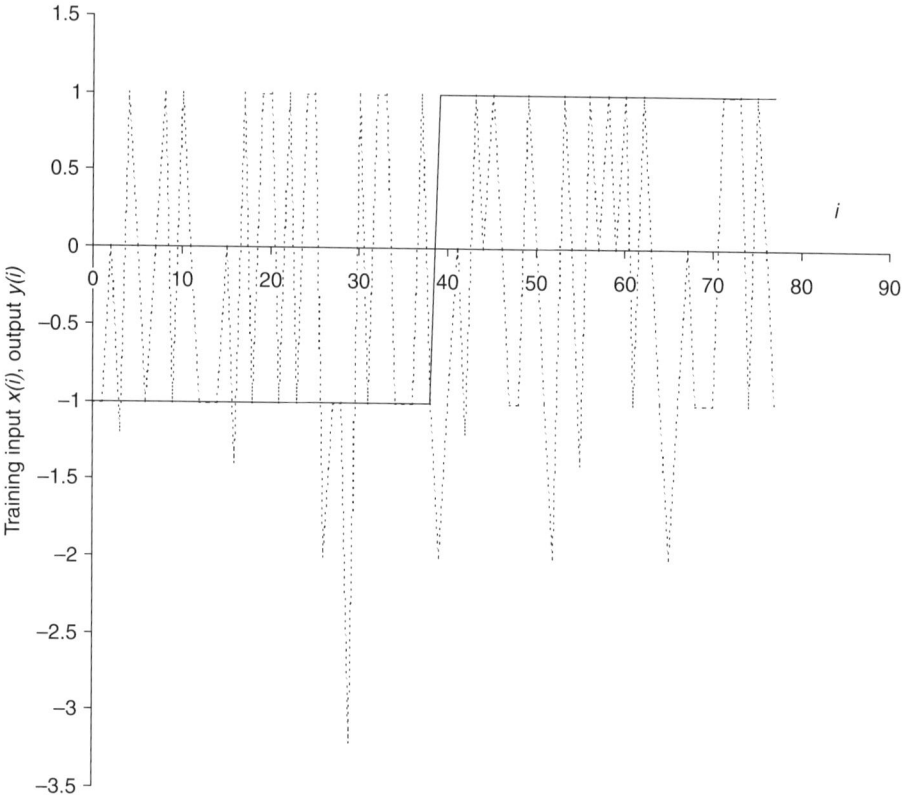

Fig. 1. Training input $x(i)$ (dotted line) formed by splicing together the numerically-encoded biomarkers (**Table 1**, right column) from the first three "failed outcome" (F) cases and the first three "survivor outcome" (S) cases. The biomarkers used were the 13 whose average values differed between the three F and the three S cases. Training output $y(i)$ (solid line) defined as -1 over the "failed outcome" portions of the training input and 1 over the "survivor outcome" portions. The training input and output were used to identify a parallel cascade model of the form in **Fig. 2**.

systems, suggested a parallel *LNL* cascade model in which the static nonlinearities were exponential and logarithmic functions rather than the polynomials used here *(9)*.

Parallel cascade identification is then used to identify the model directly from the training input and output. Briefly, a first cascade of a dynamic linear element followed by a static nonlinearity is found to approximate the defined input/output relation. The residual, i.e., the difference between the cascade output and the training output, is treated as the output of a new nonlinear system, and a second cascade is found to approximate the latter system. The new residual is then computed,

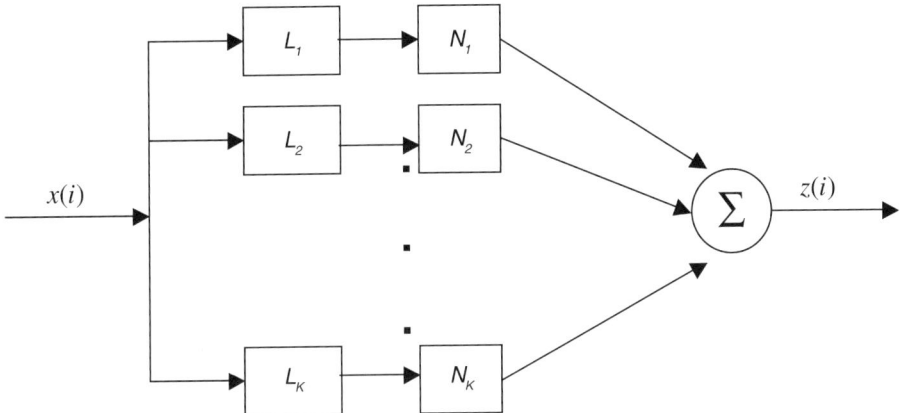

Fig. 2. Parallel cascade model used to predict follicular lymphoma overall survival. Each L is a dynamic linear element; each N is a polynomial static nonlinearity.

a third cascade is found to improve the approximation, and so on. Under broad conditions, the original nonlinear system can be approximated to an arbitrary degree of accuracy by a sum of a sufficient number of these cascades, which have been found individually. A detailed description of PCI is given in **ref. 9**.

To identify a parallel cascade model, several architectural parameter settings have to be determined:

1. Memory length of each linear element L.
2. The degree of each polynomial static nonlinearity N.
3. The maximum number of cascades in the model.
4. A threshold regulating the reduction in mean-square error required to admit a candidate cascade into the model.

Several PCI models, corresponding to different trial settings of these parameters, were identified from the training input and output (**Fig. 1**), then their accuracy was compared in classifying the remaining 44 cases associated with known outcomes. It was quickly found that using lower degree polynomials, especially first-degree, for the static nonlinearities resulted in more accurate classifiers. This is not unexpected because imprecise, qualitative assessments underlay much of the scoring, e.g., MIB1 proliferation scored as $-1, -2, -3$. Higher degree polynomials could have tended to accentuate small differences in biomarker values, e.g., overemphasizing the difference between a little less than and a little more than 50% proliferation (*see* **Note 3**).

Using first degree polynomials for the static nonlinearities also simplified the determination of other parameter settings. The parallel cascade is then equivalent to a single dynamic linear element (with the same memory length as in a cascade) plus a constant, no matter how many cascade paths are in the PCI model (*see* **Note 4**). Thus, provided that the memory length was not excessive (*see* **Note 5**), there was no danger of introducing more variables into the model than output points used for training and it was not necessary to restrict the number of cascades in the model. Hence, when the static nonlinearities were first-degree polynomials, memory length of the dynamic linear elements was the only architectural parameter setting that had to be determined (*see* **Note 6**).

The latter was chosen by trial and error, exploring a range of memory lengths, as well as small variations in the encoding scheme, and checking the resulting classifier accuracy over the remaining 44 cases of the TMA not used to form the training input, for which the outcomes were known. A memory length of nine samples appeared to produce an effective classifier, when the scoring system of **Table 1** was employed (*see* **Subheading 3.5.**).

However, because the memory length, polynomial degree, and encoding scheme had been chosen for the particular training exemplars to enhance classification accuracy over the remaining cases with known outcomes, this does not mean that the resulting predictor will perform well on novel cases from the TMA. To gauge whether these parameter settings and encoding scheme could be effective for classifiers trained on different exemplars, the next three cases corresponding to failed outcomes (denoted F4–F6), and the next three for successful outcomes (denoted S4–S6), were instead used to construct a new training input. This time, 18 biomarkers (all except BCL2, CD4/8[arch]) were found to differ on average between the 3 F and 3 S training cases, so that a 108-point training input resulted. This produced a second predictor that was then tested on the remaining 44 cases from the TMA with known outcomes. A third predictor was trained using the next three cases from each class (denoted F7–F9, S7–S9), tested on remaining known outcome cases, and so on. Each time the same PCI architecture parameter settings and encoding scheme from **Table 1** were used, and Fisher's exact test was employed to measure the effectiveness of the resulting classifier. Although 29 cases were associated with successful outcomes, there were only 21 cases for failed outcomes, so that 7 outcome predictors in total were produced.

Only a predictor statistically significant over known outcome cases, not used for the training input, was allowed to predict over the reserved subgroup with masked outcomes. A one-tailed test was used to determine which predictors reached significance over the known subgroup of cases. This is because, owing to the way each model had been trained (–1 denoted failure and 1 denoted successful outcome), it was expected to have negative output for failed outcomes and positive output for successful outcomes. Indeed, any predictor whose predicted outcome negatively correlated with actual outcome, no matter how strong the correlation, was regarded as performing insignificantly and rejected. Kaplan–Meier survival analysis *(10)* was used to evaluate the predicted outcome over the reserved subgroup.

3.4. Using a Classifier Model to Predict Outcome

1. The novel case to be classified was first converted to an input signal by using the right column of **Table 1** to numerically encode those biomarkers that were used by the predictor. The resulting values were then appended in the order the biomarkers appear in **Table 1**. For the first predictor, this produced a 13-point input signal, and for the second predictor, an 18-point signal.

2. The input signal was fed through the classifier model and, once the memory length was reached, the resulting output signal was averaged. For example, each of the 7 predictors had memory length of 9, so for the first predictor, output points 9 to 13 were averaged. For the second predictor, output points 9 to 18 were averaged.

3. If the average output value was negative, then a failed outcome was predicted, and otherwise a successful outcome was predicted.

3.5. Results

3.5.1. Subgroup for Which Outcomes Were Labeled

For the results in **Table 2**, note that the 44 test cases (18F, 26S) are not exactly the same for the 7 predictors because each predictor was evaluated on all but the 6 cases used to construct its training input. Only the first two predictors were significant on Fisher's exact test, so only these were chosen to predict outcome over the subgroup with masked outcomes. The sixth predictor correlated quite strongly with outcome, but *negatively*, and hence was treated as not significant and rejected on the one-tail test.

On this subgroup, the first predictor performs best; however, for its particular training input, the PCI architectural parameter values and the encoding scheme had been tailored to enhance accuracy. No further searching for good parameter values was conducted to build the remaining predictors from their respective training inputs: they were simply identified from their training data after adopting the same architectural settings and encoding scheme as used for the first predictor. So one might expect that the second predictor is less likely to have its accuracy inflated over this subgroup than the first predictor. The second predictor made 14 errors, and was much more accurate recognizing S than F profiles.

One point of interest is how the second predictor performs over the low-intensity CD68 lymphomas, forming a low-macrophage subgroup. These are the cases whose LAM score is less than 15 cells per high-power field. For the training input of this predictor, all 3 S cases were low-macrophage (LAM scores: 10, 8, 9), as were 2 of the 3 F ones (LAM scores: 12, 20, 7). Over the 24 low-macrophage S cases not used in the training input for this predictor, 22 were correctly classified. Over the 14 low-macrophage F cases not used in the training input, 5 were correctly classified. Thus over these low-macrophage cases, Matthews' correlation *(11)* coefficient r equalled .34, $p < 0.05$, one-tail, on Fisher's exact test.

Table 2
Predictor Performance Over Subgroup With Known Outcomes

Predictor	Training cases	Number of biomarkers used	No. of 18 F cases correct	No. of 26 S cases correct	Correlation with outcome	p-value (one-tailed)
1	F1–F3, S1–S3	13	13	18	0.41	0.00775
2	F4–F6, S4–S6	18	8	22	0.32	0.038
3	F7–F9, S7–S9	11	4	16	−0.17	ns
4	F10–F12, S10–S12	15	8	19	0.18	0.189
5	F13–F15, S13–S15	12	14	3	−0.14	ns
6	F16–F18, S16–S18	16	2	13	−0.4	ns
7	F19–F21, S19–S21	14	8	11	−0.13	ns

ns, not significant.

3.5.2. Subgroup for Which Outcomes Were Masked

Of the 43 cases in this subgroup, the first predictor classified 17 as S and 26 as F. The researcher who did this analysis was not told which predictions were correct, but that there were a total of 18 errors, with 12 actual S and 6 actual F misclassified. The approx 58% success rate here is considerably lower than the accuracy of about 70% observed over the subgroup with known outcomes. The disparity seems a result of having tailored the model architectural settings and encoding scheme for the particular training input, to enhance accuracy over the known subgroup.

This supposition is supported by the fact that the second predictor, which did not have any readjustments for its training input, had very similar accuracy over the masked subgroup of cases as it did over the known subgroup. Again, it made 14 errors in total, and was much more accurate recognizing S than F profiles: 3 actual S and 11 actual F were misclassified. **Figure 3** shows the overall survival comparing the predicted successful group of 31 patients with the predicted failure group of 12 patients. On Kaplan–Meier survival analysis, the difference between the groups is highly significant at $p = 0.0056$.

The next question was whether the second predictor could distinguish failures from successful outcomes over the low-macrophage patients, as it had

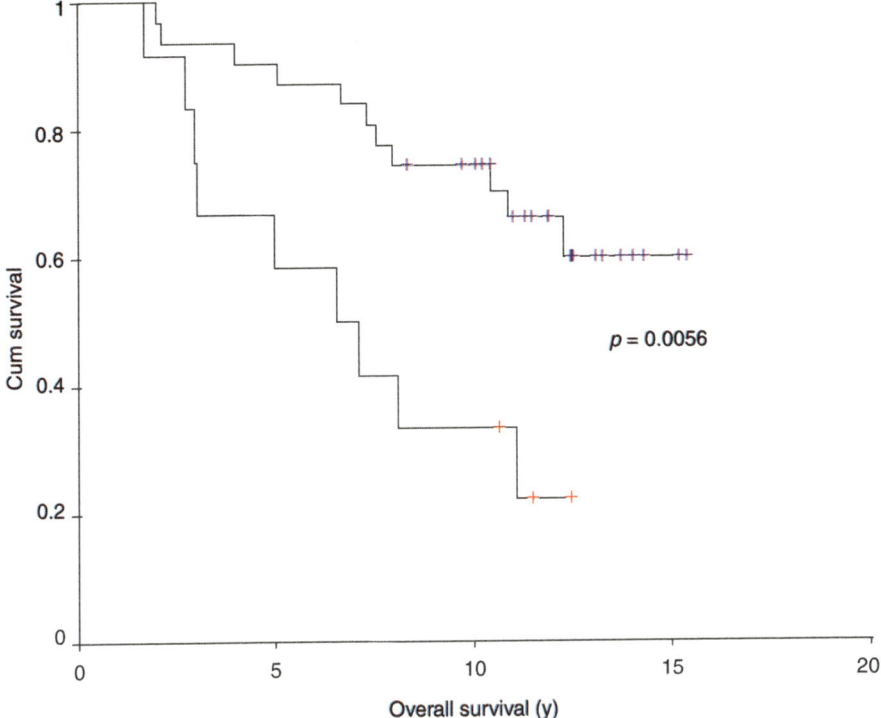

Fig. 3. Overall survival of the reserved subgroup based on second predictor. The top curve represents 31 predicted successful outcome patients; the bottom curve, 12 predicted failed outcome patients.

done in the known subgroup. This is a much harder distinction, where the LAM score alone would predict all to survive. Over the reserved subgroup, Kaplan–Meier analysis showed a clear trend for the predicted successful outcome patients to survive longer than the predicted failed outcome patients, but the difference did not reach statistical significance.

3.5.3. Both Subgroups

The survival difference between predicted successful and predicted failed outcomes for low-macrophage cases becomes much clearer by looking at all such cases not used in forming the second predictor's training input (**Fig. 4**). The difference between the 62 predicted successful outcomes and the 14 predicted failures is significant at $p = 0.014$. This conclusion requires confirmation when a larger group of patients with masked outcomes becomes available.

Finally, **Fig. 5** shows survival differences between the 63 predicted successful and the 24 predicted failures for all cases not used to form the second predictor's

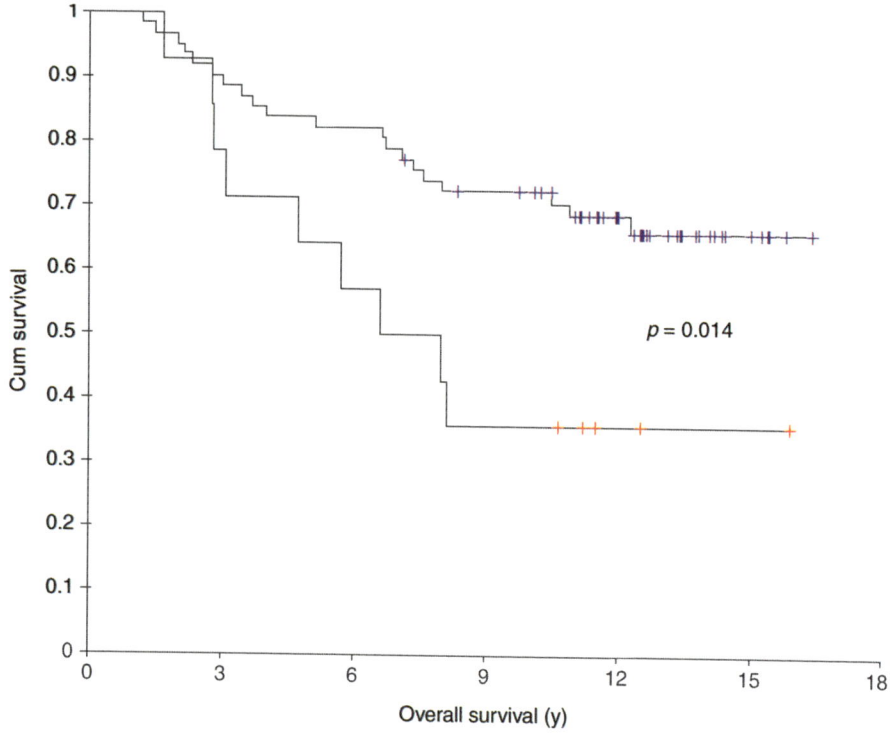

Fig. 4. Overall survival of low-macrophage cases based upon the second predictor, excluding cases used to form the training input. The top curve represents 62 predicted successful outcome patients; the bottom curve, 14 predicted failed outcome patients.

training input. The difference is highly significant at $p = 0.0007$, and corroborates the result in **Fig. 3** for the reserved subgroup. Moreover, in a Cox multivariate model (12), the predictor was an independent variable distinct from the LAM score, and its introduction into the model improved the level of significance from 0.003 to 0.001. However, this finding, and the strength of the result in **Fig. 5**, require confirmation on a larger set of masked outcomes. The evidence herein does suggest that a multibiomarker predictor can improve prediction of overall survival in follicular lymphoma.

4. Notes

1. Typically, a white input is advantageous for nonlinear system identification. This is an input whose autocovariance equals zero except at zero lag (a δ-function), but such an input is an idealization. As a working compromise, it is helpful to use an input with considerable variability in its values. One way of increasing the variability is by varying the order of appending the biomarker values so that the resulting training input autocovariance becomes closest to a δ-function (5).

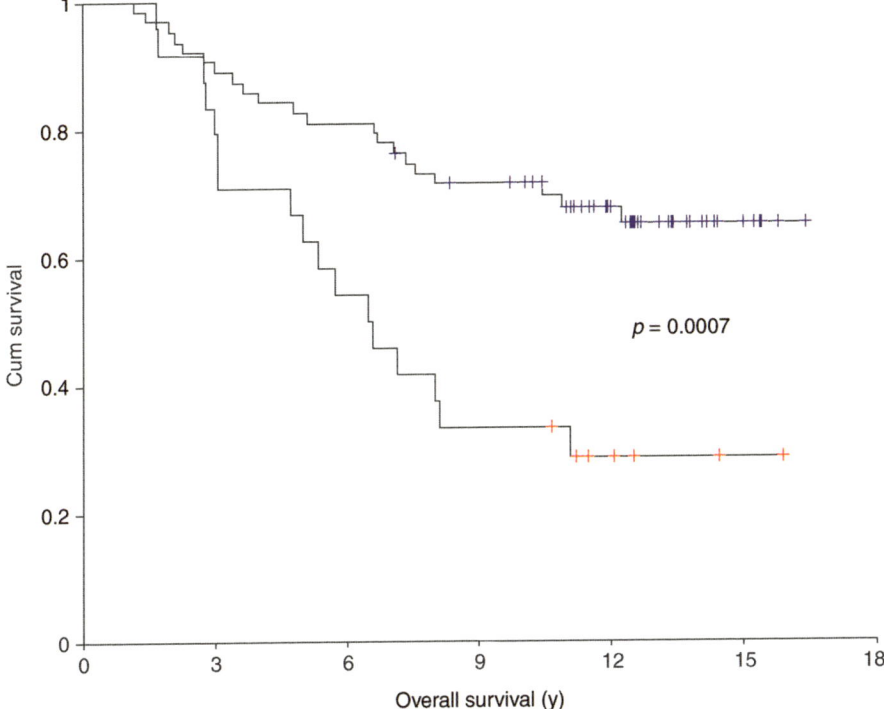

Fig. 5. Overall survival based upon the second predictor for all cases not used to form its training input. The top curve represents 63 predicted successful outcome patients; the bottom curve, 24 predicted failed outcome patients.

2. In most applications, far more than three cases from each class to be distinguished may be needed to form an effective training input. In the present application, the number of cases used to make the training input was deliberately limited to leave a large number of test cases with known outcomes.

3. Higher degree polynomials tend to emphasize small differences in their input values. They can also be harder to fit accurately. Inherent subjectivity in the qualitative scoring of biomarkers causes a lack of precision and does not justify use of higher degree polynomials; generally more accurate predictors will result from using polynomials of first degree. For gene expression data, there is greater precision, and using higher degree polynomials typically improves classification accuracy *(5,6)*.

4. When all the static nonlinearities are first-degree polynomials, the parallel cascade can be collapsed into an equivalent linear system plus a constant. For example, if each dynamic linear element *L* in **Fig. 2** has memory length of 9, then so does the equivalent linear system, and hence a total of 10 variables are introduced into the model (counting the constant) and have to be determined. This is true no matter how many cascades are permitted in the model.

5. Suppose a predictor that uses first-degree polynomials has memory length 9 and its training input is based on 13 biomarkers. This allows those output points corresponding to points 9–13 of each of the 6 training input segments, hence 30 points in total, to be used to determine the 10 variables. If instead the training input is based on 18 biomarkers, then training output points corresponding to points 9–18 of each training input segment can be used in the identification; hence 60 output points in total are available to determine the 10 variables.

6. When there is no downside to allowing more cascade paths, a threshold of zero, admitting every candidate cascade, can be used. Here 100 cascades were added because the mean-square of the residual did not decline significantly thereafter.

References

1. Glas, A. M., Kersten, M. J., Delahaye, L. J. M. J., et al. (2005) Gene expression profiling in follicular lymphoma to assess clinical aggressiveness and to guide the choice of treatment. *Blood* **105**, 301–307.
2. Horning, S. J. and Rosenberg, S. A. (1984) The natural history of initially untreated low-grade non-Hodgkin's lymphomas. *N. Engl. J. Med.* **311**, 1471–1475.
3. Solal-Celigny, P., Roy, P., Colombat, P., et al. (2004) Follicular lymphoma international prognostic index. *Blood* **104**, 1258–1265.
4. Farinha, P., Masoudi, H., Skinnider, B. F., et al. (2005) Analysis of multiple biomarkers shows that lymphoma-associated macrophage (LAM) content is an independent predictor of survival in follicular lymphoma (FL). *Blood* **106**, 2169–2174.
5. Korenberg, M. J. (2002) Prediction of treatment response using gene expression profiles. *J. Proteome Res.* **1**, 55–61.
6. Korenberg, M. J. (2003) Gene expression monitoring accurately predicts medulloblastoma positive and negative clinical outcomes. *FEBS Lett.* **533**, 110–114.
7. Kirkpatrick, P. (2002) Look into the future. *Nature Rev. Drug Discovery* **1**, 334.
8. Palm, G. (1979) On representation and approximation of nonlinear systems. Part II: Discrete time. *Biol. Cybern.* **34**, 49–52.
9. Korenberg, M. J. (1991) Parallel cascade identification and kernel estimation for nonlinear systems. *Ann. Biomed. Eng.* **19**, 429–455.
10. Kaplan, E. L. and Meier, P. (1958) Nonparametric estimation for incomplete observations. *Am. J. Stat. Assoc.* **53**, 457–481.
11. Matthews, B. W. (1975) Comparison of the predicted and observed secondary structure of T4 phage lysozyme. *Biochim. Biophys. Acta* **405**, 442–451.
12. Cox, D. R. (1972) Regression models and life tables. *J. R. Stat. Soc.* **B34**, 187–220.

Index